Sustainable Development in Chemical Engineering

Sustainable Development in Chemical Engineering Innovative Technologies

Editors

VINCENZO PIEMONTE

University Campus Bio-Medico of Rome, Italy

MARCELLO DE FALCO

University Campus Bio-Medico of Rome, Italy

ANGELO BASILE

ITM-CNR, Rende (CS) Italy

WILEY

This edition first published 2013
2013 John Wiley & Sons Ltd

Registered office: John Wiley & Sons Ltd, The Atrium, Southern Gate, Chichester, West Sussex, PO19 8SQ, United Kingdom

For details of our global editorial offices, for customer services and for information about how to apply for permission to reuse the copyright material in this book please see our website at www.wiley.com.

Library of Congress Cataloging-in-Publication Data applied for.

A catalogue record for this book is available from the British Library.

ISBN: 978-1-119-95352-4

Set in 10/12pt Times by Laserwords Private Limited, Chennai, India
Printed and bound in Malaysia by Vivar Printing Sdn Bhd

Contents

List of Contributors

Gianna Allegrone, Department of Chemical, Food, Pharmaceutical and Pharmacological Sciences (DiSCAFF), Università del Piemonte Orientale "Amedeo Avogadro", Italy

D. Lawrence Arockiasamy, King Abdullah Institute for Nanotechnology, King Saud University, Saudi Arabia

Angelo Basile, Institute of Membrane Technology, Italian National Research Council (ITM-CNR), c/o University of Calabria, Italy

Chiranjib Bhattacharjee, Department of Chemical Engineering, Jadavpur University, India

Massimo Cavallo, Department of Chemical, Food, Pharmaceutical and Pharmacological Sciences (DiSCAFF), Università del Piemonte Orientale "Amedeo Avogadro", Italy

Sudip Chakraborty, Department of Chemical Engineering, Jadavpur University, West Bengal, India *and* Department of Chemical Engineering and Materials, CNR-ITM, University of Calabria, Italy

Giuseppe Chidichimo, Department of Chemistry, University of Calabria, Italy

Daniela Cupelli, Department of Pharmaceutical Sciences, University of Calabria, Italy

Stefano Curcio, Department of Engineering Modeling, University of Calabria, Italy

Ranjana Das Mondal, Department of Chemical Engineering, Jadavpur University, India

Professor Yi Ding, Centre for Electric Technology, Department of Electrical Engineering, Technical University of Denmark, Denmark

Marcello de Falco, Faculty of Engineering, University Campus Bio-Medico of Rome, Italy

Giovanni De Filpo, Department of Chemistry, University of Calabria, Italy

Patrizia Formoso, Department of Pharmaceutical Sciences, University of Calabria, Italy

Letizia Fracchia, Department of Chemical, Food, Pharmaceutical and Pharmacological Sciences (DiSCAFF), Università del Piemonte Orientale "Amedeo Avogadro", Italy

Alessandro Franco, Department of Energy and System Engineering (DESE), Università di Pisa, Italy

Adolfo Iulianelli, Institute of Membrane Technology, Italian National Research Council (ITM-CNR), c/o University of Calabria, Italy

Simona Liguori, Institute of Membrane Technology, Italian National Research Council (ITM-CNR), c/o University of Calabria, Italy

Mrinal Kanti Mandal, Chemical Engineering Department, National Institute of Technology Durgapur, India

Maria Giovanna Martinotti, Department of Chemical, Food, Pharmaceutical and Pharmacological Sciences (DiSCAFF), Università del Piemonte Orientale "Amedeo Avogadro", Italy

Salvador Pineda Morente, Centre for Electric Technology, Department of Electrical Engineering, Technical University of Denmark, Denmark

Debolina Mukherjee, Department of Geological Sciences, University of Calabria, Italy

Fiore Pasquale Nicoletta, Department of Pharmaceutical Sciences, University of Calabria, Italy

Jacob Østergaard, Centre for Electric Technology, Department of Electrical Engineering, Technical University of Denmark, Denmark

Alessandro Piccolo, Dipartimento di Science del Suolo, della Pianta dell'Ambiente e delle Produzioni Animali, Università di Napoli Federico II, Italy

Vincenzo Piemonte, Faculty of Engineering, University Campus Bio-Medico of Rome, Italy

Filomena Sannino, Dipartimento di Science del Suolo, della Pianta dell'Ambiente e delle Produzioni Animali, Università di Napoli Federico II, Italy

Jaya Sikder, Chemical Engineering Department, National Institute of Technology Durgapur, India

Qiuwei Wu, Centre for Electric Technology, Department of Electrical Engineering, Technical University of Denmark, Denmark

Preface

This book aims to examine the newest technologies for sustainable development, through a careful analysis not only of the technical aspects but also on the possible fields of industrial development. In other words, the book aims to shed light, giving a broad but very detailed view on the latest technologies aimed at sustainable development, through a point of view typical of an industrial engineer.

The book is divided in four sections (*Energy*, *Process Intensification*, *Bio-Based Platform for Biomolecule Production* and *Soil and Water Remediaton*) in order to provide a powerful and organic tool to the readers.

The first chapter (by Piemonte, Basile, De Falco) is devoted to an overview of the main arguments in the book and to provide a useful key lecture to the reader for a more easy understanding of the topics analysed in further chapters.

In the second chapter (De Falco), Concentrated Solar Power (CSP) technology is presented and a particular application, that is, the cogenerative production of electricity and pure hydrogen by means of a steam reforming reactor is studied in depth and assessed in order to make clear the huge potentialities of CSP plants in the industrial sector.

The third chapter (Franco) analyses some aspects in connection with the problem of new renewable energy penetration. The case of Italian energy production is considered as a meaningful reference due to its characteristic size and the complexity. The various energy scenarios are evaluated with the aid of multipurpose software, taking into account the interconnections between different energy uses.

The last chapter (Ding, Østergaard, Morente, and Wu) in the *Energy* section discusses the smart grid as response for integrating Distributed Generation to provide a balancing capacity for mitigating the high volatility of renewable energy resources in the future.

The second section opens with a chapter on Process Intensification (PI) in the chemical industry. In this chapter (Curcio) a description of some process units designed on the basis of PI concepts has been presented, pointing out their major features, the advantages determined by the exploitation of these PI units and, in some cases, on the existing barriers that are currently limiting their spread on an industrial scale.

The sixth chapter (Basile, Iulianelli, Liguori) is devoted to summarizing the importance of PI in the chemical and petrochemical industries focusing on the membrane reactor (MR) role as a new technology. In particular, it illustrates how integration of MRs in the industrial field could constitutes a good solution to the reduction of the reaction/separation/purification steps, thus allowing a reduction in plant size and improving overall process performance.

The first chapter (Chakraborty, Das Mondal, Mukherjee, Bhattacharjee) in the section on the bio-based platform for biomolecule production deals with a wide and detailed review of the science and technology for sustainable biofuel production. In particular, the production processes of bioethanol and biodiesel are analysed deeply, paying attention also to the sustainability of biofuel use issue.

The eighth chapter (Piemonte) depicts the complex world of bioplastics through the analysis of the bioplastics concept and the description of the most important production processes of bioplastics. Particular attention has been paid to the bioplastic footprint on the environment by analysing the environmental impact of two of the most important bioplastics in the world (PLA and Mater-Bi) in comparison with some petroleum-based plastics (PET and PE) in order to answer, if possible, the most important reader's question: *how green are bioplastics?*

The ninth chapter (Martinotti, Allegrone, Cavallo, and Fracchia) focuses on the most recent results obtained in the field of production, optimization, recovery, and applications of biosurfactants. The chapter spans environmental to biomedical applications of biosurfactants, covering agricultural, biotechnological and nanotechnological applications.

The first chapter (Chakraborty, Sikder, Mukherjee, Mandal, and Arockiasamy) of the soil and water remediaton section presents a state-of-the-art report on the past and existing knowledge of water remediation technologies for the environmentalist who evaluates the quality of environment, implements and evaluates the remediation alternatives at a given contaminated site. The chapter provides a basic understanding of the bioremediation technologies for water recycling to the reader.

The fourth section continues with a chapter (Sannino and Piccolo) on soil remediation, which reviews innovative sustainable strategies that can be applied to remediate soil contaminated by organic pollutants and based on biological, physical and advanced chemical processes. These approaches are illustrated together with the related technical, environmental and economic aspects which should be considered when selecting the most useful remediation method for given soil conditions.

The book concludes with the last chapter (Chidichimo, Cupelli, De Filpo, Formoso, and Fiore) in the soil and water remediaton section, which reports on recent progress in remediation by nanomaterials, describing synthesis and properties of different classes of nanoparticles. The main physico-chemical principles and advantages of using nanoparticles in remediation of wastewaters contaminated by dyes, heavy metals and organic pollutants are discussed. Special attention is given to the modification of nanoparticle surface properties in order to increase efficiency and selectivity. Advances in some particular nanosystems, and perspectives on environment and health impacts by massive use of nanodevices are also reported.

Finally, let us conclude this preface by thanking all the authors who have contributed to the realization of this book, without whom this project would never have been born. We wish to thank them for their participation and patience during the preparation of this book. We are also grateful that they have entrusted us with editing their contributions as per the requirements of each chapter. We hope that readers will find this book useful.

Powerpoint slides of figures in this book for teaching purposes can be downloaded from http://booksupport.wiley.com by entering the book title, author or ISBN.

Vincenzo Piemonte
Marcello De Falco
Angelo Basile

Italy
December 2012

1

Sustainable Development Strategies: An Overview

Vincenzo Piemonte[1,*], Marcello De Falco[1], and Angelo Basile[2]

[1]Faculty of Engineering, University Campus Bio-Medico of Rome, 00128 Rome, Italy
[2]CNR-ITM, c/o University of Calabria87030 Rende (CS), Italy

1.1 Renewable Energies: State of the Art and Diffusion

Energy is a crucial challenge that scientific and technological communities face with more to come in the future. The environmental impact of fossil fuels, their cost fluctuations due both to economical/political reasons and their reducing availability boost research toward the development of new processes and technologies, which are more sustainable and renewable, such as solar energy, wind, biomass and geothermal.

Governments have facilitated renewable energy production diffusion by means of incentive schemes as the feed-in tariff (FIT) and Green Certificates (GCs), achieving unforeseeable success. In fact, the change in the world energy politics is substantially modifying the energy production network. The European Union target to increase the share of renewable energy sources (RES) in its gross final consumption of energy to 20% by 2020 from the 9.2% in 2006, which seemed unlikely up until recently, is now almost there thanks mainly to the strong increase of wind power, photovoltaics and plant biomass installations, together with the implementation of more efficient energy-consuming technologies in domestic, industrial and transport sectors, able to reduce global energy consumption.

The following charts in Figures 1.1–1.3 report wind power, photovoltaic and biomass-fired power station (by wood, municipal solid wastes and bio-gas) electrical energy production trends in recent years in EU-27 (Ruska and Kiviluoma, 2011): it is a worthy

Sustainable Development in Chemical Engineering – Innovative Technologies, First Edition.
Edited by Vincenzo Piemonte, Marcello De Falco and Angelo Basile.
© 2013 John Wiley & Sons, Ltd. Published 2013 by John Wiley & Sons, Ltd.

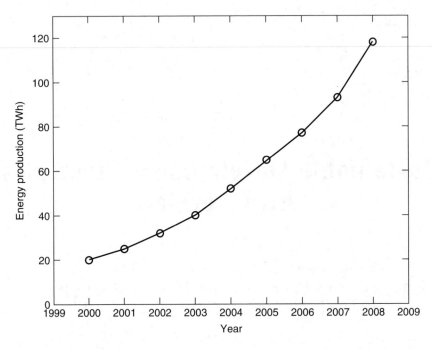

Figure 1.1   *Wind energy production in EU-27 (2000–2008)*

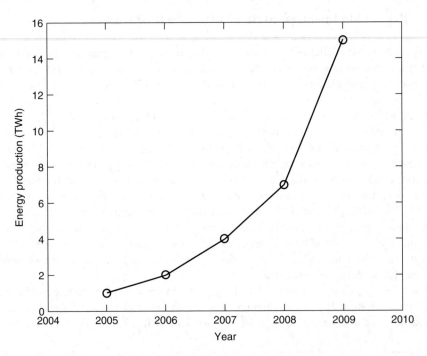

Figure 1.2   *PV energy production in EU-27 (2000–2008)*

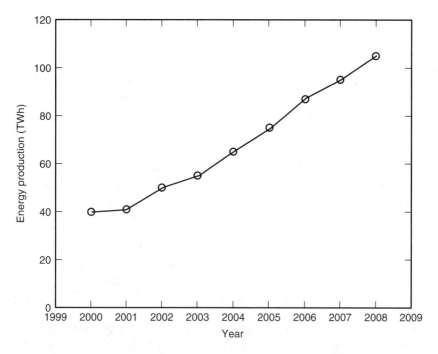

Figure 1.3　*Biomass plant energy production in EU-27 (2000–2008)*

assessment that the diffusion of such technologies follows an exponential profile. The total renewable installed capacity (hydropower, wind, biomass-fired power stations, geothermal plants, photovoltaics) was 200 GW in 2008 and it is continuously increasing.

The International Energy Outlook (*Bloomberg*, 2009) estimates that more than 42% of the new electrical power capacity to be installed up to 2020 will be based on renewable energies, with an average annual growth rates of 4.1%. By 2020 it is foreseen that US$150 bn will be invested worldwide on renewable energies. In Europe, €35 bn has been devoted to clean energy investment in 2008 (http://www.newenergyfinance.com, 2019–2013), and capital expenditure needed to achieve the EU objectives will be approximately €70 bn per year until 2020 in order to reach the 20% target.

From all these data, it is clear how the renewable energy market is becoming mainstream both from technical and financial points of view. Surely, public incentives must be one of the main reasons for renewable penetration in the energy sector, since they have allowed convenient investment when the technologies were not competitive. The increase of investors' interest on this market has pushed industrial production, with the effect of a strong reduction in prices. Taking the PV sector as an example, polycrystalline modules had a cost of about 3000 €/kW in 2009, while now the average price is 700–800 €/kW in 2011 thanks to the development of numerous modular manufacturing industries in Europe and China.

But, concerning the perspectives of renewable energies market in the next years, two crucial aspects have to be considered:

- The economic crisis is stimulating a debate about renewable energy public incentives, which have an increasing effect on the energy bills. The next target is the 'grid parity', that is, the point at which generating electricity from alternative energy produces power at a levelled cost equal to or less than the price of purchasing power from the grid.

- The penetration of renewable energy and the increase in its contribution to total electricity input in the grid lead to the problem of electricity network overload due to clean energy production fluctuations. PV and wind energy production depends on environmental conditions: during sunny and windy days renewable production could invoke serious problems for the grid. This problem stimulates the development of smart grid technologies, able to control and manage grid overloading and electricity storage systems.

Solving both these problems, which have the potential stop renewable energy use, is the main scientific and technological challenge for the future. In this context, proposing, developing and implementing new technologies able to reduce installation costs reaching grid parity and managing energy production is absolutely necessary in order to assure a clean energy future and further enhance its share in energy total production.

The EU assists innovative technology research and development process by allocating many resources to renewable energy projects funding. Figure 1.4 summarizes the organization of the RES financing programmes within the EU (ECOFYS project, 2011) for a total funding amount devoted to energy projects equal to about €4 bn for the next two years.

Thanks to EU support and to the expertise and creativity of worldwide scientific community, the next issues of renewable energy sector can be suitably overcome, allowing the implementation of a 100% clean energy system and achieving the objective of total decarbonation of economies and industries.

1.2   Process Intensification

Following Gorak and Stankiewicz (2011), process intensification (PI) is commonly considered to be one of the most promising development paths for the chemical process industry and one of the most important progress areas for chemical engineering research nowadays.

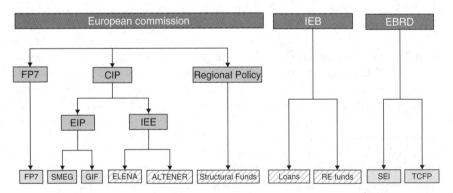

Figure 1.4   *Financial organization of renewable energy in Europe*

When introduced in the 1970s as a general approach, PI suggested a design strategy which aspired to the reduction in processing size of existing technology without any reduction in process output and quality. From that time, PI meaning has been changed several times and many definitions have been proposed, which, despite to their common point of view on innovation, were often different in substance. In 2009, Gerven and Stankiewicz (2009) defined the fundamentals of PI, suggesting that PI should follow a function oriented approach distinguishing four main principles:

- maximize the effectiveness of intra and intermolecular events;
- give each molecule the same processing experience;
- optimize the driving forces at every scale and maximize the specific surface area to which these forces apply;
- maximize the synergistic effects from partial processes.

In particular, the PI principles refer to all scales existing in chemical processes, from molecular to meso- and macro-scales and represent the targets that an intensified process aims to reach.

By applying these principles the PI offers, to an industrial company, many opportunities which can be summarized using only four words: smaller, cheaper, safer and slicker. Indeed, PI leads to the reduction of both investment (reduced equipment or integrated processing units) and operating costs (raw materials and utilities) and less waste. Moreover, by reducing the size of process equipment and the amount of raw material it is possible to ensure a safety benefit, especially in the nuclear/oil industry.

Generally, the PI can be divided in two domains: (1) *process intensifying equipment*, which considers equipment for both carrying out chemical reactions and not involving chemical reactions; and (2) *process intensifying methods*, which takes into account unit operations and is classified furthermore into four different areas (Stankiewicz and Moulijn, 2000).

1.2.1   Process Intensifying Equipment

As mentioned previously, this domain includes both equipment for carrying out the reaction such as the spinning disk reactor, spinning mixer reactor, static mixer catalyst, microreactors and heat exchange reactors, and equipment for non-reactive operations such as the static mixer, compact heat exchangers, rotor/stator mixers and so on.

As a classic example of process intensifying equipment already used in industrial processes, the static mixer reactors must be mentioned, due to their capability in combining mixing and intensive heat removal/supply (Thakur *et al.*, 2003). Moreover, they require less space, low equipment cost and good mixing at low shear rates. On the contrary, one of the most important drawbacks is their sensitivity to clogging by solids. It must also be said that this problem can be partially avoided by developing an open-crossflow-structure catalyst, a structured packing with good static-mixing properties and at the same time, used as catalytic support. The best known of this family is the so-called Katapak, commercialized by Sulzer, and characterized by both good mixing and radial heat-transfer (Stringaro *et al.*, 1998; Irandoust *et al.*, 1998). Usually, Katapak can be applied in catalytic distillation as well as in some gas-phase exothermic oxidation.

Heterogeneous catalytic processes can be intensified by using monolithic catalysts (Kapteijn *et al.*, 1999). Among their many features, it is possible to distinguish some very important benefits such as low pressure drop, high mass transfer area, a low space requirement, low cost and better safety. Another interesting example of process intensifying equipment is the microreactor, used for highly exothermic reactions or for toxic or explosive reactants/products. This device is a small reactor characterized by a structure that has a considerable number of layers with micro-channels. The layers perform various functions such as: mixing, catalytic reaction, heat exchange, or separation (Charpentier, 2007).

1.2.2   Process Intensifying Methods

Process intensifying methods can be divided into four areas: multifunctional reactors, hybrid separation, alternative energy sources and other methods. In the first two categories, the PI is expressed by the novelty of the processing methods in which two or more operations are combined, such as reaction/separation or separation/heat exchange and so on.

A well-known example of a multifunctional reactor is the membrane reactor, in which separation and reaction take place in the same tool. This alternative device represents a real model of intensification showing a higher efficiency compared to both conventional separation and reaction operations. An extensive discussion on these membrane reactors will be given in Chapter 6.

Another example of multifunctional reactors widely studied is the reverse-flow reactor, in which the reaction is combined with the heat transfer in only one unit operation (Matros and Bunimovich, 1996). The idea is to couple indirectly the energy necessary for endothermic reactions and energy released by exothermic reactions, without mixing both endothermic and exothermic reactants in closed-loop reverse flow operation. Usually, this reactor is used for SO_2 oxidation, total oxidation of hydrocarbons and NO_x reduction (Matros and Bunimovich, 1995).

Reactive distillation is another one of the best known examples of reaction and separation combination used commercially (De Garmo *et al.*, 1992). In this case, the reactor consists of a distillation column filled with catalyst. The aim of the distillation column is to separate the reaction products by fractionation or to remove impurities or undesired species. The main benefits of reactive distillation are reduced energy requirements and lower capital investment. Moreover, the continuous removal of reaction products allows us to obtain higher yields compared to conventional systems (Stadig, 1987). Nowadays, this device has been used on a commercial scale even if the potential of this technique has not yet completely exploited.

Reactive extraction is the combination of processes such as reaction and solvent extraction. The main benefit of this integration results in fewer process steps overall, thereby reducing capital cost. Moreover, this combination allows the enhancement of both selectivity and yields of desired products consequently reducing recycle flows and waste formation (Krishna, 2002).

Multifunctional reactors can also combine reaction and phase transition, and the reactive extrusion represents an example of such combination. Currently, this reactor is used in polymer industries, which enables the processing of highly viscous materials without requiring large amounts of solvents (Minotti *et al.*, 1998; Samant and Ng, 1999). Also hybrid separations are characterized by coupling of two or more different unit operations, which lead to a sustainable increase in the process performances owing to the synergy effects among the operations. The most important category in this area is represented by the combination of membranes with another separation unit operation.

Membrane distillation is probably the best known of hybrid separation (Lawson *et al.*, 1997; Godino *et al.*, 1996). It consists of the permeation of a volatile component contained in a liquid stream through a porous membrane as a vapour and condensing on the other side into a permeate liquid. In this process, the driving force is represented by the temperature difference. This technique is widely considered as an alternative to reverse osmosis and evaporation. In comparison with other separation operations, membrane distillation shows very important benefits, such as a complete rejection of colloids, macro-molecules and non-volatile species, lower operating temperature and pressure, and therefore lower risk and low equipment cost, and less membrane fouling due to larger pore size (Tomaszewska, 2000).

Other examples of hybrid separation are membrane absorption and stripping, in which the membrane serves as a permeable barrier between the gas and liquid phases (Jansen *et al.*, 1995; Poddar *et al.*, 1996).

Adsorptive distillation represents a hybrid separation process not involving membranes (Yu *et al.*, 1996). In this technique, a selective adsorbent is added to a distillation mixture which allows us to increase separation ability. Adsorptive distillation can be used for the removal of trace impurities in the manufacturing of fine chemicals or it can present an attractive option in separation of azeotropes or close boiling components. Also alternative energy sources can be considered as an example of PI. Indeed, for instance, alternative forms of energy, such as microwaves, can accelerate chemical processes by hundreds of times compared to the conventional unit operation.

However, other techniques not belonging to the three aforementioned areas can also be considered as intensified processes, such as supercritical fluids and cryogenic techniques. In particular, supercritical fluids are currently applied in mass transfer operations, such as extraction (McHugh and Krukonis, 1994) and for chemical reactions (Savage *et al.*, 1995; Hyde *et al.*, 2001) owing to their high diffusion coefficient; instead, the cryogenic technique, combining distillation with adsorption, is used for industrial gas production but it can present a future option for separation operations in fine chemical industries (Jain and Tseng, 1997; Stankiewicz, 2003).

Anyhow, despite the benefits arising by application of PI principles and by considering that some PI technologies have already been implemented, PI industrial applications on a large scale are faced with several barriers. These obstacles are represented by an insufficient PI knowledge and know-how among process technologists, no pilot plant or possibility to use an existing pilot line, both technical and financial risk in the development of first industrial prototype and the implementation of PI modules in existing production plants and low awareness of potential benefits of PI technologies at the management level.

Only a broad action plan including not only technical factors (technological R&D, upscaling and industrial implementation), but also social and economic factors, can ensure the fast and successful implementation of PI.

1.3   Concept and Potentialities of Bio-based Platforms for Biomolecule Production

Around the world significant steps are being taken to move from today's fossil based economy to a more sustainable economy based on biomass. The transition to a bio-based economy has multiple drivers:

- the need to develop an environmentally, economically and socially sustainable global economy;
- the anticipation that oil, gas, coal and phosphorus will reach peak production in the not-too-distant future and that prices will climb;
- the desire of many countries to reduce an over dependency on fossil fuel imports, so the need for countries to diversify their energy sources;
- the global issue of climate change and the need to reduce atmospheric greenhouse gases (GHG) emissions.

The production of bio-based chemicals is not new, nor is it an historic artefact. Current global bio-based chemical and polymer production (excluding biofuels) is estimated to be around 50 000 000 tonnes (Higson, 2011). Notably, examples of bio-based chemicals include non-food starch, cellulose fibres and cellulose derivatives, tall oils, fatty acids and fermentation products such as ethanol and citric acid. However, the majority of organic chemicals and polymers are still derived from fossil based feedstocks, predominantly oil and gas.

Historically, bio-based chemical producers have targeted high value fine or speciality chemical markets, often where specific functionality played an important role. The low price of crude oil acted as barrier to bio-based commodity chemical production and producers focussed on the specific attributes of bio-based chemicals, such as their complex structure, to justify production costs.

The recent climb in oil prices, the consumer demand for environmentally friendly products, population growth and limited supplies of non-renewable resources have now opened new windows of opportunity for bio-based chemicals and polymers. Bio-based products (chemicals, materials) can be produced in single product processes; however, the production in integrated biorefinery processes producing both bio-based products and secondary energy carriers (fuels, power, heat), in analogy with oil refineries, probably is a more efficient approach for the sustainable valorization of biomass resources in a future bio-based economy (Kamm, 2006; World Economic Forum, 2010).

However, the main driver for the development and implementation of biorefinery processes today is the transportation sector. Significant amounts of renewable fuels are necessary in the short and midterm to meet policy regulations both in- and outside Europe. A very promising approach to reduce biofuel production costs is to use so called biofuel-driven biorefineries for the co-production of both value-added products (chemicals, materials, food, feed) and biofuels from biomass resources in a very efficient integrated approach.

From an overall point of view, a key factor in the realization of a successful biobased economy will be the development of biorefinery systems that are well integrated into the existing infrastructure. Through biorefinery development, highly efficient and cost effective processing of biological raw materials into a range of bio-based products

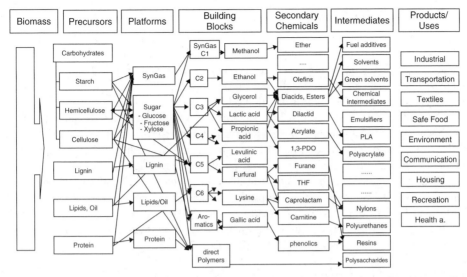

Figure 1.5 *Biorefinery system scheme (Kamm et al., 2006). Copyright Wiley-VCH Verlag GmbH & Co. KGaA. Reproduced with permission.*

can be achieved. On a global scale, the production of bio-based chemicals could generate US$10–15 bn of revenue for the global chemical industry (World Economic Forum, 2010).

Biorefineries can be classified on the basis of a number of their key characteristics (see Figure 1.5). Major feedstocks include perennial grasses, starch crops (e.g. wheat and maize), sugar crops (e.g. beet and cane), lignocellulosic crops (e.g. managed forest, short rotation coppice, switchgrass), lignocellulosic residues (e.g. stover and straw), oil crops (e.g. palm and oilseed rape), aquatic biomass (e.g. algae and seaweeds), and organic residues (e.g. industrial, commercial and post-consumer waste). These feedstocks can be processed to a range of biorefinery streams termed platforms. The platforms include single carbon molecules such as biogas and syngas, five- and six-carbon carbohydrates from starch, sucrose or cellulose; a mixed five- and six-carbon carbohydrate stream derived from hemicelluloses, lignin, oils (plant-based or algal), organic solutions from grasses, pyrolytic liquids. These primary platforms can be converted to wide range of marketable products using combinations of thermal, biological and chemical processes.

1.3.1   Biogas Platform

Currently, biogas production is mainly based on the anaerobic digestion of 'high moisture content biomass' such as manure, waste streams from food processing plants or biosolids from municipal effluent treatment systems. Biogas production from energy crops will also increase and will have to be based on a wide range of crops that are grown in versatile, sustainable crop rotations. Biogas production can be part of sustainable biochemical and biofuel-based biorefinery concepts as it can derive value from wet streams. Value can be increased by optimizing methane yield and economic efficiency of biogas production (Bauer *et al.*, 2007), and deriving nutrient value from the digestate streams (De Jong *et al.*, 2011).

1.3.2   Sugar Platform

Six-carbon sugar platforms can be accessed from sucrose or through the hydrolysis of starch or cellulose to give glucose. Glucose serves as feedstock for (biological) fermentation processes providing access to a variety of important chemical building blocks. Glucose can also be converted by chemical processing to useful chemical building blocks. Mixed six- and five-carbon platforms are produced from the hydrolysis of hemicelluloses. The fermentation of these carbohydrate streams can in theory produce the same products as six-carbon sugar streams; however, technical, biological and economic barriers need to be overcome before these opportunities can be exploited. Chemical manipulation of these streams can provide a range of useful molecules (see Figure 1.6).

Six- and five-carbon carbohydrates can undergo selective dehydration, hydrogenation and oxidation reactions to give useful products, such as: sorbitol, furfural, glucaric acid, hydroxymethylfurfural (HMF), and levulinic acid. Over 1 000 000 tonnes of sorbitol is produced per year as a food ingredient, a personal care ingredient (e.g. toothpaste), and for industrial use (Vlachos *et al.*, 2010, ERRMA, 2011).

1.3.3   Vegetable Oil Platform

Global oleochemical production in 2009 amounted to 7.7 million tonnes of fatty acids and 2.0 million tonnes of fatty alcohols (ICIS Chemical Business, 2010). The majority of fatty acid derivatives are used as surface active agents in soaps, detergents and personal care products (Taylor *et al.*, 2011).

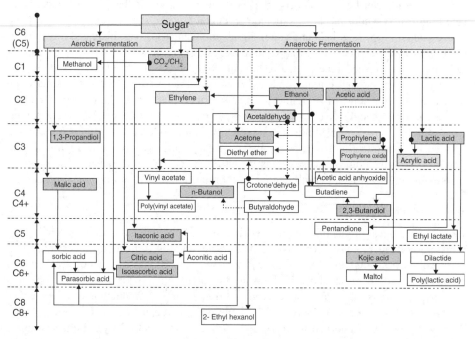

Figure 1.6   *Sugar platform scheme (Kamm et al., 2006). Copyright Wiley-VCH Verlag GmbH & Co. KGaA. Reproduced with permission.*

Major sources for these applications are coconut, palm and palm kernel oil, which are rich in C12–C18 saturated and monounsaturated fatty acids. Important products of unsaturated oils, such as soybean, sunflower and linseed oil, include alkyd resins, linoleum and epoxidized oils. Rapeseed oil, high in oleic acid, is a favoured source for biolubricants. Commercialized biofunctional building blocks for bio-based plastics include sebacic acid and 11-aminoundecanoic acid, both from castor oil, and azelaic acid derived from oleic acid. Dimerized fatty acids are primarily used for polyamide resins and polyamide hot melt adhesives.

Biodiesel production has increased significantly in recent years with a large percentage being derived from palm, rapeseed and soy oils. In 2009 biodiesel production was around 14 million tonnes; this quantity of biodiesel co-produces around 1.4 million tonnes of glycerol. Glycerol is an important co-product of fatty acid/alcohol production. The glycerol market demand in 2009 was 1.8 million tonnes (ICIS Chemical Business, 2010). Glycerol is also an important co-product of fatty acid methyl ester (FAME) biodiesel production. It can be purified and sold for a variety of uses (De Jong *et al.*, 2011).

1.3.4   Algae Oil Platform

Algae biomass can be a sustainable renewable resource for chemicals and energy. The major advantages of using microalgae as a renewable resource are:

- Compared to plants, algae have a higher productivity. This is mostly due to the fact that the entire biomass can be used in contrast to plants which have roots, stems and leafs. For example, the oil productivity per land surface can be up to 10 times higher than palm oil.
- Microalgae can be cultivated in seawater or brackish water on non-arable land, and do not compete for resources with conventional agriculture.
- The essential elements for growth are sunlight, water, CO_2 (a greenhouse gas), and inorganic nutrients such as nitrogen and phosphorous which can be found in residual streams.
- The biomass can be harvested during all seasons and is homogenous and free of lignocellulose.

The main components of microalgae are species dependent but can contain a high protein content, quantities can be up to 50% of dry weight in growing cultures with all 20 amino acids present. Carbohydrates as storage products are also present and some species are rich in storage and functional lipids, they can accumulate up to 50% lipids, and in very specific cases up to 80% (the green algae *Botryococcus*) which accumulates long chain hydrocarbons. Other valuable compounds include: pigments, antioxidants, fatty acids, vitamins, anti-fungal, -microbial, -viral toxins and sterols (De Jong *et al.*, 2011).

1.3.5   Lignin Platform

Up to now the vast majority of industrial applications have been developed for lignosulfonates. These sulfonates are isolated from acid sulfite pulping and are used in a wide range of lower value applications. Around 67.5% of world consumption of lignosulfonates in 2008 was for dispersant applications followed by binder and adhesive applications at 32.5%. Major end-use markets include construction, mining, animal feeds and agriculture uses.

Besides lignosulfonates, Kraft lignin is produced as commercial product at about 60 kton per year. New extraction technologies developed in Sweden will lead to an increase in Kraft lignin production at the mill-side for use as an external energy source and for production of value added applications (Öhman *et al.*, 2009) (see Figure 1.7). The production of bioethanol from lignocellulosic feedstocks could result in new forms of higher quality lignin becoming available for chemical applications. The production of more value added chemicals from lignin (e.g. resins, composites and polymers, aromatic compounds, carbon fibres) is viewed as a medium to long term opportunity which depends on the quality and functionality of the lignin that can be obtained. The potential of catalytic conversions of lignin (degradation products) has been recently reviewed (Zakzeski, 2010).

1.3.6   Opportunities and Growth Predictions

The potential for chemical and polymer production from biomass has been comprehensively assessed in several reports and papers (Shen *et al.*, 2009; US Department of Agriculture, 2008; Patel *et al.*, 2006; Bozell and Petersen, 2010; Werpy and Petersen, 2004; Nexant ChemSystems, 2008).

An international study (Patel *et al.*, 2006) found that with favourable market conditions the production of bulk chemicals from renewable resources could reach 113 million tonnes by 2050, representing 38% of all organic chemical production. Under more

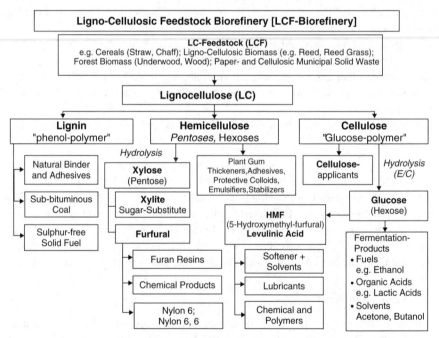

Figure 1.7   *Lignin platform scheme (Kamm et al., 2006). Copyright Wiley-VCH Verlag GmbH & Co. KGaA. Reproduced with permission.*

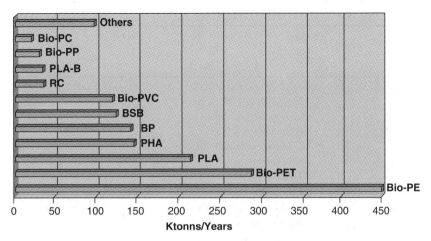

Figure 1.8 *Plastics Europe anticipated biopolymer production capacity (in tonnes/year) by 2015*

conservative market conditions the market could still be a significant 26 million tonnes representing 17.5% of organic chemical production.

Recently Plastemart (Plastemart, 2011) gave an overview of the biorenewable market which was estimated to be worth US$2.4 bn globally in 2010. This steadily growing market has experienced a compounded annual growth rate (CAGR) of 14.8%, a growth trend that is going to increase as the world resumes a more normal production pace and new bio-based chemicals such as bio-ethylene come to market (see Figure 1.8).

The platform biorenewable chemicals glycerine and lactic acid make up the bulk of biorenewable chemicals being sold in 2010, accounting for 79.2% of the market. There is a large range in market maturity for platform biochemicals, ranging from mature markets such as lactic acid to nascent markets for chemicals such as succinic acid.

Currently, commercialized biopolymers (i.e. PLA, PHA, thermoplastic starch) are demonstrating strong market growth. Market analysis shows growth per annum to be in the 10–30% range (Pira, 2010; SRI Consulting, 2012; Helmut Kaiser Consultancy, 2012). Bio-based polymer markets have been dominated by biodegradable food packaging and food service applications. It can be rationalized that the production of more stable, stronger and longer lasting biopolymers will lead to CO_2 being sequestered for longer periods and leads to (thermochemical) recycling rather than composting where the carbon is released very quickly without any energy benefits (De Jong *et al.*, 2011).

1.4 Soil and Water Remediation

One of the most urgent problems being faced worldwide is contamination of soil and water due to domestic and industrial activity. Large polluted areas have lost their eco-functionality and also often present a serious risk to human health. A policy of restoration of natural resources is thus a priority. In recent years, a range of technologies for the remediation of contaminated sites have been developed. Treatment methods are

Table 1.1   *Overview of soil remediation methods*

SOIL Remediation Techniques	Degree of Effectiveness	References
Biological treatments	Removal efficiencies >80% for mineral oil & >95% for monocyclic aromatic hydrocarbons (HC) • HC biodegradation rates under biopile–44,000 g/m3/day	Hoeppel and Hinchee (1994), Tsang *et al.* (1994), Lei *et al.* (1994), Alcade *et al.* (2006), Adekunle (2011)
Chemical treatments Remediation using actinide chelators	Potential solubility problems, stability, and pH requirements that could limit the use of chelators exist	Gopalan *et al.* (1993), Henry and Warner (2003)
Chemical immobilization	Tests results demonstrated that with chemical treatment, heavy metals mobility was drastically reduced; 82–95% metals confined	Czupyrna *et al.* (1989), Khan and Husain (2007)
Critical fluid extraction	Extraction efficiencies between 90–98% demonstrated using PCB-laden sediments • Process is complex; predicting the efficiency of such extraction is difficult	Bellandi (1995)
Oxidation • Chlorine dioxide CD • Hydrogen Peroxide HP • Photolysis P • Reductive Dechlorination RD	• Oxidation can reduce or eliminate volume & toxicity • CD best applied to aqueous phase chemicals • HP easily oxidizes organic and compound rings • UV shown to degrade PCB, dioxins, PAHs	Bellandi (1995), Aristov and Habekost (2010), Fontaine and Piccolo (2011)
In situ catalysed peroxide remediation	• Injection pressure and injection depth influenced decontamination efficiency • Nitrobenzene was reduced in concentration by >50% over 15 days	Ho *et al.* (1995), Gates and Siegrist (1994)
Photodegradation with uranium recovery	• Overall rate of photodegradation faster at pH 3.5 than 6.0	Dodge and Francis (1994), Alvarez *et al.* (2007), Marin *et al.* (2012)

Table 1.1 *(continued)*

SOIL Remediation Techniques	Degree of Effectiveness	References
	• Absence of O_2, excess citric acid, and intermediate degradation products prevented uranium precipitation	
Physical treatments	• Interim stabilization successful at Hanford	Bellandi (1995), Henry and Warner (2003)
Capping	• If uncontrolled, methane gas that migrates within cover system can balloon and possibly combust	
Cementitious waste forms sulfur polymer cement concrete (SPCC) Sulfur polymer cement (SPC)	NRC requires only 500 psi compressive strength and SPCC avg. is 4000 psi • SPC resists attack by most acids and salts, less permeable than concrete	Smith and Hayward (1993), Darnell (1994)
Electrokinetic remediation	• Results indicate optimum moisture content for soil between 14–18 weight % exists • Possibility of inducing greater flow thru fine-grained soils	Lindgren *et al.* (1994), Swartzbaugh *et al.* (1990), Krukowski (1993), Park *et al.* (2007)
Incineration technologies • Rotary kiln RK • Infrared conveyor furnaces ICF • Liquid injection LI • Plasma arc PA • Fluidized bed FB • Multiple hearth MH Feed rates (FR) Destruction and removal efficiency (DRE)	• Incineration reduces volume & toxicity • RK solid waste feed rate 160–170 g/s • ICF DRE 99.9999% for PCBs • PA efficiency at high temperatures is exceedingly high • FB enhances efficiency (larger particles remain suspended until combustion) • MH FR 9–16g/m2/s	Bellandi (1988 and 1995), Aronne *et al.* (2012)

(continued overleaf)

Table 1.1   *(continued)*

SOIL Remediation Techniques	Degree of Effectiveness	References
In situ grouting ISG	• ISG of shallow landfills has been used to effectively control inflow of surface water into hazardous and radioactive waste sites • Chemical grouts – high penetration potential	Spence and Tamura (1989), Karol (1990)
In situ vitrification ISV	• Reduces toxicity, mobility, and volume of waste, and residual product rendered relatively innocuous • Volume reduction of soil matrix ~20–40%	Bellandi (1995), Oma (1994), Tixier *et al.* (1992), Luey *et al.* (1992), Spalding (1994)
Soil washing	• Knowing distribution of contaminants among various particle-size fractions is key to predicting effectiveness • Mobile washing system capacity of 2–4 tons/hr demonstrated reduction in lead concentrations by factor of ~20 • Contaminants must be Hydrophobic	Bellandi (1988 and 1995), Scholz and Milanowski (1983), Masters *et al.* (1991), Griffiths (1995), Gombert (1994), Kim (1993), Wilson and Clarke (1994a,b), Berselli *et al.* (2004),Conte *et al.* (2005)
Sorting methods	Potentially contaminated soil is processed at a rate of 750 m3/ week with a volume reduction between 95–99%	Bramlitt (1990)
Stabilization/ solidification S/S	Polybutadiene resin used in S/S is durable, resists corrosion, and is impermeable to leachates	Conner (1994), Unger *et al.* (1989)
Thermal desorption	Advantage over incineration–reduced amount of gases produced, thereby reducing the size of the off-gas handling system	Bellandi (1995), Ayen *et al.* (1994), Wilson and Clarke (1994a,b)
Vapour stripping	• Large volumes can be readily treated, cleanup times short, toxic material removed and destroyed • Environmental impacts are low	Bellandi (1995), Wilson and Clarke (1994a,b), Thompson (1996)

Table 1.2   *Overview of surface and groundwater remediation methods*

WATER Remediation Techniques	Degree of Effectiveness	References
Biological treatments	• Specific environment governs success of process • Aerobic fluidized-bed had higher chlorophenol loading rates and better quality effluent than those reported	Bellandi (1989 and 1995), Wilson and Clarke (1994a,b), Jarvinen *et al.* (1994), Okonko and Shittu (2007), Cicek (2003)
Chemical treatments Electron-beam irradiation	• Low dose rates of electrons more efficient • E-beam technology has shown removal efficiencies up to 99.99% in full-scale Operation	Rosocha *et al.* (1994), Nyer (1992)
Mercury extraction	Microemulsion containing a cation exchanger reduces mercury content of aqueous phase from 500 ppm to 0.3 ppm, a 40-fold improvement over equilibrium extraction	Larson and Wiencek (1993 and 1994)
Radiocolloid treatment	In situ colloid remediation process using polyelectrolyte capture successful in laboratory column tests	Nuttall and Kale (1994), Nuttall *et al.* (1992)
Removal by sorption to organo-oxides	Advantages: can be regenerated in situ; selective removal achieved if specific surfactant that sorbs contaminant selectively is used; solute removed can be Recovered	Park and Jaffe (1994)
Physical treatments Air sparging/Air stripping In situ air sparging IAS Pump & treat P&T	• Efficiency with which O_2 is transferred to groundwater must be addressed if IAS to be proven effective • Air and steam stripping technologies most effective with VOCs and ammonia	Bellandi (1995), Wilson and Clarke (1994a,b), Cartwright (1991), Hinchee (1994), Isherwood (1993), Johnson (1994), Looney *et al.* (1991)

divided into those for soil remediation and for surface and groundwater remediation. Further categorization results in the consideration of biological (also called bioremediation), chemical, and physical treatment techniques.

The majority of techniques are categorized as physical treatments with only one process, and treatment based on biodegradation falling in the biological treatment category. Chemical treatments involve the application of agents to promote extraction of the hazardous substance, and physical treatments involve removal of the hazard through physical means. The relative benefit of the various remediation methods has dependence in large-scale applicability as well as overall cost. Techniques such as in situ vitrification can be applied only to finite areas in each application, however, because of the associated expense, multiple applications in different areas of the same waste site increase the method's cost-effectiveness.

In the following we try to give an overview of the most important processes for water and soil remediation in a schematic way, through Tables 1.1 and 1.2, which report the remediation processes applied to water and soil remediation along with the relative degree of effectiveness as well as the reference papers that the readers can refer for detailed information on each method.

As for the newest technologies now available for water and soil remediation, the reader can directly refer to the dedicated chapters (Chapters 10 and 11) in this book.

1.4.1   Soil Remediation

In the following table the most used methods for soil remediation along with their degree of effectiveness are reported. For detailed information about each method the reader can refer to the original articles (see Table 1.1).

1.4.2   Water Remediation

In the following table the most used methods for water remediation along with their degree of effectiveness are reported. For detailed information about each method the reader can refer to the original articles (see Table 1.2).

Acknowledgement

The authors thank IEA Bioenergy for their useful contribution.

References

I. M. Adekunle, 2011, Bioremediation of soils contaminated with Nigerian petroleum products using composted municipal wastes, *Bioremediation Journal*, **15**, 230–241.

M. Alcade, M. Ferrer, F. J. Plou and A. Ballesteros, 2006, Environmental biocatalysis: from remediation with enzymes to novel green processes, *Trends Biotechnol.*, **24**, 281–287.

M. Alvarez, T. Lopez, J. A. Odriozola, *et al.* 2007, 2,4-Dichlorophenoxyacetic acid (2,4-D) photodegradation using an Mn+/ZrO2 photocatalyst: XPS, UV–vis, XRD characterization, *Appl. Catal. B-Environ.*, **73**, 34–41.

N. Aristov and A. Habekost, 2010, Heterogeneous dehalogenation of PCBs with iron/toluene or iron/quicklime, *Chemosphere*, **80**, 113–115.

A. Aronne, F. Sannino, S. R. Bonavolontà, *et al.*, 2012, Use of a new hybrid sol–gel zirconia matrix in the removal of the herbicide MCPA: a sorption/degradation process, *Environ. Sci. Technol.*, **46**, 1755–1763.

R. J. Ayen, C. R. Palmer and C. P. Swanstrom, 1994, Thermal desorption. In: *Hazardous Waste Site Soil Remediation: Theory and Application of Innovative Technologies*. D. J. Wilson and A. N. Clarke (eds). New York: Marcel Dekker.

A. Bauer, A. Hrbek, B. Amon, *et al.* 2007. Potential of biogas production in sustainable biorefinery concepts. (Available online at http://www.nas.boku.ac.at/uploads /media/OD7.1_Berlin.pdf, accessed 4 February, 2013).

R. Bellandi (ed.), 1988, *Hazardous Waste Site Remediation: The Engineer's Perspective*. New York: Van Nostrand Reinhold.

R. Bellandi (ed), 1995, *Innovative Engineering Technologies for Hazardous Waste Remediation*. New York: Van Nostrand Reinhold.

S. Berselli, G. Milone, P. Canepa, *et al.*, 2004, Effects of cyclodextrins, humic substances, and rhamnolipids on the washing of a historically contaminated soil and on the aerobic bioremediation of the resulting effluents, *Biotechnol. Bioeng.*, **88**, 111–120.

Bloomberg New Energy Finance, 2013, International Energy Outlook 2009, May 2009 (available online at http://www.newenergyfinance.com, accessed 4 February, 2013).

J. J. Bozell and G. R. Petersen 2010. Technology development for the production of biobased products from biorefinery carbohydrates – the US Department of Energy's 'Top 10' revisited. *Green Chemistry*, **12**, 539–554.

E. T. Bramlitt, 1990, Clean soil at Eniwetok and Johnston Atolls. *Transactions of the American Nuclear Society*. **62**, 70–71.

G. C. Cartwright, 1991, Limitations of pump and treat technology. *Pollution Engineering*, **23**, 64–68.

J. C. Charpentier, 2007, In the frame of globalisation and sustainability, Process Intensification, a path to the future of chemical and process engineering (molecules into money), *Chem. Eng. J.*, **134**, 84–92.

N. Cicek, 2003, A review of membrane bioreactors and their potential application in the treatment of agricultural wastewater. *Canadian Biosystems Engineering, Canada*, **45**, 6.37–6.49.

J. R. Conner, 1994, Chemical stabilization of contaminated soils. In: *Hazardous Waste Site Soil Remediation: Theory and Application of Innovative Technologies*. D. J. Wilson and A. N. Clark (eds). New York: M. Dekker.

P. Conte, A. Agretto, R. Spaccini and A. Piccolo, 2005, Soil remediation: humic acids as natural surfactants in the washings of highly contaminated soils, *Environ. Pollut.*, **135**, 515–522.

G. Czupyrna, R. D. Levy, A. I. MacLean and H. Gold, 1989, *In Situ Immobilization of Heavy-Metal Contaminated Soils*. Park Ridge, NJ: Noyes Data Corporation.

G. R. Darnell, 1994, Sulfur polymer cement as a final waste form for radioactive hazardous wastes. In: *Emerging Technologies in Hazardous Waste Management IV*. D. W. Tedder and F. G. Pohland, (eds). Washington, DC: American Chemical Society.

J. L. De Garmo, V. N. Parulekar and V. Pinjala, 1992, Consider reactive distillation, *Chem. Eng. Prog.*, **88**, 43–50.

E. De Jong, A. Higson, P. Walsh and M. Wellisch, 2011, Bio-based Chemicals Value Added Products from Biorefineries, IEA Bioenergy, Task 42 Biorefinery.

C. J. Dodge and A. J. Francis, 1994, Photodegradation of uranium-citrate complex with uranium recovery. *Environmental Science & Technology*, **28**, 1300–1306.

ECOFYS project, *Financing Renewable Energy in the European Energy Market*, Final Report, Jan. 2011.

ERRMA, 2011, EU-Public/Private Innovation Partnership *Building the Bio-economy by 2020*.

B. Fontaine and A. Piccolo, 2011, Co-polymerization of penta-halogenated phenols in humic substances by catalytic oxidation using biomimetic catalysis, *Environ. Sci. Pollut. Res.*, DOI 10.1007/s11356-011-0626-x.

D. D. Gates and R. L. Siegrist, 1994, In-situ chemical oxidation of trichloroethylene using hydrogen peroxide. *Journal of Environmental Engineering*, **121**(9), 639–644.

T. Gerven and A. Stankiewicz, 2009, Structure, energy, synergy, time: the fundamentals of process intensification, *Ind. Eng. Chem. Res.*, **48**, 2465–2475.

P. Godino, L. Peña and J. I. Mengual, 1996, Membrane distillation: theory and experiments, *J. Membr. Sci.*, **121**, 89–93.

D. Gombert, 1994, Soil washing and radioactive contamination. *Environmental Progress*, **13**, 138–142.

A. Gopalan, O. Zincircioglu and P. Smith, 1993, Minimization and remediation of DOE nuclear waste problems using high selectivity actinide chelators. *Radioactive Waste Management and the Nuclear Fuel Cycle*, **17**, 161–175.

A. Gorak and A. Stankiewicz, 2011, "Intensified reaction and separation systems", *Annu. Rev. Chem. Biomol. Eng.*, **2**, 431–451.

R. A. Griffiths, 1995, Soil-washing technology and practice. *Journal of Hazardous Materials*, **40**, 175–189.

Helmut Kaiser Consultancy, 2012 *Bioplastics Market Worldwide 2007–2025*. (Available online at http://www.hkc22.com/bioplastics.html, accessed 4 February, 2013.

S. M. Henry and S. D. Warner, 2003, Chlorinated solvent and DNAPL remediation: An overview of physical, chemical, and biological processes. In *Chlorinated Solvent and DNAPL Remediation: Innovative Strategies for Subsurface Cleanup*, S.M. Henry, C.H. Hardcastle and S.D. Warner (eds) *ACS Symposium Series*, **837**, 1–20.

A. Higson. 2011, NNFCC. *Estimate of chemicals and polymers from renewable resources*. 2010. NNFCC. *Estimate of fermentation products*. 2010. Personal communication.

R. E. Hinchee, 1994, Air sparging state of the art. In: *Air Sparging for Site Remediation*. Boca Raton, FL: Lewis Publishers.

C. L. Ho, M. A. Shebl and R. J. Watts, 1995, Development of an injection system for in situ catalyzed peroxide remediation of contaminated soil. *Hazardous Waste & Hazardous Materials* **12**, 15–25.

R. E. Hoeppel and R. E. Hinchee, 1994, Enhanced biodegradation for on-site remediation of contaminated soils and groundwater. In: *Hazardous Waste Site Soil Remediation: Theory and Application of Innovative Technologies*. D. J. Wilson and A. N. Clarke (eds). New York, NY: Marcel Dekker.

J. R. Hyde, P. Licence, D. Carter and M. Poliakoff, 2001, "Continuous catalytic reactions in supercritical fluids", *Appl. Catal. A: Gen.*, **222**, 119–131.

ICIS Chemical Business. 2010, Soaps & Detergents Oleochemicals. *ICIS Chemical Business*. 25 January–7 February.

S. Irandoust, A. Cybulski and J. A. Moulijn, 1998, The use of monolithic catalysts for three-phase reactions. In *Structured Catalysts and Reactors*, A. Cybulski and J. A. Moulijn (eds), Marcel Dekker, New York, 239–265.

W. F. Isherwood, D. Rice and J. Ziagos, 1993, 'Smart' pump and treat. *Journal of Hazardous Materials*, **35**, 413–426.

R. Jain and J. T. Tseng, 1997, Production of High Purity Gases by Cryogenic Adsorption, presented at *AIChE Ann. Mtg.*, Los Angeles.

A. E. Jansen, R. Klaassen and P. H. M. Feron., 1995, Membrane gas absorption – a new tool in sustainable technology development, *Proc. 1st Int. Conf. Proc. Intensif for the Chem. Ind.*, 18, BHR Group, London, 145–153.

K. T. Jarvinen, E. S. Melin and J. A. Puhakka, 1994, High-rate bioremediation of chlorophenolcontaminated groundwater at low temperatures. *Environmental Science & Technology*, **28**, 2387–2392.

R. L. Johnson, 1994, Enhancing biodegradation with in situ air sparging: a conception model. In: *Air Sparging for Site Remediation*. Boca Raton, FL: Lewis Publishers.

Kamm B., Gruber P. and Kamm M.. (eds). 2006, *Biorefineries: Industrial Processes and Products*. Weinheim: Wiley-VCH Verlag. ISBN-13 978-3-527-31027-2.

F. Kapteijn, J. J. Heiszwolf, T. A. Nijhuis and J. A. Moulijin, 1999, Monoliths in multiphase catalytic processes-aspects and prospects, *Cattech.*, **3**, 24–41.

R. H. Karol, 1990, *Chemical Grouting* 2nd edn New York, NY: Marcel Dekker, Inc.

A. A. Khan and Q. Husain, 2007, Decolorization and removal of textile and non-textile dyes from polluted wastewater and dyeing effluent by using potato (*Solanum tuberosum*) soluble and immobilized polyphenol oxidase, *Biores. Technol.*, **98**, 1012–1019.

I. Kim, 1993, Mobile soil-washing system. *Chemical Engineering*, **100**, 104.

R. Krishna, 2002, Reactive separations: more ways to skin a cat, *Chem. Eng. Sci.*, **57**, 1491–1504.

J. Krukowski, 1993, Electrokinetics: old technology generates new interest. *Pollution Engineering*, **25**(6), 16.

K. A. Larson and J. M. Wiencek, 1993, Kinetics of mercury extraction using oleic acid. *Industrial and Engineering Chemistry Research*, **32**(11), 2854–2862.

K. A. Larson and J. M. Wiencek., 1994, Extraction of mercury from wastewater using microemulsion liquid membranes: kinetics of extraction. In: *Emerging Technologies in Hazardous Waste Management IV*. W. Tedder and F.G. Pohland (eds). Washington, DC: ACS Symposium Series 554.

K. W. Lawson and D. R. Lloyd, 1997, Membrane distillation: a review, *J. Membr. Sci.*, **124**, 1–25.

J. Lei, J. I. Sansregret and B. Cyr, 1994, Biopiles and biofilters combined for soil cleanup. *Pollution Engineering*, **26**(6), 56–58.

E. R. Lindgren, E. D. Mattson and M. W. Kozak, 1994, Electrokinetic remediation of unsaturated soils. In: *Emerging Technologies in Hazardous Waste Management IV*. Washington, DC: American Chemical Society.

B. B. Looney, T. C. Hazen, D. S. Kaback and C. A. Eddy., 1991, Full scale field test of the in-situ air stripping process at the Savannah River Integrated Demonstration Test Site. Report No. WSRC-91–22. Westinghouse Savannah River Company, Aiken, SC.

J. K. Luey, C. H. Kindle and R. G. Winkelman, 1992, In-situ vitrification of the 116-B-6A crib: large-scale demonstration results. *Transactions of the American Nuclear Society*, **65**, 28–29.

M. L. Marin, L. Santos-Juanes, A. Arques, *et al.*, 2012 Organic photocatalysts for the oxidation of pollutants and model compounds, *Chem. Rev.*, **112**, 1710–1750.

H. Masters, B. Rubin, R. Gaire and P. Cardenas., 1991, EPA's mobile volume reduction unit for soil washing; remedial action, treatment, and disposal of hazardous wastes. Washington, DC: EPA Report No. EPA/600/9-91/002.

Y. S. Matros and G. A. Bunimovich, 1995, Control of volatile organic compounds by the catalytic reverse process, *Ind. Eng. Chem. Res.*, **34**, 1630–1640.

Y. S. Matros and G. A. Bunimovich, 1996, Reverse-flow operation in fixed-bed catalytic reactors, *Catal. Rev. Sci. Eng.*, **38**, 1–68.

M. A. McHugh and V. J. Krukonis, 1994, *Supercritical Fluid Extraction*, Butterworth-Heinemann, Boston.

M. Minotti, M. F. Doherty and M. F. Malone, 1998, Design for simultaneous reaction and liquid–liquid extraction,' *Ind. Eng. Chem. Res.*, **37**, 4746–4755.

Nexant ChemSystems, 2008, *Biochemical Opportunities in the United Kingdom*. York: NNFCC.

H. E. Nuttall and R. Kale, 1994, Remediation of toxic particles from groundwater. *Journal of Hazardous Materials*, **37**, 41–48.

H. E. Nuttall, S. Rao, R. Jain, *et al.*, 1992, Colloid remediation in groundwater by polyelectrolyte capture. In: *Transport and Remediation of Subsurface Contaminants: Colloidal, Interfacial and Surfactant Phenomena*. D. A. Sabatini and R. C. Knox (eds). Washington, DC: American Chemical Society.

E. K. Nyer, 1992, *Groundwater Treatment Technology 2nd edn New York*, NY: Van Nostrand Reinhold.

F. Öhman, H. Theliander, P. Tomani and P. Axegard. 2009. *A method for separating lignin from black liquor, a lignin product, and use of a lignin product for the production of fuels or materials.* WO104995 (International Patent Application PCT/SE 2008/000142).

I. O. Okonko and O. B. Shittu, 2007, Bioremediation of waste water and municipal water treatment using latex exudate from *Calotropis procera* (Sodom apple). *EJEAF Chem.*, **6**(3), 1890–1904.

R. K. Oma, 1994, In-situ vitrification. In: *Hazardous Waste Site Soil Remediation: Theory and Application of Innovative Technologies*. D. J. Wilson and A. N. Clarke (eds). New York, NY: Marcel Dekker.

J. W. Park and P. R. Jaffe, 1994, Removal of nonionic organic pollutants from water by sorption to organo-oxides. In: *Emerging Technologies in Hazardous Waste Management IV*. W. Tedder and F. G. Pohland (eds). Washington, DC: ACS Symposium Series 554.

J. Y. Park, H. H. Lee, S. J. Kim, *et al.*, 2007, Surfactant-enhanced electrokinetic removal of phenanthrene from kaolinite, *J. Hazard. Mater.*, **140**, 230–236.

M. Patel, M. Crank, V. Dornburg., *et al.*, 2006. *Medium and long-term opportunities and risks of the biotechnological production of bulk chemicals from renewable resources – The BREW Project*. (Available online at http://www.projects.science.uu.nl /brew/programme.html, accessed 4 February, 2013).

Pira, 2010, *The Future of Bioplastics for Packaging to 2020*. s.l.: Pira.

Plastemart.com 30/05/2011. Biorenewable chemicals market to be worth US$6.8 bln by 2015 at a CAGR of 22.8%

T. K. Poddar, S. Majumdar and K. K. Sirkar, 1996, Removal of VOCs from air by membrane-based absorption and stripping, *J. Membr. Sci.*, **120**, 221–237.

L. A. Rosocha, D. A. Secker, D. A. Smith and J. D. Smith., 1994, Kinetic modeling of trichloroethylene and carbon tetrachloride removal from water by electron-beam irradiation. In: *Emerging Technologies in Hazardous Waste Management IV*. W. Tedder and F. G. Pohland (eds). Washington, DC: ACS Symposium Series 554.

M. Ruska and J. Kiviluoma., 2011, Renewable electricity in Europe. Current state, drivers, and scenarios for 2020, *VTT Research Note 2584*.

K. D. Samant and K. M. Ng 1999, Systematic development of extractive reaction process, *Chem. Eng. Technol.*, **22**, 877–880.

P. E. Savage, S. Gopalan, T. I. Mizan, *et al*., 1995, Reactions at supercritical conditions: applications and fundamentals, *AIChE J.*, **41**, 1723–1178.

R. Scholz and J. Milanowski., 1983, *Mobile system for extracting spilled hazardous materials from excavated soils*. Washington, DC: EPA Report No. EPA-600/2-83-100.

L. Shen, J. Haufe and M. K. Patel 2009, *Product overview and market projection of emerging bio-based plastics*. s.l.: Utrecht University.

D. L. Smith and W. M. Hayward, 1993, Decommissioning of a resource conservation and recovery act treatment, storage, and disposal facility: a case study of the interim stabilization of the 216-A-29 ditch at the Hanford Site. *Waste Management*, **13**, 109–116.

B. P. Spalding, 1994, Volatilization of cesium-137 from soil with chloride amendments during heating and vitrification. *Environmental Science & Technology*, **28**, 1116–1123.

R. D. Spence and T. Tamura., 1989, In situ grouting of shallow landfill radioactive waste trenches. In: *Environmental Aspects of Stabilization and Solidification of Hazardous and Radioactive Wastes*. P. Cote and M. Gilliam (eds). *American Society for Testing and Materials*. Philadelphia, PA: STP 1033.

SRI Consulting (M. Malveda and K. Yokose), 2012, *Biodegradable Polymers*. (Available online at http://chemical.ihs.com/CEH/Public/Reports/580.0280 accessed 4 February, 2013).

W. P. Stadig, 1987, Catalytic distillation: combining chemical reaction with product separation Chem. *Proc.*, **50**, 27–32.

A. Stankiewicz, 2003, Reactive separations for process intensification: an industrial perspective, *Chem. Eng. Prog.*, **42**, 137–144.

A. Stankiewicz and J. A. Moulijn, 2000, Process intensification: transforming chemical engineering, *Chem. Eng. Prog.*, **96**, 22–34.

J. P. Stringaro, P. Collins and O. Bailer, 1998, Open cross-flow-channel catalysts and catalysts supports. In *Structured Catalysts and Reactors*, A. Cybulski and J. A. Moulijn (eds), New York: Marcel Dekker, 393–416.

J. T. Swartzbaugh, A. Weisman and D. Gabrera-Guzman, 1990, The use of electrokinetics for hazardous waste site remediation. *Journal of Air and Waste Management Association*, **40**(12), 1670–1677.

D. C. Taylor, M. A. Smith, P. Fobert., *et al*. 2011 Plant systems – Metabolic engineering of higher plants to produce bio-industrial oils. In *Comprehensive Biotechnology*, 2nd edn, Vol. 4, M. Moo-Young (ed.), pp. 67–85. Amsterdam: Elsevier.

R. K. Thakur, Ch. Vial, K.D.P. Nigam, *et al*., 2003, Static mixer in the process industries – a review, *Trans. I Chem. E*, **81**, 787–826.

W. Thompson, 1996, Encouraging innovative methods at Superfund sites. *Water Environment and Technology*, **8**(2), 24.

J. S. Tixier, G. K. Jacobs, B. P. Spalding and T. D. Powell, 1992, In situ vitrification of a simulated seepage trench: a radioactive field test at ORNL. *Transactions of the American Nuclear Society*, **65**, 27–28.

M. Tomaszewska, 2000, Membrane distillation – examples of applications in technology and environmental protection. *Polish J. Env.*, **9**, 27–36.

K. W. Tsang, P. R. Dugan and R. M. Pfister., 1994, Mobilization of Bi, Cd, Pb, Th, and U ions from contaminated soil and the influence of bacteria on the process. In: *Emerging Technologies in Hazardous Waste Management IV*. Washington, DC: American Chemical Society.

US Department of Agriculture, 2008, U.S. Biobased Products, Market Potential and Projections Through 2025. s.l.: U.S. Department of Agriculture.

S. L. Unger, R. W. Telles and R. H. Lubowitz., 1989, Surface encapsulation process for stabilizing intractable contaminants. In: *Environmental Aspects of Stabilization and Solidification of Hazardous and Radioactive Wastes*. P. Cote and M. Gilliam (eds). American Society for Testing and Materials. Philadelphia, PA: STP 1033.

D. G. Vlachos, J. G. Chen, R. J. Gorte *et al.* 2010, Catalysis Center for Energy Innovation for Biomass Processing: Research Strategies and Goals. *Catal. Lett.*, **140**, 77–84.

T. Werpy, and G. Petersen., 2004, *Top Value Added Chemicals from Biomass*, Volume 1 Results of Screening for Potential Candidates from Sugars and Synthesis Gas. (Available at http://www1.eere.energy.gov/biomass/pdfs/35523.pdf accessed 4 February, 2013).

D. J. Wilson and A. N. Clarke, 1994a, Soil surfactant flushing/washing. In: *Hazardous Waste Site Soil Remediation: Theory and Application of Innovative Technologies*. D. J. Wilson and A. N. Clarke (eds). New York, NY: Marcel Dekker.

D. J. Wilson and A. N. Clarke, 1994b, Soil vapor stripping. In: *Hazardous Waste Site Soil Remediation: Theory and Application of Innovative Technologies*. D. J. Wilson and A. N. Clarke (eds). New York, NY: Marcel Dekker.

World Economic Forum, 2010, *The Future of Industrial Biorefineries*. s.l.: World Economic Forum.

K. T. Yu, M. Zhou and C. J. Xu., 1996, A novel separation process: distillation accompanied by adsorption, *Proc. 5th World Congress of Chem. Eng.*, San Diego, 347–352.

J. Zakzeski, P. C. A. Bruijnincx, A. L. Jongerius and B. M. Weckhuysen, 2010, The catalytic valorization of lignin for the production of renewable chemicals. *Chemical Reviews*, **110** (6), 3552–3599.

2

Innovative Solar Technology: CSP Plants for Combined Production of Hydrogen and Electricity

Marcello De Falco
University Campus Bio-Medico of Rome, Italy

2.1 Principles

The scientific and technological communities will have to face an extremely important challenge in future: developing a new and more sustainable production model. The pillars of such a model will be the application of efficient technologies able to respect production specifics but consuming a lower amount of energy, materials, water and so on, and the exploitation of renewable energy as solar energy, wind, biomasses and hydro-power, respecting the base concepts of economic/financial convenience. Today, many efforts have already been made in this direction, but more and more has to be devoted to these issues in the coming years before a real sustainable and competitive new industrial model will be implemented worldwide.

Concentrating Solar Power (CSP) is a promising technology perfectly coherent with the needs of developing renewable energy-driven plants respecting financial parameters and, therefore, making investment competitive. The functioning logic is simple: a series of parabolic mirrors concentrate the Sun's rays on their focal line, where a tube (the receiver) is placed. Across the tube a heat transfer fluid (diathermic oil, molten salts) removes high temperature solar heat from the receiver and afterwards it is collected in an insulated heat storage tank, to be pumped on demand to heat users where it releases

Sustainable Development in Chemical Engineering – Innovative Technologies, First Edition.
Edited by Vincenzo Piemonte, Marcello De Falco and Angelo Basile.
© 2013 John Wiley & Sons, Ltd. Published 2013 by John Wiley & Sons, Ltd.

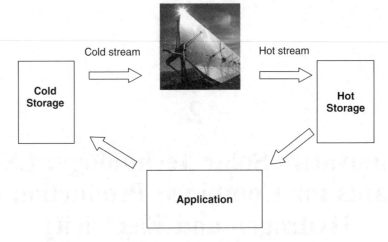

Figure 2.1   *CSP plant layout*

sensible heat. Finally, the heat carrier fluid is stored in a lower temperature tank ready to restart the solar heat collection loop. A proper dimensioning of the heat storage system and the application of backup energy sources (as biomass burners) to relied on when solar energy is not available, allows continuous driving of the process. Figure 2.1 shows a layout of CSP plant.

The heat stored in hot storage tanks is used to produce clean electricity, by generating steam for feeding a steam–turbine system.

Among solar energy technologies for electricity production, CSP is the most competitive for two main reasons:

1. The CSP plant does not produce directly electricity as photovoltaic (PV), but the output is a hot stream able to be stored in proper storage system. By this way, the plant can produce continuously, working for up to 8000 hours/year, versus 1000–1600 hours/year of PV.
2. Diathermic oil-driven CSP plant already costs <1000 \$/kW, versus 2000 \$/kW of PV plant. Moreover, the new molten salt CSP technology, by which a molten salt steam is used as thermal fluid reaching an output temperature of 550 °C versus 390 °C of mainstream technology, has the potential to pull down the plant cost.

However, the hot steam can be used also for many other tasks, such as heat supply for production processes, district heating (providing a hot stream to the domestic market), and for any other process requiring heat (water desalination, industrial processes, etc.).

In this chapter, attention is focused on molten-salt based CSP technology known as *Archimede*. The use of molten salt instead of oils as thermal fluid is competitive for a variety of reasons. Molten salts can operate at higher temperatures than oils (up to 550 °C instead of 390 °C), increasing efficiency and power output of the plant. With the higher temperature heat storage allowed by the direct use of salts, the plant can operate up to 24 hour a day for several days in the absence of Sun. Safety and environmental concerns related to the use of oils are eliminated, since molten salts are cheap, non-toxic common fertilizers and do not catch fire, as opposed to synthetic oils currently used in CSP plants around the world. Moreover, the higher temperatures reached by the molten salts enable the use of steam turbines at the standard pressure/temperature parameters as used in most common gas-cycle fossil power plants. This means that conventional power plants can be integrated – or, in perspective, replaced – with this technology without expensive retrofits to the existing assets [1]. Last but not least, the higher temperature allows us to couple the CSP plant with many chemical processes which are supported at the thermal level <500 °C. Figure 2.2 shows the plant schemes for the oil-based and molten salts-based CSP plants [2]: it must be noted that if molten salt is used, the hot storage tanks are smaller and directly connected with the steam generation unit.

For all these reasons, the molten salt-driven CSP (MS–CSP) technology has the potentiality to be a disruptive industrial technology, able not only to produce competitively 100% clean electricity, but to supply thermal energy to a variety of industrial and production processes. The MS–CSP soundness has been tested in 2010 thanks to the

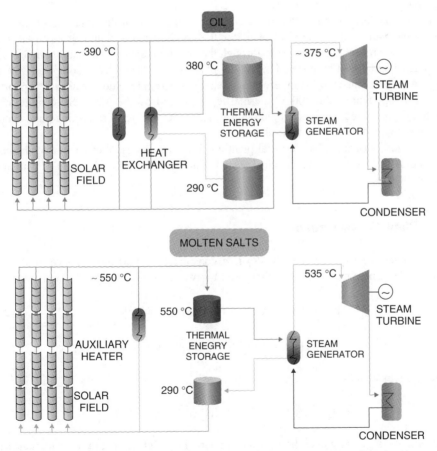

Figure 2.2 *Oil configuration versus molten salt configuration [2]. Reproduced with permission from Archimede Solar Energy © 2012.*

Figure 2.3   *The 5 MW MS–CSP plant located at Priolo, Italy [2]. Reproduced with permission from Archimede Solar Energy © 2012.*

first 5-MW plant installed at Priolo (Italy) by the Italian utility ENEL and the Research Centre ENEA (see Figure 2.3). After this experience, the interest in such a revolutionary technology has risen continuously.

In this chapter, an innovative plant configuration based on the coupling of the MS-CSP solar plant with a chemical plant for the combined production of electricity and hydrogen, is presented and assessed through the development of a simulation model. The hydrogen is produced by a steam reforming plant, whose heat duty is completely fed by the molten salt. Since the steam reforming reaction is strongly endothermic and supported at high temperatures, the 500 °C thermal level assured by the MS–CSP configuration has to be considered a low threshold, which means that the steam reforming plant cannot be linked with an oil-based CSP plant.

In the next section, solar-chemical plant configuration is described; then a mathematical model of the solar-driven chemical reactor is developed, so thus the plant performance can be evaluated.

2.2   Plant Configurations

The methane steam reforming (MSR) is a consolidated hydrogen massive production process. The chemical process is composed by two main reactions:

$$CH_4 + H_2O \leftrightarrow CO + 3H_2 \quad \Delta H^0_{298K} = 206 \frac{kJ}{mol} \tag{2.1}$$

$$CO + H_2O \leftrightarrow CO_2 + H_2 \quad \Delta H^0_{298K} = -41 \frac{kJ}{mol} \tag{2.2}$$

which together yield:

$$CH_4 + 2H_2O \leftrightarrow CO_2 + 4H_2 \quad \Delta H^0_{298K} = 165 \frac{kJ}{mol} \tag{2.3}$$

The process is globally endothermic and supported by high temperatures. The industrial process is led by placing catalytic tubular reactors inside furnaces where much high operating temperatures ($>1000\,°C$) are reached, allowing a methane conversion $>95\%$.

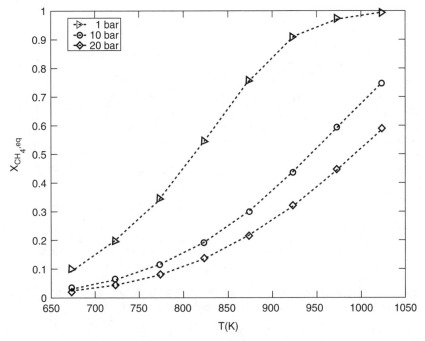

Figure 2.4 *Methane equilibrium conversion versus temperature and pressure* $(H_2O/CH_4 = 3\,mol/mol)$

If the chemical plant is coupled with the MS–CSP, the available temperature level is $<550\,°C$, which would limit the methane conversion to values lower than 30%, as shown in Figure 2.4.

But two strategies can be implemented, as described in the following:

1. The catalytic tubular reactors are replaced with a membrane reactor (MR), by which a much higher methane conversion is achieved at lower temperatures thanks to the removal of hydrogen from the reaction environment [3–6].
2. The final chemical plant product is not hydrogen, but a mixture composed by methane and hydrogen (10–30% vol.), called Enriched Methane (EM) and exploitable as fuel in methane Internal Combustion Engines (ICE), with benefits both in terms of engine efficiency and of environmental protection [7, 8].

In the next sections, both plant configurations are described in detail.

2.2.1 Solar Membrane Reactor Steam Reforming

In a membrane reactor, one or more chemical reactions, generally catalytically promoted, are carried out in the presence of a membrane selectively permeated by one of the reaction products, so that equilibrium conditions are never achieved. In this way, in the case of endothermic reactions, a lower operating temperature is required allowing the possibility of a better heat integration. Moreover, a pure product stream is directly obtained from the reactor, and the chemical plant is much more compact and easy to integrate.

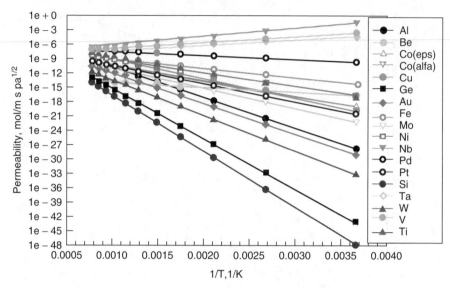

Figure 2.5   *Hydrogen permeability through different metals [10]. Reproduced with permission from International Journal of Hydrogen Energy (IJHE) © 2007*

Taking as an example the steam reforming reaction, many studies attest that conversions >60% can be achieved at operating temperatures of <500 °C [5, 9], making possible the integration of MSR plant with MS–CSP, able to supply heat at 550 °C. Many hydrogen selective membranes are produced, but the dense metallic membranes are the dominant technology thanks to the high selectivity and good permeability. Figure 2.5 illustrates H_2 permeability through different metals: although niobium (Nb), vanadium (V) and tantalum (Ta) offer higher hydrogen permeability than palladium in a temperature range between 0–700 °C, these metals give a stronger surface resistance to hydrogen transport than the palladium (Pd). For this reason, dense palladium membranes are used in preference.

Figure 2.6 shows a layout of a steam reforming membrane reactor, composed by two co-axial tubes: the packed bed zone located in the annular section, is called the *reaction zone* while the inner section is the *permeation zone* where a sweeping gas is fed to carry out the hydrogen stream produced on the catalytic bed. The inner tube is the selective membrane itself. The heat is supplied though the external tube. This is the

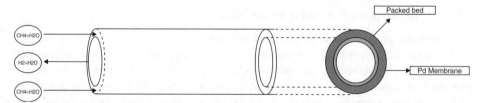

Figure 2.6   *Steam reforming membrane reactor draft [11]. Reproduced with permission from Nova Science Publishers Inc., © 2009.*

best configuration for endothermic reactions, that is, when a high reaction temperature is required while the selective membrane has to be protected from over-heating.

Figure 2.7 illustrates a membrane steam reforming plant coupled with the MS–CSP. It must be noticed that the plant output streams are:

- A pure hydrogen stream, obtained directly from the membrane reactor after a separation unit needed to separate the hydrogen from the sweeping gas. If the sweeping gas is steam water, the separation step is a condensation.
- The unreacted methane and unpermeated hydrogen stream is purified from CO (by a Water Gas Shift unit and a PROX reactor) and CO_2 (by a MDEA unit) and separated from water. This stream is characterized by a high heating value, and therefore can be sent to solar plant steam generator for electricity generation and it can be used as a backup energy source during night or overcast days.

The molten salt stream, after supplying the required heat duty to the process, is sent to steam generator to generate the steam to be fed to the steam-turbine for electricity production. All the steam needed for the chemical process, that is, both the high pressure (10–20 bar) steam for the reaction side and the low pressure (1–10 bar) steam for the permeation zone, can be taken directly from the steam generator. Of course, the water produced in the condensation steps is also sent to steam generator in a closed-loop. By this process of the molten salt stream heated up by the solar plant, both a pure hydrogen stream and clean electricity are produced exploiting solar energy.

2.2.2 Solar Enriched Methane Production

The Enriched Methane (EM) is a gas mixture composed by methane and hydrogen (10–30%.vol.). EM is a 'hybrid' fuel (Lower Heating Value – LHV = 7.84–9.23 kWh/Nm³), since it is composed by a fossil fuel (CH_4) and an energy vector as H_2, which can be produced exploiting renewable energies. Moreover, EM could constitute a first step towards the 'hydrogen economy', that is, an industrial and production system driven by energy produced by fuel cells fed by hydrogen, without emitting any pollutants or GHGs. In fact, by using EM, the main obstacles for hydrogen diffusion and exploitation are overcome:

1. EM can be stored in the traditional pressurized methane storage systems (up to 200 bar) if the H_2 content is lower than 30%.vol.;

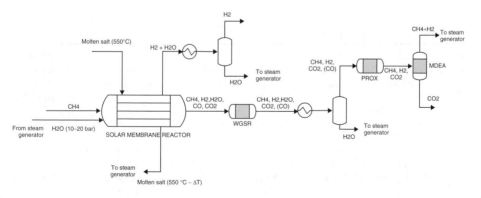

Figure 2.7 *Solar membrane reactor plant layout*

2. EM can be transported into natural gas (NG) pipelines without causing any problems to the centrifugal compressor functioning or end-users if the H_2 content is <17%.vol. [12];
3. EM can feed NG powered internal combustion engines (NG–ICE): a number of papers, appeared in the scientific literature, claim that increasing hydrogen content in the NG engine allows BSFC, BSCO2, BSCO, BSHC to be reduced, improving the engine efficiency and reducing the pollutants emissions [7, 8].

Of course, burning EM in combustion engine causes the emissions of CO_2 and pollutant, but only for the fossil fuel share (NG emissions = 0.242 kg_{CO2}/kWh), while if hydrogen is produced by renewable energy means, combustion only emits steam. Table 2.1 reports the CO_2 emission reduction burning EM at various H_2 compositions. It is worth noting that:

- for each H_2 content growth of 5%.vol. in EM composition, the CO_2 emissions reduction is 2%;
- increasing H_2 content leads to an increase of mass LHV (kWh/kg) and a reduction of volumetric LHV (kWh/m^3).

Nowadays, there is some research which has shown promising results. As an example, during the *Malmö Hydrogen and CNG/Hydrogen Filling Station and Hythane Bus* project, an EM-fuelled bus has been developed, with the following main results:

- a 28.8% unburned hydrocarbon emission reduction compared to methane engines;
- a 17.1% CO emission reduction compared to methane engines.

Other interesting projects implemented worldwide are:

1. National Renewable Energy Laboratory (NREL) and the South Coast Air Quality Management District (SCAQMD): *Development and Demonstration of Hydrogen and Compressed Natural Gas (H/CNG) Blend Transit Buses*, 15 October, 2002–30 September, 2004;
2. Brehon Energy plc., which has patented the 'Hythane' trademark (mixture of 20% H_2, 80% CH_4), proposed the project *Demonstration of the Environmental and Economic benefits for the Beijing Bus Fleet of Hythane (Hydrogen/Natural Gas Mixture) compared with Natural Gas and Diesel Fuel* in China.

Table 2.1   *CO_2 emissions reduction due to EM combustions*

H_2 content (%vol)	H_2 content (%wt)	Lower heating value (kWh/Nm^3)	Lower heating value (kWh/kg)	EM emissions (kg_{CO2}/Nm^3)	EM emissions (kg_{CO2}/kWh)	Emissions reduction compared with methane burning (%)
10	1.37	9.23	14.17	2.16	0.234	3
15	2.16	8.88	14.32	2.04	0.23	5
20	3.03	8.53	14.49	1.92	0.225	7
25	4	8.19	14.68	1.8	0.22	9
30	5.08	7.84	14.89	1.68	0.214	11

3. The FP6 European project *NaturalHy*, an integrated Project with a budget of €17.3 mn, 39 European partners, 15 gas industries and 5 years long starting from May 2004.

Moreover, some EM refuelling stations have been fulfilled in order to demonstrate the potential for EM technology:

- ALT-HY-TUDE project lead by the Research Division of Gaz de France, which intends to demonstrate the use of Hythane® as an environmentally friendly fuel, developed in two cities in France that have a strong environmental policy and wish to lead the path of innovation towards hydrogen: Dunkerque, in the north of France and Toulouse in the south-west.
- Two stations developed by ENI and GM Services in Italy (Livorno and Rome).

In this chapter, we describe the production of EM by the steam reforming process (Eqs. 2.1–2.3) heated up to 500 °C by MS–CSP. If a mean operating temperature of 500 °C is assured in the reaction environment, the equilibrium steam reforming methane conversion is 13–36% (at pressure of 10 bar and 1 bar respectively, refer to Figure 2.4), enough to reach a hydrogen composition in the methane stream of 10–30%.

Figure 2.8 shows a solar EM production plant configuration layout.

It must be noticed that:

1. Compared with the solar membrane reactor plant, only one steam stream, the reaction steam is required.
2. The reactor operating pressure can also be maintained at values lower than 10 bar, since the selective membrane is not assembled and a high pressure driving force is not required.

In the following, both the solar membrane reactor and the solar EM production reactor are modelled by a two-dimensional (2D) simulation model. Then, some simulations are performed in order to assess the plants configuration potentialities in terms of hydrogen production.

2.3 Mathematical Models

In this section, the 2D mathematical models of both the solar membrane reactor and the solar enriched methane production reactor are defined. Then, some plants simulations are performed in order to assess the plant configuration potentials.

The models are based on mass, energy and momentum balances, with the following assumptions:

- Pseudo-homogeneous condition inside the packed beds, that is, the gas and solid phases are assumed as one homogeneous phase;
- For the steam reforming process, only reactions in Equations (2.1–2.3) are considered, secondary reactions are neglected;
- Negligible axial dispersion and radial convective terms;
- Steady-state conditions;
- Ideal gas behaviour;
- In the membrane reactor, the perm-selectivity of Pd-based membrane toward hydrogen is assumed at 100%.

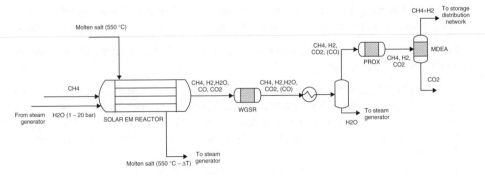

Figure 2.8 *EM production process layout*

2.3.1 Solar Enriched Methane Reactor Modelling

The reactor is a traditional tubular packed bed, externally heated up by a liquid stream (molten salts). It is assumed that the reactors are placed in a tube-tube-and shell configuration, as shown in Figure 2.9: the molten salt is sent into the shell, while the catalyst pellets are packed in four tubular reactors.

The mathematical model of such a configuration is described in [13]. Here, the model equations are reported, together with the boundary conditions to impose in order to solve the PDE (Partial Derivative Equation) set.

Mass balances:

$$\frac{\partial(\tilde{u}_z \tilde{c}_i)}{\partial \tilde{z}} = \frac{d_p \cdot L}{Pe_{mr} \cdot r_i^2} \cdot \left(\frac{\partial^2 (\tilde{u}_z \tilde{c}_i)}{\partial \tilde{r}^2} + \frac{1}{\tilde{r}} \cdot \frac{\partial(\tilde{u}_z \tilde{c}_i)}{\partial \tilde{r}} \right) - \frac{\eta \cdot \rho_{cat} \cdot (1 - \varepsilon) \cdot L}{u_{z,0} c_{CH_4,0}} \cdot R_i \quad (2.4)$$

Where r_i and L are the internal radius and the length of the reformer, $\tilde{z}$ and $\tilde{r}$ the axial and radial dimensionless coordinates, $\tilde{u}_z$ and $u_{z,0}$ the dimensionless gas velocity and its inlet value, $\tilde{c}_i$ and $c_{CH_4,0}$ the dimensionless i component concentration and the inlet methane concentration, R_i is the reaction rate of the component i according to Xu-Froment kinetics model [14], η the effectiveness factor, ρ_{cat} the catalyst density , ε the bed void fraction, d_p the catalyst particle diameter. $Pe_{mr} = \frac{u_z d_p}{D_{er}}$ is the mass effective radial Pèclet number calculated according to [15].

Energy balances: The energy balance in the reactor is defined as follows:

$$\frac{\partial \tilde{T}_R}{\partial \tilde{z}} = \frac{\lambda_{er} \cdot L}{(u_z c_{tot}) \cdot c_{p,mix} \cdot r_i^2} \cdot \left(\frac{\partial^2 \tilde{T}_R}{\partial \tilde{r}^2} + \frac{1}{\tilde{r}} \cdot \frac{\partial \tilde{T}_R}{\partial \tilde{r}} \right)$$

$$+ \frac{\rho_{cat} \cdot (1 - \varepsilon) \cdot L \cdot \sum_{j=1}^{3} \eta_j \cdot (-\Delta H_j) \cdot R_j}{(u_z c_{tot}) \cdot c_{p,mix} \cdot T_{R,0}} \quad (2.5)$$

where $\tilde{T}_R$ and $T_{R,0}$ are the dimensionless and the inlet reactor temperature, η_j, $(-\Delta H_j)$ and r_j are the effectiveness factor, the enthalpy and rate of the reaction j, $c_{p,mix}$ the

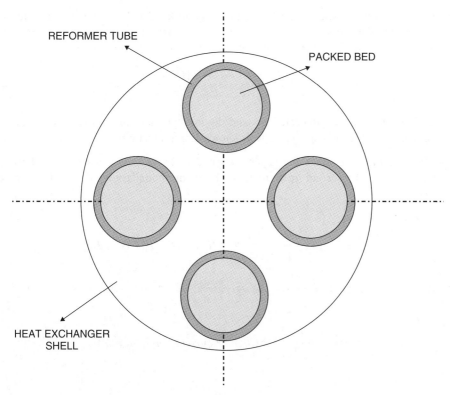

REFORMER TUBE

PACKED BED

HEAT EXCHANGER
SHELL

Figure 2.9 *Enriched methane reactor shell-and-tube configuration [13]. Reproduced with permission from International Journal of Hydrogen Energy (IJHE) © 2009*

gas mixture specific heat. The effective radial thermal conductivity λ_{er} is calculated according to [16].

The energy balance in the molten salt shell is one-dimensional because of the negligible radial temperature profiles:

$$\frac{d\tilde{T}_{MS}}{d\tilde{z}} = \pm \frac{U \cdot L}{w_{MS} \cdot c_{p,MS}} (\tilde{T}_{MS} - \tilde{T}_R) \cdot 2\pi \cdot r_i \qquad (2.6)$$

where $\tilde{T}_{MS}$, w_{MS} and $c_{p,MS}$ are the dimensionless temperature, the mass flow-rate and the specific heat of the molten salt stream respectively.

A crucial parameter is the overall heat transport coefficient U between the molten salt and the catalyst bed, calculated as follows:

$$U = \left[\frac{1}{h_{MS}} \cdot \frac{r_i}{r_o} + \frac{\delta_{met}}{k_{met}} + \frac{1}{h_W} \right]^{-1} \qquad (2.7)$$

where r_o is the external radius of the reformer tube, h_{MS} is the heat transport coefficient in the molten salt side [17], δ_{met} and k_{met} are the metal tube thickness and conductivity respectively, while the wall-to-fluid heat transport coefficient h_W is evaluated as the usual heat transport coefficient of an unmixed layer near the wall [18] according to [19].

Momentum balance: The momentum balance is imposed only in the reactor, since it is assumed that the pressure drop in the molten salt shell is negligible. Moreover, only momentum variations along the axial direction are considered (plug-flow assumption):

$$\frac{d\tilde{P}}{d\tilde{z}} = \frac{f \cdot G \cdot \mu_g \cdot L}{\rho_g \cdot d_p{}^2 \cdot P_0} \cdot \frac{(1-\varepsilon)^2}{\varepsilon^3} \tag{2.8}$$

where $\tilde{P}$ and P_0 are the dimensionless and the inlet pressure in the reaction zone. The friction factor f is calculated by the well-known Ergun equation.

Boundary conditions: Since the Equation set (2.4–2.6, 2.8) is a PDE set of the first order in $\tilde{z}$ and of the second order in $\tilde{r}$, the following boundary conditions have to be imposed:

- $\tilde{z} = 0, \forall \tilde{r}$:

$$\tilde{u}_z \tilde{c}_{CH_4} = 1; \quad \tilde{u}_z \tilde{c}_i = \frac{u_z c_i}{u_{z,0} c_{CH_4,0}} \quad (i = H_2O, H_2, CO, CO_2);$$

$$\tilde{T}_R = 1; \quad \tilde{P} = 1; \quad \tilde{T}_{MS} = \frac{T_{MS,in}}{T_{R,0}} \quad (\textit{Co-current configuration})$$

- $\tilde{z} = 1, \forall \tilde{r}$:

$$\tilde{T}_{MS} = \frac{T_{MS,in}}{T_{R,0}} (\textit{ Counter-current configuration})$$

- $\tilde{r} = 1, \forall \tilde{z}$:

$$\frac{\partial(\tilde{u}_z \tilde{c}_i)}{\partial \tilde{r}} = 0; \quad \lambda_{er} \frac{\partial \tilde{T}_R}{\partial \tilde{r}} = \frac{q_r \cdot r_i}{T_{R,0}} = \frac{U \cdot r_i}{T_{R,0}}(T_{MS} - T_{R|r_i})$$

where q_r is the heat flux from the molten salt to the reactor bed.
- $\tilde{r} = 0, \forall \tilde{z}$:

$$\frac{\partial(\tilde{u}_z \tilde{c}_i)}{\partial \tilde{r}} = 0; \quad \frac{\partial \tilde{T}_R}{\partial \tilde{r}} = 0$$

2.3.2   Membrane Reactor Modelling

As shown in Figure 2.6, a membrane reactor is composed by two co-axial tubes, which defines two different zones: the *annular reaction zone* where the catalyst is packed, and the steam reforming reactions are promoted, and the *permeation zone* where the hydrogen permeated through the selective membrane is collected and carried out by a sweeping gas (steam/water). The reactor configuration analysed in the four-tubes and shell configuration is depicted in the Figure 2.10.

Concerning the reaction zone, the mass, energy and momentum balances are the same as defined in Equations (2.4, 2.5 and 2.8). Similarly, the molten salt energy balance is defined according to Equation (2.6). In these equations, the mass and energy balances of the permeation zone have to be added. Such balances are one-dimensional, since it is assumed that in permeation zone the H_2 concentration and the temperature radial gradients are negligible. Refer to [20] for a detailed description of the solar membrane reactor modelling.

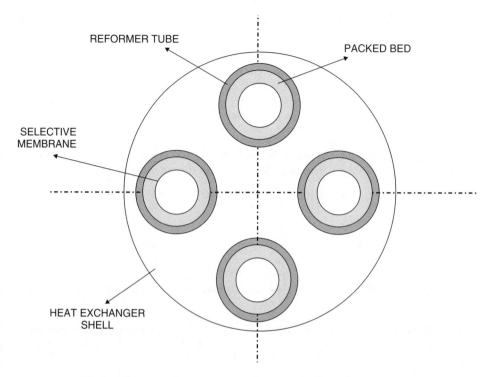

Figure 2.10 *Membrane reactor tubes-and-shell configuration [20]*

Permeation zone mass balance:

$$\frac{dY_{H_2}}{d\tilde{z}} = \pm \frac{N_{H_2}^m \cdot 2\pi \cdot r_{o,i}}{F_{CH4,in}} \qquad (2.9)$$

Where Y_{H2} is the ratio between permeated hydrogen ($F_{H_2,perm}$) and inlet methane flow rate ($F_{CH4,in}$), $r_{o,i}$ is the membrane external radius and $N_{H_2}^m$ is the hydrogen flux permeating through the membrane, calculated according to the well-known Sieverts' law:

$$N_{H_2}^m = \frac{B_H}{\delta} \cdot \left(p_{H_2,reac}^{0.5} - p_{H_2,perm}^{0.5} \right) \qquad (2.10)$$

where B_H is the hydrogen permeability, calculated according to [21] for 20 μm thick Pd-Ag membranes, δ is the membrane thickness, $p_{H2,reac}$ and $p_{H2,perm}$ the hydrogen partial pressures in reaction and permeation zone. The $+$ or $-$ sign in Equation (2.9) refers to co-current or counter-current sweeping gas configuration.

Permeation zone energy balance:

$$\frac{d\tilde{T}_P}{d\tilde{z}} = \pm \frac{L}{F_{P,tot} \cdot c_{p,perm} \cdot T_{R,in}} \cdot$$

$$\left[U_1 \cdot 2\pi \cdot r_{i,i} \cdot T_{R,in} \left(\tilde{T}_R - \tilde{T}_P \right) + N_{H_2}^m \cdot \pi \cdot r_{o,i} \cdot (h_{R,H2} - h_{P,H2}) \right] \qquad (2.11)$$

Where $\tilde{T}_P$ is the dimensionless permeation zone temperature, $F_{P,tot}$ and $c_{p,perm}$ the total molar flow rate and the permeation zone gas mixture specific heat, $h_{R,H2}$ and $h_{P,H2}$ the reaction and permeation zone hydrogen enthalpies.

The overall heat transfer coefficient U_1 between reaction and permeation zone is given by:

$$U_1 = \left[\frac{1}{h_W} + \frac{\delta}{k_{mem}} + \frac{r_{o,i}}{r_{i,i}} \cdot \frac{1}{h_{w,p}}\right]^{-1} \tag{2.12}$$

where k_{mem} is the membrane thermal conductivity and $h_{w,p}$ is the heat convective transport coefficient in the permeation zone, calculated in turbulent conditions.

Boundary conditions: The following conditions are imposed:

- $\tilde{z} = 0, \forall \tilde{r}$:

$$\tilde{u}_z \tilde{c}_{CH_4} = 1; \quad \tilde{u}_z \tilde{c}_i = \frac{u_z c_i}{u_{z,0} c_{CH_4,0}} \quad (i = H_2O, H_2, CO, CO_2);$$

$$\tilde{T}_R = 1; \quad \tilde{P} = 1; \quad \tilde{T}_{MS} = \frac{T_{MS,in}}{T_{R,0}} \quad (\text{Co-current configuration});$$

$$Y_{H_2} = 0 (\text{Co-current configuration}); \quad \tilde{T}_P = \frac{T_{P,in}}{T_{R,in}} \quad (\text{Co-current configuration})$$

- $\tilde{z} = 1, \forall \tilde{r}$:

$$\tilde{T}_{MS} = \frac{T_{MS,in}}{T_{R,0}} \quad (\text{Counter-current configuration}); \quad Y_{H_2} = 0 \quad (\text{Co-current configuration});$$

$$\tilde{T}_P = \frac{T_{P,in}}{T_{R,in}} \quad (\text{Co-current configuration})$$

- $\tilde{r} = 1, \forall \tilde{z}$:

$$\frac{\partial (\tilde{u}_z \tilde{c}_i)}{\partial \tilde{r}} = 0; \quad \lambda_{er} \frac{\partial \tilde{T}_R}{\partial \tilde{r}} = \frac{q_r \cdot r_i}{T_{R,0}} = \frac{U \cdot r_i}{T_{R,0}} (T_{MS} - T_{R|r_i})$$

- $\tilde{r} = r_{o,i}/r_i, \forall \tilde{z}$ (Conditions on membrane tube radius)

$$\frac{\partial (\tilde{u}_z \tilde{c}_i)}{\partial \tilde{r}} = 0 \quad (i = CH_4, H_2O, CO, CO_2); \quad \frac{d_p}{Pe_{mr}} \cdot \frac{u_{z,0} c_{CH_4,0}}{r_i} \frac{\partial (\tilde{u}_z \tilde{c}_{H2})}{\partial \tilde{r}} = N_{H_2}^m;$$

$$\frac{\lambda_{er}}{r_{i,0}} \cdot \frac{\partial \tilde{T}_R}{\partial \tilde{r}} = U_1 \cdot (\tilde{T}_{R|r_{o,i}} - \tilde{T}_P)$$

2.3.3   WGS, Separation Units and the Electricity Production Model

The WGS reaction system is not modelled in detail, since a traditional and consolidated technology is applied in both the plants process schemes. It is simply assumed that 99% of CO content of inlet flowrate is converted into H_2 and CO_2 according to the reaction in Equation (2.2).

The Preferential Oxidation Reactor (PROX) is assembled to eliminate the CO residue. The PROX technology has been developed in recent years as a safety system to be applied upstream PEM fuel cell stacks, which requires CO contents <20 ppm [22]. It is assumed that after the PROX reactor the gas stream does not contain CO. The PROX is inserted after the water steam condensation step since it operates at temperature lower than the steam condensation temperature at operating pressure.

The molten salt stream out-coming from the chemical plant is sent to a Rankine cycle, where its residual sensible heat is converted into electrical energy. By this type of generation cycle, conversion efficiencies up to 38% are claimed to be attainable in medium-size plants (tens of MWs) [23] while only 28–32% are achievable in small-size plants (a few MWs). In the following simulations, a conservative value of 28% is imposed for Rankine cycle efficiency.

2.4   Plant Simulations

Both the solar plant configurations are simulated here in order to evaluate the performance potentiality. Firstly, the reactors are simulated imposed optimal conditions obtained in previous works [13, 20]. Concentrations and temperatures profiles are shown to understand the reactors' behaviour and performance. Then, the entire plants are simulated, focusing on the following outputs:

- methane conversion;
- hydrogen produced (pure hydrogen for membrane reactor configuration, hydrogen in methane for EM production);
- the final MS temperature after the reactor;
- the electric power produced by the residual sensible heat of MS stream.

The MS physical and thermal properties are reported in [17]. The total mass flowrate sent in the shell (Figures 2.9 and 2.10) is $4 \, \text{kg s}^{-1}$, according to the data collected in the pre-industrial MS-CSP plant installed in Priolo.

2.4.1   EM Reactor

The following operating conditions are imposed [13]:

- GHSV (Gas Hourly Space Velocity) = $28 \, 250 \, \text{h}^{-1}$
- steam to carbon ratio, S:C = 2.5;
- operating pressure = 5 bar;
- inlet gas mixture temperature = 500 °C.

The reactor geometrical parameters are reported in Table 2.2.

The MS stream is sent co-currently with the reactants gas mixture. Figure 2.11 shows the methane conversion and the CO_2 yield (calculated as the ratio between CO_2 flowrate and inlet CH_4 flowrate) versus the axial coordinate $\tilde{z}$.

The final methane conversion is 0.17, while the CO_2 yield (X_{CO2}) is 0.152. The outlet hydrogen composition is about 17%, before of CO conversion in WGSR and PROX and CO_2 and H_2O separation. Figure 2.12 reports the temperature profiles of MS stream and

Table 2.2 *EM reactor geometrical parameters*

L (m)	2
r_i (cm)	3.81
d_p (cm)	1.1
ρ_{cat} (kg/m^3)	1016
ε	0.5
η_j	0.02

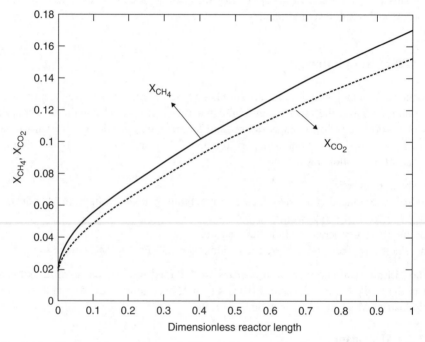

Figure 2.11 *Methane conversion and CO$_2$ yield versus axial coordinate for a solar EM production reactor*

at three different radial coordinates: axis ($\tilde{r} = 0$), $\tilde{r} = 0.5$ and at the first layer of gas near the tube wall ($\tilde{r} = 1$).

The final MS temperature after supplying the reactions heat duty is 817.9 K (544.75 °C), that is, 5.25 °C lower than inlet temperature. If the heat duty needed to generate the reactant steam is considered, the final MS temperature is 530.7 °C and the temperature drop is equal to 19.3 °C. It must be noticed that there is 'cold spot' in the first part of the reactor, where the reactions are strongly promoted since the reactants are fresh. Then, the temperatures increase since the reactions are slower (see the curve slope in Figure 2.11) while the heat supplied by MS stream is higher thanks to the greater temperature gradient between the MS and the packed bed.

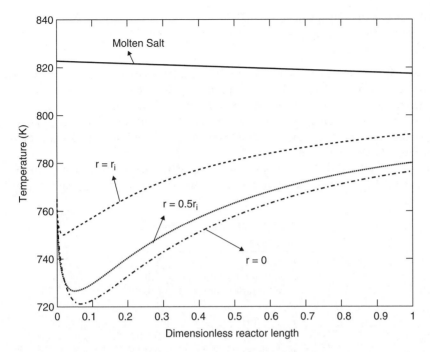

Figure 2.12 *Temperature profiles versus axial coordinate for a solar EM production reactor*

Table 2.3 *Main simulation results for a solar EM production reactor*

X_{CH4}	0.17
X_{CO2}	0.152
Molten salt outlet temperature	544.75 °C
Molten salt final temperature (after reactant steam generation)	530.7 °C
Pressure drop	1.65 bar

Concerning the pressure drops, a total pressure reduction of 1.65 bar is calculated at the end of the reactor. Table 2.3 summarizes the main results of solar EM reactor simulation.

2.4.2 Membrane Reactor

Concerning the membrane reactor simulation, the following operating conditions are imposed:

- steam to carbon ratio, S/C = 2.5;
- operating pressure = 10 bar, higher than EM reactor simulation since in a MR a high pressure difference between reaction and permeation zones is required to maximize the hydrogen flux through the membrane;

- inlet gas mixture temperature, $= 500\,°C$;
- ratio between sweeping gas and inlet methane flowrates $(F_{sweep}/F_{CH_4,in}) = 1$;
- permeation zone pressure $= 1$ bar.

Since MRs are extremely sensitive to GHSV [20], the space velocity effect on reactor performance, that is, X_{CH4} and Y_{H2}, is specifically assessed.

The reactor geometrical parameters are reported in Table 2.4.

Both MS and sweeping gas are sent co-currently with reactants gas mixture. The methane conversion X_{CH4} and the pure hydrogen permeated to inlet methane ratio Y_{H2} at different GHSV are reported in Figure 2.13.

Since a smaller GHSV corresponds to a longer residence time, both methane conversion and Y_{H2} strongly decrease increasing the gas velocity: X_{CH4} and Y_2 are 0.48

Table 2.4 MR geometrical parameters

L (m)	2
Membrane radius (cm)	3.81
Reactor radius (cm)	7.62
d_p (cm)	1.1
ρ_{cat} (kg/m³)	1016
ε	0.5
η_j	0.02

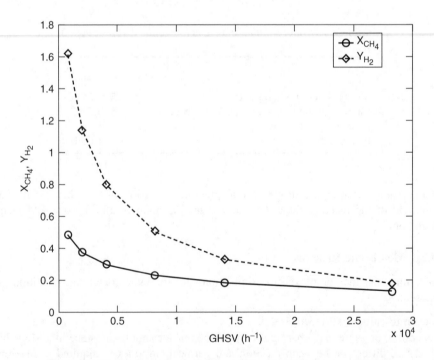

Figure 2.13 *Methane conversion and H_2 permeated versus GHSV*

and 1.62 at $819\,h^{-1}$, 0.13 and 0.175 at 28 $250\,h^{-1}$. On the other hand, decreasing the GHSV at fixed geometrical parameters means reducing the inlet methane flowrate and consequently, the total pure hydrogen produced, as shown in Figure 2.14. Practically, increasing GHSV leads to a reduction of reactor performance in terms of X_{CH4} and Y_2, but at the same time to an increase of pure hydrogen stream production.

In the following, a GHSV compromise value of $4096\,h^{-1}$ is imposed ($X_{CH4} = 0.3$, $Y_2 = 0.796$, H_2 production $= 25.5\,Nm^3/h$). It is worth noting that the reaction zone outlet gas mixture is composed of about 70% of the inlet methane and by about 33% of the hydrogen produced and not permeated. Therefore, such a stream is extremely rich and can be used as:

1. a backup energy source for heating up the MS hot storage tank during night or during rainy days;
2. A feedstock for steam generation for electricity production;
3. an EM stream to be sent to natural gas distribution network.

Figure 2.15 reports the methane conversion and CO_2 yield along the reactor axis, while Figure 2.16 shows the Y_{H2} axial profile.

Figure 2.17 shows the temperature profiles of MS stream, the gas layer near a hot wall and at the axis in the reaction zone, and of the sweeping gas + permeated H_2 in the permeation zone.

Table 2.5 summarizes the main results of solar EM reactor simulation.

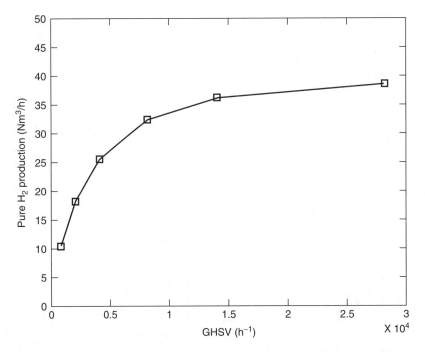

Figure 2.14 *Pure hydrogen production versus GHSV*

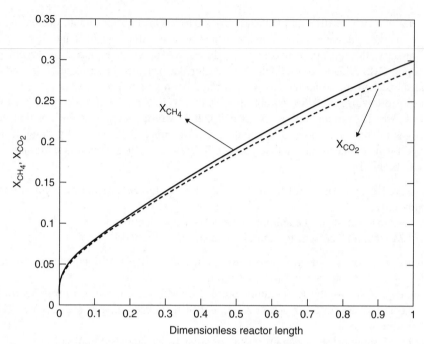

Figure 2.15 *Methane conversion and CO₂ yield along the reactor axis for a MR (GHSV = 4096 h⁻¹)*

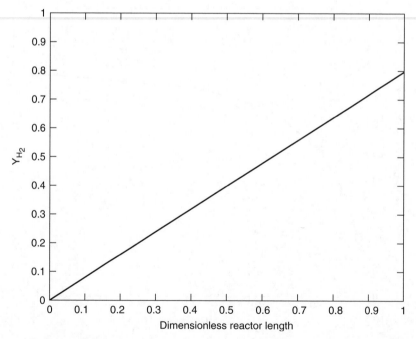

Figure 2.16 *Pure hydrogen production in terms of Y_{H2} along the reactor axis for a MR (GHSV = 4096 h⁻¹)*

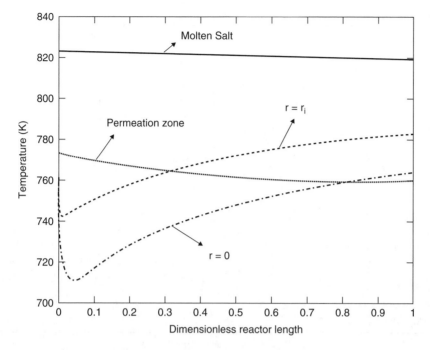

Figure 2.17 *Temperature profiles in the MR*

Table 2.5 *Main simulation results for solar MR*

X_{CH4}	0.3
X_{CO2}	0.29
Y_{H2}	0.796
Molten salt outlet temperature	546.4 °C
Molten salt final temperature (after reactant steam generation)	540.6 °C

2.4.3 Global Plant Simulations and Comparison

In the following, the entire CSP + hydrogen/EM production plants are simulated. For EM production, the reactor configuration described in Figure 2.9 and Table 2.2 at the operating conditions listed in Section 2.4.1, is imposed while for MR the configuration used is shown in Figure 2.10, and described in Table 2.4 and Section 2.4.2. The rest of the plant is simulated according to the description in Section 2.3.3. The MS stream is fixed at $4\,kg\,s^{-1}$ at 550 °C. Table 2.6 lists the main outputs of solar plant configurations.

It has to be noticed that:

1. The membrane reactor leads to a greater methane conversion, thanks to the positive effect of membrane integration in the reaction environment.
2. The EM reactor produces a greater amount of hydrogen, since the GHSV is much higher ($28\,250\,h^{-1}$ vs. $4096\,h^{-1}$). However, the H_2 stream produced in MR is 100% pure, while the H_2 composition in EM stream is maintained <=20%.vol.

Table 2.6   *Main outputs of solar plant configurations*

	EM production	Membrane reactor
X_{CH4}	0.17	0.3
X_{CO2}	0.152	0.29
Y_{H2}	0	0.796
Hydrogen production	50.2 Nm³/h in mixture with methane (H_2 composition = 45.1%vol).	25.5 Nm³/h (100% composition)
	If the H_2 composition has to be <20%, a pure CH_4 stream of 139.6 Nm³/h is added at the end of the plant.	
Final MS temperature after chemical plant	530.7 °C	540.6 °C
Electric power produced	414.5 kW	431.6 kW + 18.2 kW from the $CH_4 + H_2$ outlet mixture coming from the reaction zone.
Yearly electric energy (8000 MWh/year)	3316	3598

3. The final MS temperature is higher in MR configuration, since the hydrogen produced is lower. As a consequence, the electric power produced by MR configuration is approximately 4% greater. Moreover, if the $CH_4 + H_2$ stream coming from the reaction zone outlet is sent to steam generator + steam turbine system, another 18.2 kW can be produced.

Globally, MR configuration seems to be more desirable since it improves methane conversion, produces a pure hydrogen stream and a greater amount of clean electricity. On the other hand, the EM stream leads to a series of applicative benefits in respect to pure hydrogen (as described in Section 2.2.2) and EM configuration is much simpler and more compact.

2.5   Conclusions

This chapter has demonstrated the potentials of two innovative solar hydrogen production processes: the steam reforming solar membrane reactor and the EM production solar reactor. Both chemical processes are heated up exploiting the sensible heat of a molten salt stream coming from a CSP plant. Such a stream, after generating hydrogen by means of steam reforming process, is sent to a steam generator + steam turbine system able to produce electricity. Therefore, the plants' configurations are cogenerative, since both hydrogen and electricity are produced by solar energy.

Simulating both the processes by means of a 2D reactors mathematical model, the plant behaviour has been quantitatively assessed: the CSP + four-tube-and-shell membrane reactor configuration can produce 25.5 Nm³/h of pure hydrogen and up to 449.8 kW$_{el}$, corresponding to 3598 MWh/year of clean electricity, while the CSP + EM reactor produces 50.2 Nm³/h of H$_2$ in a mixture of methane, and 414.5 kW$_{el}$ (3316 MWh/year).

From these data, it is clear that the coupling of a chemical process with the CSP plant is possible and would lead to the production of clean hydrogen without significantly influencing the electricity production since the MS temperature after the steam reforming process is only 9.4 °C (MR) and 19.3 °C (EM reactor) lower than the initial value (550 °C).

Nomenclature

B_H	hydrogen permeability
$\tilde{c}_i$	dimensionless concentration of component i
$c_{i,0}$	inlet concentration of component i
$c_{p,mix}$	gas mixture specific heat
$c_{p,MS}$	specific heat of molten salt stream
$c_{p,perm}$	specific heat of permeation zone mixture
CSP	Concentrating Solar Power
D_{er}	effective radial mass diffusivity
d_p	equivalent particle diameter
EM	enriched methane
f	friction factor
$F_{P,tot}$	permeation zone total molar flow rate
GHSV	gas hourly space velocity
$h_{P,H2}$	hydrogen specific enthalpy in permeation zone
$h_{R,H2}$	hydrogen specific enthalpy in reaction zone
h_{MS}	heat transport coefficient in the molten salt side
$h_{w,p}$	forced convection heat transport coefficient in the permeation zone
h_W	heat transport coefficient in the first layer near the tube wall
$(-\Delta H)$	heat of reaction
k_{mem}	thermal conductivity of the membrane
k_{met}	tube wall conductivity
L	reactor length
MS	molten salt
$N_{H_2}^m$	hydrogen flux permeating through the membrane
$\tilde{P}$	dimensionless reactor pressure
$p_{H_2,perm}$	hydrogen partial pressures in the permeation zone
$p_{H_2,react}$	hydrogen partial pressures in the reaction zone
P_0	inlet pressure
Pe_{mr}	radial mass Peclét number
q_r	heat flux from the external source to the reactor packed bed

$\tilde{r}$	Dimensionless reactor radius
r_i	internal radius of catalytic reactor
r_o	external radius of catalytic reactor
$r_{o,i}$	membrane external radius
R_i	i-component reaction rate
$\tilde{T}_{MS}$	dimensionless molten salt temperature
$\tilde{T}_R$	dimensionless reactor temperature
$T_{R,0}$	inlet temperature
$\tilde{T}_p$	permeation zone temperature
U	overall heat transport coefficient between external energy source and reaction bed
U_1	overall heat transfer coefficient between reaction and permeation zone
$\tilde{u}_z$	gas superficial velocity
w_{MS}	mass flowrate of the molten salt stream
Y_{H2}	ratio between permeated hydrogen and inlet methane flowrate
$\tilde{z}$	dimensionless reactor length
Greek symbols	
δ	thickness of the membrane
δ_{met}	metal tube thickness
ε	packed bed void fraction
η	catalyst effectiveness factor
λ_{er}	effective radial thermal conductivity of the pseudo-homogeneous phase
ρ_{cat}	catalytic bed density

References

[1] http://www.opportunityenergy.org
[2] http://www.archimedesolarenergy.it
[3] M. De Falco, L. Di Paola, L. Marrelli and P. Nardella (2007), Simulation of large-scale membrane reformers by a two-dimensional model, *Chem. Eng. J.*, **128**, 115–125.
[4] E. Kikuchi (1995), Palladium/ceramic membranes for selective hydrogen permeation and their application to membrane reactor. *Catal. Today*, **25**, 333–337.
[5] F. Gallucci, L. Paturzo and A. Basile (2004), A simulation study of steam reforming of methane in a dense tubular membrane reactor, *Int. J. Hydrogen Energy*, **29**, 611–617.
[6] J. Oklany, K. Hou and R. Hughes (1998), A simulative comparison of dense and microporous membrane reactors for the steam reforming of methane, *Appl. Catal A: General*, **170**, 13–22.
[7] C.G. Bauer and T.W. Forest (2001), Effect of hydrogen addition on the performance of methane-fueled vehicles. Part I: effect of S.I. engine performance, *Int. J. Hydrogen Energy*, **26**, 55–70.

[8] S. Orhan Akansu, Z. Dulger, N. Kaharaman and T.N. Veziroglu (2004), Internal combustion engines fuelled by natural gas – hydrogen mixtures, *Int. J. Hydrogen Energy*, **29**, 1527–1539.

[9] M. De Falco, P. Nardella, L. Marrelli, *et al.* (2008), The effect of heat flux profile and of other geometric and operating variables in designing industrial membrane steam reformers, *Chem. Eng. J.*, **138**, 442–451.

[10] F. Gallucci, M. De Falco, S. Tosti, *et al.*, (2007), The effect of the hydrogen flux pressure and temperature dependence factors on the membrane reactor performances, *Int. J. Hydrogen En.*, **32**, 4052–4058.

[11] M. De Falco, G. Iaquaniello, B. Cucchiella and L Marrelli. (2009), Reformer and membrane mod-ules plant to optimize natural gas conversion to hydrogen. In: *Syngas: Production Methods, Post Treatment and Economics*, Nova Science Publishers Inc., ISBN 978-1-60741-841-2.

[12] D. Haeseldonckx and W. D'Haeseleer, (2007), The use of natural-gas pipeline infrastructure for hydrogen transport in a changing market structure, *Int. J. Hydrogen Energy*, **32**, 1381–1386.

[13] M. De Falco, A. Giaconia, L. Marrelli, *et al.* (2009), Enriched methane production using solar energy: an assessment of plant performance, *Int. J. Hydrogen Energy*, **34**, 98–109.

[14] J. Xu and G. Froment (1989), Methane steam reforming, methanation and water-gas shift: I. Intrinsic kinetics, *AIChE J.*, **35**, 88–96.

[15] B.D. Kulkarni and L.K. Doraiswamy (1980), 'Estimation of effective transport properties in packed bed reactors', *Catalysis Reviews: Science and Engineering*, **22**, 431–483.

[16] A.P. De Wasch and G.F. Froment (1972), Heat transfer in packed beds, *Chem. Eng. Science*, **27**, 567–76.

[17] A. Miliozzi, G.M. Giannuzzi, P. Tarquini and A. La Barbera. (2001) Fluido termovettore: dati di base della miscela di nitrati di sodio e potassio, ENEA Report, ENEA/SOL/RD/2001/07.

[18] E. Tsotsas and E. Schlünder (1990), Heat transfer in packed beds with fluid flow: remarks on the meaning and the calculation of a heat transfer coefficient at the wall, *Chem. Eng. Science*, **45**, 819–837.

[19] C. Li and B. Finlayson, (1977), Heat transfer in packed beds – a reevaluation, *Chem. Eng. Science*, **32**, 1055–1066.

[20] M. De Falco, A. Basile and F. Gallucci (2010), Solar membrane natural gas steam reforming process: evaluation of reactor performance, Special Issue of *Asia-Pacific Journal of Chemical Engineering on Membrane Reactors*, **5**, 179–190.

[21] J. Shu, B. Grandjean and S. Kaliaguine (1994), Methane steam reforming in asymmetric Pd and Pd-Ag porous SS membrane reactors', *Appl. Catal. A: General*, **119**, 305–325.

[22] C.D. Dudfield, R. Chen and P.L. Adcock (2001), A carbon monoxide PROX reactor for PEMFC automotive application, *Int. J. Hydrogen Energy*, **26**, 763–775.

[23] Sargent & Lundy LLC Consulting Group (2003), Assessment of Parabolic Trough and Power Tower Solar Technology Cost and Performance Forecasts. NREL Subcontractor Report, NREL/SR-550-34440, October.

3

Strategies for Increasing Electrical Energy Production from Intermittent Renewables

Alessandro Franco

Department of Energy and Systems Engineering (DESE), University of Pisa, Pisa, Italy

3.1 Introduction

Renewable energy sources (RES) could play an important role in the sustainability of energy systems, although at present their contribution is often marginal in energy balances of many countries. Since the beginning of the twenty-first century, the European Union (EU) has officially recognized the need to promote RES as a priority measure both for reduction of energetic dependence on fossil fuels and for environmental protection.

In recent years the share of renewable energy systems like Combined Heat and Power (CHP) and District Heating (DH) systems has been permanently increased (EIA, 2009). So far, penetration of new renewables and more efficient energy systems like CHP has been politically encouraged by strong financial incentives in different European countries (Campoccia *et al.*, 2009).

Renewable energy sources have several advantages. Important penetration of RESs, mainly wind and photovoltaic (PV) power, like increased decentralized production, resulted in a growing number of small and medium size producers and a lot of new CHP systems that are operated almost continuously in order to achieve economic efficiency, connected to energy networks and in particular to electricity grids. But the fluctuating nature of wind and solar energy, and the simultaneous production of heat and electricity

Sustainable Development in Chemical Engineering – Innovative Technologies, First Edition.
Edited by Vincenzo Piemonte, Marcello De Falco and Angelo Basile.
© 2013 John Wiley & Sons, Ltd. Published 2013 by John Wiley & Sons, Ltd.

in CHP systems cause sensible amounts of problems when the fluctuating supply must be adapted to electricity demand, that has a well-defined structure (Lund and Clark, 2002; Greenblatt *et al.*, 2007).

Maintaining the momentary balance between supply and demand in energy systems with the use of large quantities of fluctuating renewable energy sources appears to be a quite complex problem. Therefore, many new problems will arise, in relation to management and operation of energy transfer as well as in relation to efficient distribution of wind power and other renewable energy sources, such as PV, to the grids. In order to bring about a substantial long-term penetration of distributed energy resources in Europe, it is necessary to address the key issues related to their integration into existing energy systems, dominated by a large variety of thermoelectric power plants so that an effective Primary Energy Saving (PES) could be obtained. For this reason one of the most important future challenges for RES increase will be the management of the integration of fluctuations in the electricity production from renewable energy sources and from CHP units in complex energy systems. (Lund, 2005).

Although this has not yet incurred many technical problems, it seems necessary to wonder whether such a fragmented logic of development on a local-scale could be sustainable system-wide or not. Therefore, further penetration must be encouraged with rational plans of development that must take into account the problem in its completeness. Future electricity supply systems with high penetrations of renewable energy sources, such as wind energy and PV, and efficient conversion technologies, like CHP systems, are characterized by an increased need for energy storage.

The problem of penetration of RES has been considered in a great number of recent papers, referred to various different context. The various approaches are well summarized by Banos *et al.* (2011) in a recent paper that presents a review of the current state of the art in computational optimization methods applied to renewable and sustainable energy, offering a clear vision of the latest research advances in this field.

In this study we focus on electricity production from intermittent renewable energy sources such as wind turbines and solar PV systems. An electric system completely based on renewable energies is feasible in principle but appears not sustainable for sizing problems because it would require a very large storage capacity. In Italy for example 55 GW of peak power are required on the net and this demand is currently mainly satisfied with thermoelectric plants with fossil fuel input. To sustain the same demand with new renewable sources about 300 GW of power installed (an average value of 1500 full load-hours per annum for wind energy and an average value of 1200 full load hours per annum of PV vs 8000 full load hours for thermoelectric power plants) would be necessary with a similar number of storage systems. we also consider the growing interest in technologies like the development of innovative engines in transport systems using electricity or hydrogen (that can be used as further element for chemical storage too) for mobility, or the introduction of peak-off peak tariff of charge to consumers so as to achieve a flattening in daily power demand profiles.

Following a traditional approach, taking the increase in efficiency of single sub-systems as the main target, these technologies would definitely lose part of their potential contribution to mitigate energy problems: in the worst but not so unrealistic case, they could even have a negative impact, ending up with a kind of *rebound effect* with an increase of non-necessary energy production. In this sense, the development of fluctuant renewable

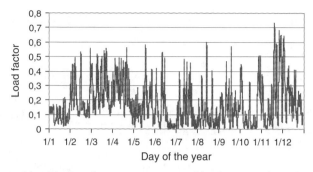

Figure 3.1 *Typical load factor of intermittent renewable: the case of wind power production during a year*

energy sources is a paradigmatic case to analyse. The problem of power grid balancing with high renewable energy penetration stimulates a lot of interesting analysis carried out from a general standpoint (Salg and Lund, 2008; Stadler, 2008) and with reference to specific sectors (Greenblatt *et al.*, 2007, Hoogwijk *et al.*, 2007, Osteergard, 2009; Fthenakis *et al.*, 2009). The main problem, well evidenced by Fig. 3.1 is that some RES are available with a load factor (defined as the power available divided the total power installed) that varies in a range between 0–0.7.

Traditionally, the *sectorial decomposition* in the three conventional energy sectors: electricity, thermal (heat) and transport is the first step in energy systems analysis. If, on one hand, this leads to the study of less complex systems, with the advantage of a smaller number of variables, on the other hand a completely unrelated view of the three reference areas threatens to become a limit to the logical planning of a problem that is getting more and more difficult because of its economic (prices of fuels), environmental (pollution, emissions) and political (energy dependence) consequences. In the near future, a new concept of *integrated analysis* of energy systems will be needed, so as to regain information about interconnections between the three sectors mentioned: these interactions will become essential for a more efficient evolution of energy systems, both on the supply side and demand side (Fig. 3.2). As stated before, in principle, the production system has to follow demand variations, which means that when neglecting international trade, the production should equal demand.

In consideration of renewable energy sources, cogeneration of heat and power for industrial (process heat) or civil (residential heating or also cooling in tri-generation systems) purposes, additional time variations are introduced. For example, the production of a CHP plant is determined by temperature variations which vary according to a daily and a seasonal cycle as well as a stochastic element. The same applies to PV-based electricity generation, as the altitude of the Sun varies during the yearly cycle and local climatic conditions influence cloud cover.

Many analyses have treated flexible energy systems, where RES and CHP plants are regulated with the perspective of optimization (Lund, 2005). This especially the case in northern Europe (mainly in Germany, Denmark and the UK). Flexible demand (Stadler, 2008; Lund and Mathisen, 2009), improved forecasting (Pinson *et al.*, 2009), inclusion of technologies such as heat pumps adding flexibility to the system (Blarke and

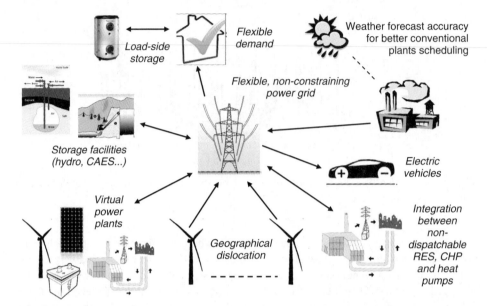

Figure 3.2   *The complex energy system integration: integration of the three energy production of electricity, thermal and transport*

Lund, 2008), use of electric vehicles (Franco and Salza, 2011a,b), increase of use of storage technologies of the various size (Salgi and Lund, 2008; Beaudin *et al.*, 2010), improved control of generating technology (Østergaard, 2010) and improved control strategies and architecture, analysed by a lot of authors in the electric systems field, are the main paths along which research developed in the last 10 years.

However, a lot of questions remain and it is difficult to use typical solutions for small size systems (like the Danish case) for more complex systems. In this work, the knowledge of the aforementioned problems related to the rational growth of energy systems characterized by a high level of penetration of intermittent energy sources is developed, analysing at first critical elements, then trying to provide useful analysis tools to efficiently plan the future development of such a kind of systems. The methodology can be extrapolated and applied to a variety of contexts, with different scale or features: from technical-economic optimization of single or virtually aggregate power plants (industrial plants, cogeneration based district heating, *virtual power plants*) to local energy planning (regional energy plans, analysis of single macrozones subjected to constraints of power flow through national transmission line) (Wille-Hausmman *et al.*, 2010). All three sectors will be taken into account, even if particular attention will be given to the electrical sector, because of its central role also in integrated analysis (consider the already cited cogeneration technology and electric traction transport systems) and to wind power, because of its currently greater potential and because a good technical literature exists, developed by those countries that have a significant penetration level of the resource. The Italian energy system is considered as a reference scenario for case study analysis.

3.2 Penetration of Renewable Energies into the Electricity Market and Issues Related to Their Development: Some Interesting Cases

With regard to the implementation of the change from fossil fuels to renewable energy, some interesting cases are available (Terna, 2011a). The examples of Denmark, Ireland, Switzerland and southern Italy (see Table 3.1 and Fig. 3.3) describe some interesting cases of countries with remarkable penetration of RES. In each case, countries are totally dependent on the import of oil, gas and coal for electricity production; Germany and Denmark are characterized by a quite high percentage of intermittent renewable with respect to the total installed power. Denmark was the first country in which in 2004–2005

Table 3.1 Summary data on renewable sources – comparison between countries (data from 31 December, 2009)

	Italy	Denmark	Germany
Total installed capacity (MW)	105 200	13 800	146 600
Total energy production (TWh)	308	39	636
Wind capacity (MW)	4898	3200	25 800
PV capacity (MW)	1142	5	10 800

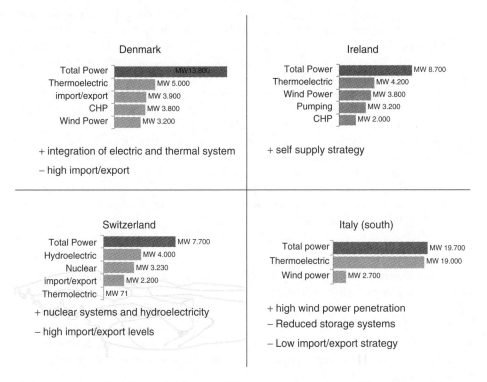

Figure 3.3 Examples of remarkable renewable energy penetration in different energy contexts

problems of excess of wind energy production with respect to the total demand have been observed.

The cases of Ireland and Switzerland are completely different, concerning the size and the distribution of RES and the remaining thermoelectric capacity. The case of southern Italy is similar to Denmark in the case of wind energy penetration, regarding the size of the whole system (19 200 MW installed and the amount of wind power installed). Almost all transport and residential heating is based on oil.

The problem of penetration of new RESs in relation to the typical rigidity of a complex energy systems based on thermoelectric base plants is well known. At modest penetration, wind power, PV power and small hydro power merely substitutes electricity generated from the various thermal power plants. In this case, they are able to provide energetic and economic benefits in terms of saved marginal fuel and operation and maintenance costs. At higher penetrations, it becomes increasingly important for the energy system to be able to operate without costly reserve capacity awaiting fluctuations in demand or intermittent renewable power generation. Existing transmission interconnections have mainly been established in order to assist in reducing the reserve capacity of thermal power systems. While indeed relevant in thermal systems, this is typically even more important in renewable energy-based systems, in which fluctuations to a large extent are less controllable (Østergaard, 2008). Main critical elements related to the growth of *non-dispatchable* renewable energy sources are the following:

- *Increase of the total number of plants*, caused by the reduced size of the plants based on renewable sources (PV, wind and mini-hydro) and by the increase of total power installed to obtain the same energy amount (if a thermoelectric plant can operate at rated powers up to 8000–8500 hours per year, a wind power plant operates at the rated power not more than 1500 hours per year);
- *Excess electricity production* risk, caused both by high level of installed renewable power plants capacity and by the intrinsic poor flexibility of thermal power plants (baseload, startup/shut off times, size of plants ...);
- *Supply/demand mismatch*, to be solved by energy storage solutions (in this case Italy can take advantage of a good amount of pumped hydro power plants), by exporting surplus, or also by the implementation of innovative utilization systems;
- An appropriate *system security* level is needed, that is hour by hour a certain percentage of power must be supplied by those plants that can provide ancillary services (i.e. load following, frequency regulating, reactive power support ...);
- Constraint of electricity flows through power grid lines among different areas of the electrical system.

Possible solutions adopted or proposed by the countries that already have to deal with problems mentioned earlier, focus on the following key points:

System design improvements:

- improvement of transmission interconnections, so as to avoid constraints of maximum load flows between contiguous areas or countries (Excess Export Strategy);

- storage facilities:

  - *centralized* storage (pumped hydro plants, large compressed air energy storage CAES ...);
  - *supply-side* storage (small CAES, hydrogen storage systems ...);
  - *load-side* electrical or thermal storage systems (so as to make the demand more flexible);

- introduction of innovative mobility solutions, like vehicles with electrical or hydrogen engines, to add load on off peak hours (especially night-time).

System operation improvements:

- improved weather forecast, so as to facilitate the scheduling of dispatchable power plants, whose output can be varied according to demand;
- virtual power plants, that is clusters of physically distributed generation units (such as combined heat and power plants CHP, wind turbines, small hydro, PV panels, storage systems) which are collectively run by a central control entity;
- integration between non-dispatchable energy sources and cogenerative district heating supported by high efficiency electrical systems like heat pumps, so as to easily balance thermal and electric energy demand and supply by switching from different technologies.

3.3   An Approach to Expansion of RES and Efficiency Policy in an Integrated Energy System

The key functionality of a 'virtuous' energy system is to control the produced power by clustering different plants so that it is possible to balance fluctuating generation with controllable generators, and replacing conventional power plants with renewable power plants. By increasing storage capacities, the degrees of freedom for operation management can be increased. The problem of penetration of RES appears to be like the solution of an optimization problem with multiple objectives.

According to the schematization of Fig. 3.4, a complex energy system can be considered to be composed of a series of plants that operates according to a well-defined behaviour. Concerning the electricity production and the link between electricity and thermal sector, the following systems can be considered:

- Thermoelectric base plants (TE base), including all the plants with reduced regulation capacity that operates approximately 8000 h or more for years (high size thermoelectric power plants with coal and natural gas as fuels, nuclear plants, etc.);
- Thermoelectric modular plants (TE modular), including combined cycle power plants, turbogas and internal combustion engine power plants, operating less than 5000–6000 h per year;
- Cogeneration and Combined Heat and Power Plants (CHP plants);
- Hydro Power Plants (Hydro): they are plants with storage and regulation capacity;

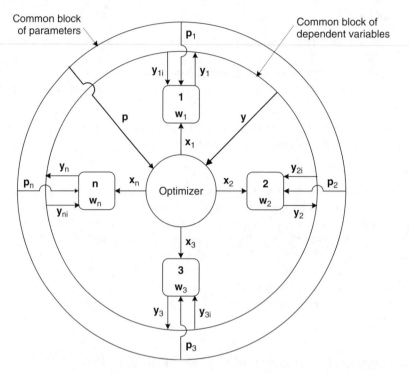

Figure 3.4   *The modular analysis concept for energy system optimization*

- Geothermal Plants (GEO);
- Wind Power Plants (Wind): plants that operates for a number of equivalent hours ranging from 1000–2000;
- PV Plants: plants that operates for a number of equivalent hours ranging from 800–1500;
- River Hydro (RH);
- Electricity imported from abroad (abroad);
- Pumping and storage Plants like Hydro and CAES (synthetically identified as PUMP).

Each of those groups can be considered operating according to a well-defined load factor that can be fixed and equal to one (plants that operates at maximum load during the whole year like TE base plants), programmable and with possibility of regulation (like TE modular) intermittent but programmable according to the period (like PV or RH), completely intermittent (like wind power plants) and completely programmable (like storage systems). The complexity of the system can be extended to consider other components of the energy systems, like vehicles, heating systems and so on, when possible connections between various energy sectors are possible.

These multi-criteria problems place great demands on mathematical formulation and methods. In the literature different approaches for solving those multi-criteria problems can be found and two basis approaches are distinguished (e.g. Lazzaretto *et al.*, 2010; Moura and De Almeida, 2010).

On one hand there are aggregation methods and on the other hand there are the so-called Pareto based solution techniques. In particular, selecting the first approach, the problem of integration of new renewable energies into the context of a complex energy system in which thermal and electric field can be structurally connected (see, for example, the case of CHP plants in which electric and thermal production are joined). So this appears to be an operational optimization problem, in which the objective is minimization of primary energy consumption for a given energy need with a series of constraints related to the operating mode of the various power plants. In this sense, optimization algorithms constitute a suitable tool for solving complex problems in the field of renewable energy systems. The development of an energy model for the energy system considered is necessary.

Some authors have reviewed different types of models such as renewable energy models, emission reduction models, energy planning models, energy supply–demand models, forecasting models, and control models using optimization methods, but many researchers are continuously proposing and applying new methods in the field of renewable energy. Many different computer tools can be used. Connolly *et al.* (2010a,b) reviewed about 68 computer tools available. Given a complex energy system a lot of questions can be analysed. Given the energy need, what is the best type of energy system to be used and what is the best system configuration (components and their interconnections)? What are the best technical characteristics of each component (dimensions, material, capacity, etc.)? Otherwise, when a number of plants are available to serve a certain region, what is the best operating point of the system at each instant of time, how should the operation and maintenance of each plant be scheduled in time? Many different objective functions can be considered: for example, maximization efficiency, minimization of fuel consumption, minimization of exergy destruction, maximization of the net power density or minimization of emitted pollutants. The decomposition of an energy system into subsystems of reduced complexity, to be optimized separately, but in a way compatible with the optimum of the global system, has been recognized as a viable solution to the problem of the design optimization of highly integrated, complex energy systems. (Lazzaretto *et al.*, 2010). In general terms it is a synthesis-design-operation optimization of complex systems under time-varying operating conditions. The solution of this problem involves a large number of variables and constraints. The optimization problem may become intractable even by the most capable software and hardware available without a simplification. The modular approach helps a lot in these cases (Frangopoulos *et al.*, 2002). The system is considered as composed of the various modules. A module may consist of group of homogeneous plants with similar characteristics (eg. Thermoelectric plants, wind power plants, PV plants).

3.3.1 Optimization Problems

The problem of introduction of RES in a bounded system can be considered as an optimum design problem articulated at three different levels (synthesis, design and operation) subject to constraints.

Given a well-defined load data (constraint), the optimal system is the one that minimizes the fossil fuel consumption with constraints on the upper level of RES penetration. The objective function of the complete optimization problem (i.e. synthesis, design, and operation) can be written in the general form:

$$\min_{X,Y,W} [f(X, Y, W)] \tag{3.1}$$

subject to equality and inequality constraints;

$$h_i(X) = 0 \qquad i = 1, 2, ..., I \tag{3.2}$$

$$g_j(X) \leq 0 \qquad j = 1, 2, ..., J \tag{3.3}$$

Where:

X is the set of independent variables for operation optimization (for example typical load factors of the various plants and in particular of RES plants),

Y is the set of independent variables for design optimization (for example nominal capacities of plants, etc.),

W is the set of independent variables for synthesis optimization; there is only one variable of this type for each plant, indicating whether the plant exists in the optimal configuration or not; it may be a binary (0 or 1), an integer or a continuous variable such as the rated power of a component, with a zero value indicating the non-existence of a component in the final configuration.

$h_i(X)$ equality constraint functions, which constitute the simulation model of the system and are derived by an analysis of the same system (for example the power required on the grid),

$g_j(X)$ inequality constraint functions corresponding to design and operation limits, state regulations, safety requirements, and so on.

Several objectives pertinent to energy systems can be written in the form of Eq. (3.1). For example, f can be the primary energy consumption from fossil fuel, the exergy destruction, the energy losses in the whole energy system analysed and so on. The system described can be the electric system or otherwise the complex energy system in which thermal energy system, electric grid and transport systems are taken into account. In the first case we will consider as constraint the fuel consumption for satisfying the electric energy required on the grid or otherwise the total primary fuel consumption. A constraint on the Exportable Excess Electricity Production (EEEP) can be considered too.

Considering a complex energy system such as a national energy system, main inputs are energy demands, renewable energy sources, traditional or CHP plants capacities, costs (only needed in economic analysis) and different regulation strategies emphasizing import/export and excess electricity production. Outputs consist of hourly energy balances and resulting annual values, fuel consumption, carbon dioxide emission and total costs.

The possibility to work on hour by hour steps is of particular interest, because of being able to valuate effects of the fluctuations of intermittent renewable energies as well as the seasonal periodicity in electricity and heat demand or in natural supply to hydro power plants with storage facilities. Table 3.2 provides a possible version of the variables.

According to the representation of Fig. 3.4, after the user defines modules and parameters, the *optimizer* compares p and y (i.e. demand, production, storage level), defines x_n for different time steps, on the basis of defined control strategies (that is, the priority of use of various modules), minimizing the load factors of plants that use fossil fuels: this

Table 3.2   *Synthetic description of the symbols in Fig. 3.4 for the case study developed*

Symbol	Mathematical signification	Real problem signification
$1,2,\ldots,n$	1st, 2nd, ..., *n-th* module	Technologies take in account (type, capacity, efficiency, primary source availability...)
p_i	Parameters	Hourly energy demand, non-dispatchable generation
x_r	Set of *independent variable* of *n*-th module	Hourly capacity factors
y_{ri}	Set of *input dependent variables* (output from other modules) in *n*-th module	Hourly energy inputs
$y_r = Y(x_r, y_{ri})$	Set of *output dependent variables* from *n*-th module	Hourly energy outputs, storage state of charge
$w_r = W(x_r, y_{ri})$	Set of *dependent variables inside* *n*-th module	Fuel consumptions, emissions

is reflected on a minimization of fuel consumption and emissions, which are the sum of various w_n (calculated by appropriate W functions). To create and compare alternative scenarios, the user can act varying parameters and modules.

3.3.2   Operational Limits and Constraints

The long term generation expansion planning problem is subject to different constraints. The basic constraint is represented by the demand. The electricity grid power balance represents the basic constraint of the system. The total power output of generators must always be equal to the sum of the power demands and the grid losses

$$\sum_{i=1}^{m} P_i = P_D + P_L \tag{3.4}$$

where P_D is the power demand for dispatcher hour; P_L is the transmission losses and m the number of dispatched generators divided in homogeneous groups. Therefore, other constraints are imposed by safety and operability requirements as:

- hourly electricity demand;
- fixed value of electricity imported from abroad;
- condensing (baseload and modulating) and cogenerative (CHP and district heating) thermoelectric generation;
- renewable energy production: geothermal, wind, PV, RH, hydro with pump and storage systems;
- residential air conditioning and heating energy demand (including thermal solar);
- industrial energy demand and cogeneration production;
- transport energy demand.

A quite low constraint on the EEEP is considered too so that it is not possible to consider the export excess approach as a valid general strategy for increasing the share of renewables.

3.3.3   Software Tools for Analysis

According to these kind of analysis, many ad hoc computer tools can be used for the optimum operation of a complex energy system: among them, various software can be used to support the analysis of different scenarios (Lund, 2009; Connolly *et al.*, 2010a,b). A very helpful tool that can be used to support such a kind of analysis is EnergyPLAN. It considers electricity, heat and transport; it is primarily used for simulating national energy-systems; it considers a large range of different technologies so that the logic of the system can be precisely followed the idea of the modular approach described in Fig. 3.4.

EnergyPLAN operates as follows. Main inputs are energy demands, renewable energy sources, plants capacities, costs (only needed in economic analysis) and different regulation strategies emphasizing import/export and excess electricity production. Outputs consist of hourly energy balances and resulting annual values, fuel consumption, carbon dioxide emission and total costs. According to the representation of Fig. 3.3, after the user defines modules and parameters, the *optimizer* compares p and y (i.e. demand, production, storage level), defines x_n for different time steps, on the basis of defined control strategies (that is, the priority of use of various modules), minimizing the load factors of plants that use fossil fuels: this is reflected by a minimization of fuel consumption and emissions, which are the sum of various w_n (calculated by appropriate W functions). To create and compare alternative scenarios, the user can act by varying parameters and modules according to the schematic representation of Fig. 3.4 with the symbols explained in Table 3.2.

3.4   Analysis of Possible Interesting Scenarios for Increasing Penetration of RES

As discussed in the previous sections, to understand the upper limit to the penetration of RES, several measures can be imposed to the control of the energy system of technical and economic nature as for example, the increase of generation flexibility, the increase of CHP systems (that introduce connection between thermal energy and electric energy production), introduction of electric vehicles (that introduce connection between electricity production and mobility) and the introduction of peak-off tariff charge, aimed to reduce the distance between maximum and minimum level of daily production curve.

Each single measure appears to be not efficient enough to produce beneficial effects mainly for complex energy systems. For this reason, to understand the perspective of upper increase of RES, analysis can be performed based on the development of specific scenarios, regarding high-scale integration of wind and PV power analysing alternative energy paths for the energy-system under analysis, in which a combination of the various measures is proposed.

The focus will be on effects on the electrical system and on the whole energy-system as well, and on the interactions between fluctuating renewable energy sources and other innovative energy production and utilization technologies. Five different scenarios of renewable energies expansion are considered. For reference data on the electricity produced and the installed power plants, see Table 3.3.

Table 3.3   *Scenarios descriptions and their influence on reference areas*

Scenario description	Main sector of interest
Renewable energies expansion in a reference scenario	Electrical (system 'as it is')
Increasing thermoelectric generation flexibility	Electrical
Introducing peak-off peak tariff of charge	Electrical
Introducing electric traction in transport sector	Transport
Increasing industrial CHP electricity production	Thermal
Increasing industrial CHP electricity production + Increasing thermoelectric generation flexibility + Introducing electric traction in transport sector	Electrical, Thermal, Transport
Introducing the concept of Virtual Power Plants	Electrical and Thermal

3.4.1   Renewable Energy Expansion in a Reference Scenario

The first possibility is to analyse the effect of an increase in wind and PV installed power capacity for immediate integration onto the energy-system without any major redevelopment. In this case, all the remaining power systems are kept under current conditions, for different values of *minimum grid stabilizing production* (i.e. the share of power, hour by hour, produced by plants capable of providing ancillary services such as frequency regulation services, reserve, reactive power, voltage regulation and network restart).

3.4.2   Increasing Thermoelectric Generation Flexibility

Another possibility for increasing the penetration of intermittent renewable energy is to create scenarios that progressively reducing thermoelectric 'baseload power': the existing thermoelectric power capacity is in fact modelled splitting the capacity itself in a *base capacity* (a kind of technical minimum) not reducible by the software, and a *variable capacity* directly managed by the software during calculation. A high value of PV capacity, equivalent to about 20% of the average power and a minimum grid stabilizing production of 75% are set to perform the analysis. The analysis can be developed to understand the trends of percentage increase of the production from intermittent renewable sources ΔE_{IR} and of the reduction in fuel consumption ΔE_{TP}, for a value of i GW of wind power, due to the reduction of the $j\%$ of base power. These two indicators can be found from:

$$(\Delta E_{IR})_{ij} = \frac{(E_{IR})_{ij} - (E_{IR})_{i0}}{(E_{IR})_{i0}} \cdot 100 \qquad (3.5)$$

$$(\Delta E_{TP})_{ij} = (E_{TP})_{ij} - (E_{TP})_{i0} \qquad (3.6)$$

For high values of wind power, the substitution of a certain share of baseload power plant with flexible power positively influences the use of fluctuant sources, significantly increasing the produced energy by the same installed capacity. This causes a reduction in primary energy consumption due not only to the gain in renewable production, but also to the decrease of excess electricity production and of the need for pumping.

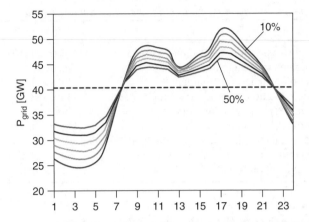

Figure 3.5   *Parametric variation of hourly demand with reference to Italian load*

3.4.3   Effects of Introducing the Peak/Off-Peak Charge Tariff

Another way to increase penetration of renewable energies is the introduction of peak/off-peak shifting policy. This measure is a typical one for influencing the energy demand. Several electricity suppliers offer to customers low prices to encourage consumption during off-peak hours, thus trying to reduce the daily peaks in demand profiles. In this scenario, the analysis are carried out for increasing level in *peak flattening*, reducing the difference between the daily maximum and minimum load from 10% to 50%. The effect of the various parametric reduction of the electricity demand on the grid is shown in Fig. 3.5 with reference to a particular day using as reference scenario the Italian grid.

Starting from a difference of about 25 GW between the peak and the minimum level, the difference can be reduced up to 9 GW in the case of maximum reduction. The effect of such a kind of policy appears to be quite effective in case of presence of Thermoelectric Baseload power, while for a system with high RES penetration it appears to be less advantageous.

3.4.4   Introducing Electric Traction in the Transport Sector: Connection between Electricity and Transport Systems

An interesting perspective that requires special investigation into promoting increase of RES based power plants is to partially shift the fuel used from oil to electricity in the transport sector, so it introduces a remarkable amount of electric mobility. The introduction of electric vehicles is considered in the event of a constant *recharge from the electric grid* at night (from 10 pm to 7 am), and no recharge during the rest of the day. The various reductions can be defined as:

$$(\Delta E_{elv,n})_j = s_j \cdot \left(\frac{1}{C_{ICE}} - \frac{1}{C_{elv} \cdot \eta_E} \right) \tag{3.7}$$

$$(\Delta_{elv})_{ij} = (\Delta E_{elv,tot})_{ij} - (\Delta E_{elv,n})_j \tag{3.8}$$

$$(\Delta E_{elv,tot})_{ij} = (E_{TP} + E_{tr})_{i0} - (E_{TP} + E_{tr})_{ij} \tag{3.9}$$

where s_j represents the kilometres of traffic by internal combustion engine vehicles substituted by electric vehicles, and $\Delta E_{elv,tot}$ the overall energy saving in thermal plant E_{TP} and transport E_{tr} consumption calculated using the defined energy-system model for a wind power capacity of i GW and a j% percentage of electric vehicles. Note how the increased penetration of electric vehicles in the transport sector results in significant gains (up to 15% in case of high values of installed wind capacity) in the annual production: adding a night load reduces the need for modulating to avoid excess production during the daytime. The reduction in consumption benefits from the combined action of the two technologies, too: increasing the share of electric vehicles increases not only the renewable energy production, but also the possibility of a direct use of the electricity without the need for pumping storage.

3.4.5 Increasing Industrial CHP Electricity Production

According to recent directives developed by the European Commission (EC) in 2004, a growing inclusion of the CHP units in the regulation (CHP – often referred to as 'cogeneration' in the EU) aims to reduce energy demand, and is considered as one way to achieve security of energy supply in Europe, contributing towards the EU's carbon-saving targets, playing a key part in reducing demand. A possible strategy for further increase of the new renewables and CHP is to run CHP units to meet both electricity demand variation and the fluctuations in the intermittent renewables (Lund and Clark, 2002). The following scenarios are created by setting 1000 MW installed with a PV capacity and a minimum grid stabilizing production of 70%: since baseload power plants production is not changed (in a first analysis), higher values of PV capacity would leave little scope for reducing variable thermal power plant generation, while higher values of instantaneous stabilizing power production would make the system too constrained. Industrial cogeneration electricity production is progressively increased from present value to twice present value. It is useful to highlight how industrial cogeneration is modelled in this work: it has *priority dispatching* (produced power is always fed into the grid without any possibility of reducing modulation), it is associated with industries with *continuous production* cycles (constant production during working days, no production in holidays), and it has *no ability to stabilize* the electrical system. It appears clear how negative interferences can emerge from parallel growth in the utilization of intermittent renewable energy sources and industrial cogeneration, mainly due to their common non-schedulable nature and inability to stabilize the electric grid. The following parameters are defined for the quantification of these effects: the percentage *decrease* of electricity from intermittent renewable source ΔE_{IR} for the same amount of i GW of installed capacity, following the j% increase in industrial cogeneration, and the difference Δ_{CHP} between the maximum saving of primary energy used by conventional thermal plants, which would occur if all the E_{CHP} cogenerated electricity were used optimally (i.e. without increasing pumping and export), and the actually achieved saving. The two quantities can be found from:

$$(\Delta E_{IR})_{ij} = \frac{(E_{IR})_{i0} - (E_{IR})_{ij}}{(E_{IR})_{i0}} \cdot 100 \tag{3.10}$$

$$(\Delta_{CHP})_{ij} = \frac{(E_{CHP})_j - (E_{CHP})_0}{\eta_E} - ((E_{TP})_{ij} - (E_{TP})_{i0}) \tag{3.11}$$

3.4.6   Developing the Concept of 'Virtual Power Plants'

The development of the concept of 'virtual power plants' is another step towards increasing RESs. It means that a complex system is articulated in a series of sub-systems that are clusters of physically distributed generation units (such as combined heat and power plants CHP, wind turbines, small hydro, PV panels, storage systems) which are collectively run by a central control entity. So, different to the situation considered in Section 3.3, the single clusters represented in Fig. 3.4 are considered as a group of heterogeneous plants. The concept of Virtual Power Plants has been already been developed in the 'Regenerative Combined Power Plant' project: a pioneering project, carried out by the companies Enercon GmbH, SolarWorld AG and Schmack Biogas AG, together with the Institute for Solar Energy Supply Technology, University of Kassel. The combined power plant consists of three wind parks (12.6 MW), 20 solar power plants (5.5 MW), four biogas systems (4.0 MW) and the pump storage Goldisthal (Output: 1.060 MW; Storage: 80 h, i.e. 8480 MWh). The development of the concept of Virtual Power Plants is considered important in order to increase the level of penetration of intermittent RES, even if up unti today the idea has been developed mainly in the field of economic incentives without a real technical analysis of the problem (the promotion policy known as *Renewable Portfolio Standard* (RPS) in the US can be considered in this field).

3.5   Analysis of a Meaningful Case Study: The Italian Scenario

To explain the meaning of the various strategies defined in the previous section, the application of a particular meaningful case is shown: the Italian electric system. Italy generates electricity with high shares of base load power stations like combined plants and coal power stations (conventional thermal). Italy has a population of approximately 60 million people. The level of primary energy consumption is estimated to be 170 Mtoe (Millions of Tonnes of Oil Equivalent) approximately, distributed by three main energy sectors (electricity, thermal and transport).

In 2010, the electricity demand was approximately 330 TWh with a peak load of 55 000 MW and an average peak load of 49 000 MW. A single operator (Terna) operates today on the electricity distribution. A good presence of large hydroelectric power plants (a lot of them storage plants) can be evidenced (about 300 plants) contributing to a good 'elasticity' to the electric system. The transport sector is almost completely powered by oil, including 295 TWh of diesel, 145 TWh of petrol, and 50 TWh of jet fuel and 20 TWh of gaseous fuels. The energy demand in the residential sector is approximately of 207 TWh mostly dominated by single boilers fuelled by natural gas.

The Italian energy system, like a lot of other systems in the world, is segregated; that is, the electricity, heat and transport sectors are very independent of one another but a meaningful increase of CHP plants permits a first significant interaction between the electricity and thermal sectors (estimated in approximately 20 TWh).

Production with renewable energy (geothermal, wind, PV and mini-hydro) plays a marginal role even though the growth of wind power and PV power installed was remarkable (mainly in the last five years). Moreover, pumped hydro power stations employ a long-established method for storing mechanical potential energy by using surplus power

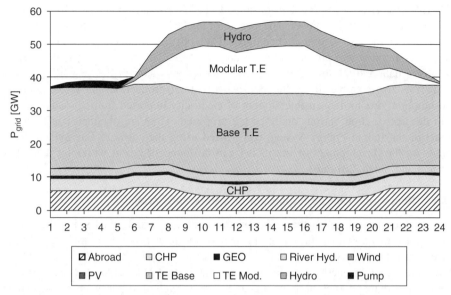

Figure 3.6 *Hourly load of gross domestic electricity consumption in Italy on a typical peak load day*

for pumping water for generating dispatchable peak or control power. But, for the future, there are technical constraints: the development of wind farms was mainly in the south of Italy so quite distant from the main hydro power plants. This makes the development of a wind-water model for the control of intermittent power difficult. Fig. 3.6 provides the hourly load of gross domestic electricity consumption in Italy on a typical peak load day of June (Terna, 2011b).

Observing the data reported in Fig. 3.6 and the data reported in the electricity sector balance of Table 3.4 it is possible to understand how the load demand on the Italian grid is satisfied by various thermoelectric plants (about 70%), hydroelectric power plants

Table 3.4 *Balance of electricity sector of Italian country in the year 2010 (the value inside parenthesis represent estimated power)*

	Production (GWh)	Installed power (MW)	Available peak power (MW)
Hydro	54 407	21 520	13 300
Thermoelectric	231 248	74 976	53 100
Geothermal electricity	5376	728	600
Wind + PV	11 032	9264	2300
Gross production	302 062		
Auxiliary services consumption	11 314		
Net production	290 748		
Import	44 160		5500*
Pumping Consumption	4464		
Total energy requirements	330 455	106 488	69 300 (74 800)

(12–13%), electricity imported from abroad (11–12%) and about 5% (16.5 TWh) by geothermal, wind and PV power. The growth of the last two renewable energies has been remarkable in the last 10 years (when only 5 TWh of geothermal energy production from the Larderello area was present in the production data). This occurred by both analysing the global data and the load data referred to the various days of the year. It is possible to analyse, for example, the difference between total production and gross electricity consumption, due to the sensible amount of electricity imported from abroad.

In the next section various analyses are proposed. They are obtained constructing a model of the Italian Energy System constructed using the software EnergyPlan. Major details on the development of the model can be found in recent works by this author (Franco and Salza, 2011a,b).

3.5.1   Renewable Energy Expansion in a Reference Scenario

The first step is to analyse the effect of an increase in wind and PV installed power capacity for the case that the remaining part of the power system is maintained under current conditions, for different values of *minimum grid stabilizing production*. Figure 3.7 shows the trends in energy production from intermittent renewable sources, in relation to the total demand (net of pumping and export) and of primary energy consumed by thermal power plants. Ideal curves simply describe a proportional relation between installed capacity and produced energy (thermal plants consumptions consider constraints in minimal base production).

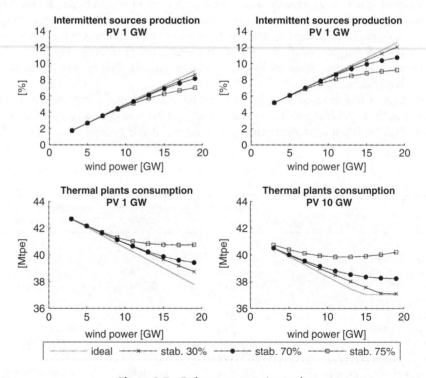

Figure 3.7   *Reference scenario results*

This was obtained for different values of the minimum grid stabilizing production, that is, the minimum percentage or power coming from plants with possibility of stabilization of the grid. The remaining capacity is represented by RH, wind, PV and CHP plants.

According to the present Italian situation a value of 70–75 can be considered realistic in the case of a meaningful increase n intermittent RES. The case of minimum grid stabilizing production of 30% is an idealized one typical of countries in which a great penetration of RES is observed (i.e. Denmark, Portugal and Ireland have already maintained stable grids with 40–60% of power from stabilizing production). As can be seen, production curves tend to significantly diverge from ideal curves for high values of installed wind power capacity and grid stability, because of reductions to avoid *excess production*. Similarly, real reduction in thermal power plants consumption is at first smaller than ideal reduction, then consumption tends to increase: in a very constrained grid, in fact, an increase in production from fossil fuels is necessary to allow an increase in generation from renewable sources, with the consequential increase in both the energy required for *pumping* and in *exportable excess production*. The analysis of the data shows the introduction of a combination of at least 10 GW of Wind Power Plants with 10 GW of PV solar is sufficient to produce some problematic effect on the control system and it is a level of penetration sufficient for non-appreciating the beneficial effect of intermittent RES penetration in term of primary energy saving. In all the cases considered the Electric Energy production increase in defined as:

$$\Delta Elec_{IR} = \Delta Elec_{PV} + \Delta Elec_{WIND} - \Delta Elec_{RH} \tag{3.12}$$

3.5.2 Increasing Thermoelectric Generation Flexibility

In the second case, scenarios are created progressively reducing the load capacity of thermoelectric 'base' power plants: the thermoelectric power capacity is modelled splitting it in a *base capacity* (a kind of technical minimum) not reducible by the software, and a *variable capacity*. This means, considering Fig. 3.6, an increase of the amount of modular thermoelectricity with respect to base thermoelectricity. A high value of PV capacity, equivalent to 10 000 MW, and a minimum grid stabilizing production of 75% are set to perform the analysis. The results of Table 3.5 show the reduction in fuel consumption ΔE_{TP} of conventional thermoelectric power plants, for a value of i GW of wind power, due to the reduction of the baseload power, with respect to the reference case, identified with i0 according to the definition given with Eqs. (3.5) and (3.6).

For high values of wind power (above 10 000 MW), the substitution of a certain share of base power with flexible power positively influences the use of fluctuant sources, significantly increasing the produced energy by the same installed capacity. This causes a Primary Energy Saving (in the range 0–2 Mtoe) due not only to the gain in renewable production, but also to the decrease of excess electricity production and of the need for pumping.

3.5.3 Effects of Introducing a Peak/Off-Peak Charge Tariff

The perspectives of introducing a peak/off-peak shifting policy are analysed from the standpoint of RES. More and more electricity suppliers in the last 10 years offered customers low prices to encourage consumption during off-peak hours, thus trying to

Table 3.5 *Effects of a higher flexibility of thermal power plants: Thermal power plants consumption reduction* ΔE_{TP} *(Mtoe)*

	P_w (GW)								
BP_r (%)	3	5	7	9	11	13	15	17	19
0	0.00	0.00	0.00	0.00	0.00	0.00	0.00	0.00	0.00
5	0.25	0.25	0.26	0.27	0.29	0.33	0.36	0.42	0.50
10	0.47	0.47	0.48	0.50	0.54	0.61	0.68	0.76	0.90
15	0.66	0.66	0.66	0.69	0.75	0.84	0.93	1.04	1.21
20	0.83	0.83	0.84	0.86	0.92	1.03	1.14	1.28	1.47
25	0.95	0.95	0.96	0.98	1.05	1.16	1.28	1.44	1.65
30	0.98	0.99	0.99	1.03	1.11	1.23	1.36	1.54	1.76

reduce daily peaks in demand profiles. In this scenario, the analysis are carried out for increasing level in *peak flattening*, reducing the difference between the daily maximum and minimum load from 10–50%. The effect of this parametric reduction of the electricity demand on the grid is shown in Fig. 3.5 with reference to a particular winter day. Two different configurations are compared: one with a 1 000 MW and one with 10 000 MW total PV capacity with a growing level of wind power. In both cases the minimum grid stabilizing production is set to 75%. Analysis carried out shows a modest increase in the production from intermittent sources for the same installed power, especially (of course) in the case of high value of PV capacity installed, which operates just during the central part of the day (in this case, for low wind power values there is even a slight lowering effect on the production). The analysis of the data shows that the effect of this particular energy policy action scarcely aids the penetration of intermittent RES. An important penetration of RES (10 GW and more that 15% of wind power) does not guarantee a primary energy saving higher than 1 MToe. This analysis confirms the reduced effect of this kind of policy in order to increase the share of intermittent RES.

3.5.4   Introduction of a Connection between Electricity and Transport Systems: The Increase in Electric Cars

The introduction of electric vehicles is considered in the event of a constant *recharge from the electric grid* in particular during the night-time (from 10 pm to 7 am), and no recharge during the rest of the day. In this way the difference between the maximum value of the load and the minimum one (like explained in Section 3.4.3 and Fig. 3.5) can be reduced. As previously, 10 GW of PV installed capacity and 75% grid stabilizing production are set. The graphs in Fig. 3.8 represent, besides the increase of the production from fluctuant renewable sources defined similarly as earlier, *the further* reduction in consumption of primary energy ΔE_{elv} *in addition* to the natural reduction $\Delta E_{elv,n}$ due to the greater overall efficiency of the electric traction system compared to internal combustion engines according to the definition given with Eqs. (3.7–3.9).

The penetration of electric vehicles is estimated on the base of the actual mobility system configuration. In Italy the transport sector can be estimated in terms of total km (about $690 \cdot 10^9$); so a 20% of electric vehicles penetration means that about $138 * 10^9$ km are covered with electric vehicles and this corresponds to an increase of electricity

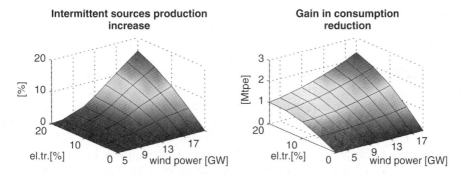

Figure 3.8   *Effects of the introduction of electric vehicles*

consumption of about 27.6 TWh (1 kWh/5 km), with a corresponding oil primary energy consumption reduction of 1 kWh/1.5 km of fuel. Considering the results depicted in Fig. 3.8, we can observe how the increased penetration of electric traction in the transport sector results in significant gains (up to 15% in case of high values of installed wind capacity) in the annual production: adding a night load reduces the need for modulating to avoid excess production. The reduction in consumptions benefits from the combined action of the two technologies, too: increasing the percentage contribution of electric vehicles (up to the 20%, i.e. corresponding to about 100 TWh of Primary Energy), increases not only the renewable energy production, but also the possibility of a direct use of electricity, without the need for pumping storage.

3.5.5   Increasing Industrial CHP Electricity Production

In this section the author refers to industrial CHP as 'non-programmable' like the intermittent renewables. So the increase of CHP creates additional problems due to the not-stabilizing load introduced by CHP.

The following scenarios are created by setting a 1000 MW installed PV capacity and a minimum grid stabilizing production of 70%: since base power plant production is not changed (in a first analysis), higher values of PV capacity would leave little scope for reducing variable thermal power plant generation, while higher values of instantaneous stabilizing power production would make the system too constrained. Industrial cogenerative electricity production is progressively increased from present value to twice present value.

It's useful to highlight how industrial cogeneration is modelled in this work: it has *priority dispatching* (produced power is always fed into the grid without the possibility of reducing modulation); it is associated with industries with *continuous production* cycles (constant production during working days, no production in holidays); it has *no ability to stabilize* electrical system. It is clear how negative interferences can emerge from the parallel growth in the utilization of fluctuant renewable energy sources and industrial cogeneration, mainly due to their common non-schedulable nature and inability to stabilize the electric grid.

The following parameters are defined for the quantification of these effects as discussed in Section 3.4.5: the percentage *decrease* of electricity from intermittent renewable source

ΔE_{IR} for the same amount of *i GW* of installed capacity, following the *j%* increase in industrial cogeneration and the difference Δ_{CHP} between the maximum saving of primary energy used by conventional thermal plants, which would occur if all the $Elec_{CHP}$ cogenerated electricity were used optimally (i.e. without increasing pumping and export) and the actually achieved saving. It is therefore clear that, for a rational use of the proposed technologies, the modelled energy system does not turn out to be adequate. An analysis is now proposed, that compares the development of industrial cogeneration in two scenarios: a *base scenario*, that is, the one just proposed and an *alternative scenario*, in which a 30% decrease of thermoelectric base power and a 20% share of electric traction transport are introduced. The benefits arising from the adoption of the alternative scenario compared to the reference case are clear (see Fig. 3.9, where the loss of energy saving obtained combining renewable energy sources and industrial cogeration increase is shown): production from unprogrammable renewable sources suffers far less from the additional, not-dispatchable and not-stabilizing load introduced by CHP.

In addition to the gain of electricity from wind power, the decrease in need for pumping and excess energy production can even halve the loss in primary energy savings compared with an optimized utilization of cogeneration.

The increase of CHP plants has an effect not only on the electric power system, but also on the possibility of an energy saving in the heat sector in industry. At present it is estimated that the heat energy demand is about 305 TWh and approximately 20 TWh are covered by means of CHP plants. To be defined is the link between thermal energy demand E_{ind}, and energy production with conventional plants E_t and with CHP E_{CHP}.

Reasoning in terms of total primary energy consumption in industry E_{ind}, transport E_{tr} and power generation E_{TP} sectors, where E_{ind} can be expressed in function of E_{CHP}:

$$E_{ind} = \frac{E_t}{\eta_t} + E_{CHP} \cdot \left(\frac{\eta_t - \eta_{t,CHP}}{\eta_t \eta_{E,CHP}} \right) \tag{3.13}$$

E_{tnd} being the industrial thermal energy demand, η_t the conventional thermal energy production efficiency, $\eta_{E,CHP}$ and $\eta_{t,CHP}$ the CHP electrical and thermal efficiencies. Table 3.6 provides the possible reduction of total primary energy consumption in the

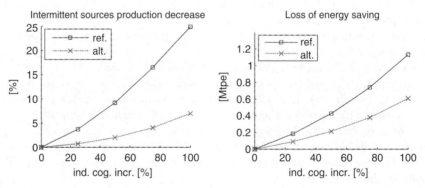

Figure 3.9 *Development of industrial cogeneration in reference scenario and alternative scenario (wind power installed at 17 GW)*

Table 3.6 *Development of industrial cogeneration in reference and alternative scenarios: results of total primary energy consumption*

CHP_{inc} (%)	P_W (GW)								
	3	5	7	9	11	13	15	17	19
0	113.93	113.40	112.88	112.38	111.90	111.48	111.12	110.86	110.68
25	113.70	113.18	112.68	112.20	111.76	111.40	111.13	110.94	110.84
50	113.50	112.99	112.51	112.07	111.71	111.44	111.26	111.16	111.12
75	113.33	112.84	112.40	112.05	111.80	111.63	111.54	111.53	111.56
100	113.20	112.76	112.43	112.20	112.07	112.01	112.02	112.08	112.15

*Total energy consumption – Reference scenario (Mtoe)

CHP_{inc} (%)	P_W (GW)								
	3	5	7	9	11	13	15	17	19
0	110.42	109.81	109.20	108.61	108.03	107.49	107.02	106.61	106.25
25	110.08	109.47	108.87	108.29	107.74	107.26	106.84	106.47	106.16
50	109.75	109.14	108.56	108.01	107.52	107.10	106.73	106.41	106.14
75	109.42	108.84	108.29	107.81	107.40	107.04	106.72	106.45	106.22
100	109.13	108.60	108.13	107.73	107.39	107.10	106.84	106.63	106.45

*Total energy consumption – Alternative scenario (Mtoe)

three sectors given by the sum:

$$E_{tot} = E_{TP} + E_{ind} + E_{tr} \tag{3.14}$$

It is interesting to observe not so much the absolute values (which may be affected by approximations of the model, and which are obviously lower in the alternative scenario) but the *shape* of the two curves of Fig. 3.10. Notice that, in the first case, for high penetration levels of wind power (in some cases over 10 GW), the curve shows a reversal trend. At first global total energy consumption decreases with increasing share

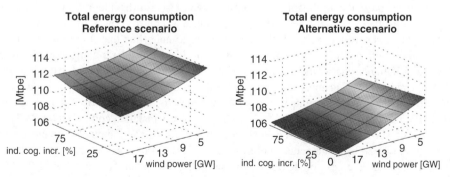

Figure 3.10 *Development of industrial cogeneration and large wind energy penetration: possible Primary Energy Saving (PES)*

of CHP production, then there is a second phase characterized by a point of minimum consumption, finally consumption monotonically increases with increasing penetration of industrial CHP.

This can be explained as follows. In a very constrained energy system such the one modelled in the reference scenario, for high values of installed wind capacity a thermal generation from CHP, which is more efficient but is neither stabilizing nor programmable, involves a significant increase in excess electricity production and in pumping requirement. These effects are not present, or at least much downsized, in the alternative scenario. From the same table it is possible to observe a negative impact on the total consumption due a too high increase of industrial cogeneration.

The combination of CHP plants increase (up to 50% with respect to the present value, that is, approximately 30 TWh of thermal energy production) and intermittent RES (Wind energy up to 15–17 GW) seems to be a promising strategy in order to obtain a meaningful primary energy saving. An analysis is finally proposed, that compares the development of industrial cogeneration in two different scenarios: a *base scenario*, that is, the one just proposed, and the *alternative scenario*, (30% decrease of thermoelectric base power and a 20% share of electric traction transport).

3.6   Analysis and Discussion

Evaluating recent data from Italy in this section, how a sustainable electrical energy supply based on a relevant penetration of wind power and pumped storage based on hydroelectric plants (it can represent a classical 'wind and water' model) might look from the standpoint of fuel saving is discussed. The operation of such an energy scheme, while being feasible in principle, shows with actual demand history the presence of an upper limit at about the 20% of the required power.

For higher penetration levels in a complex energy system with many sources of energy, many interdependencies between sources, demands and conversion systems must be developed. This makes interconnected systems an interesting option for integrating electricity produced from such energy sources. This means that increasing the geographical extension of the area in which fluctuating wind power is being exploited reduces the average need for reserve capacity in the form of power plants operating in condensing mode. In any case, the argument that an upper limit exists for the integration of renewable energies into the electricity network due to a missing control power from conventional power plants is surely not true. But it is important to underline that it is very difficult to pursue the primary energy consumption reduction. If we consider a comparison between the actual Italian scenario and an alternative scenario with an important contribution from RESs, a reduction in Primary Energy Consumption can be observed. In particular, with the introduction of an amount of renewable energies of 27 GW (17 of Wind power and 10 of PV meaning this to reach a level of 30% of Power installed composed by RES), an increase of 100% of industrial cogeneration (from 20–40 TWh annually produced), the substitution of 20% of the actual mobility park with a corresponding amount of electric vehicles and the increase of flexibility of thermal power plants, a reduction of 7 MToe of total energy and of 9 MToe of fossil fuel consumption could be pursued, maintaining the actual consumption levels (see Table 3.7).

Table 3.7   *Total energy, fossil fuel consumption and CO_2 emissions comparison*

Scenario	Total energy consumption	Fossil fuel consumption	CO_2 emissions
Italian present energy system	169.8 Mtoe	157.8 Mtoe	434 Mt
Wind 17 GW + PV 10 GW + double industrial cogeneration + 30% thermal plant flexibility + 20% electric vehicles	162.8 Mtoe	148.8 Mtoe	403 Mt
% reduction	4.12 %	5.7%	7.1%

3.7 Conclusions

The objectives of security energy supply, competitiveness and environmental protection for EU requires a further introduction of RES in connection with eco-efficiency policies. Such issues are much more complicated than usually depicted and cannot be closed confined to the socio-economic space. Many elements (mainly technical) concerning the operation of the various Energy Systems need to be considered and any option can be more or less valid according to the context in order to obtain an effective PES.

In this study the dynamics of electricity production from intermittent renewable sources like wind and solar PV for electricity production are discussed with the perspective of increasing their share. From this perspective the aim of this chapter has been to show potential of an integrated analysis in studying and planning large scale energy systems, especially focusing on different scenarios for promoting high penetration of non-dispatchable RES.

Possible upper limits are considered operating:

- on *electricity* sector only: improvements on thermal power plants operational flexibility, flattening of daily power demand profile;
- on *transport* sector only;
- on *thermal* sector only;
- on *all the three* sectors at the same time: improvements on thermal power plants operational flexibility, introduction of electric vehicles, increase of industrial cogeneration.

From the analysis, how further increase of intermittent RESs cannot be carried out considering just the peculiarities of electric energy system clearly appears. The joined consideration of CHP, electric vehicles and higher flexibility of thermoelectric power plants provide a possibility for ensuring the balance between electricity production and demand and for designing of optimal energy source mixing in a strongly constrained scenario. The analysis proposed outlines that no unequivocal answer can be found to the question of how to design an optimal energy system in the perspective of a strong penetration of fluctuating renewable energy systems. However, a non-sectorial vision of the problem of renewable energy penetrations increase within energy-systems could make it possible an optimal expansion of them. The optimization could be implemented according to techno-operational objectives and not only basing on economic perspectives.

The whole Italian energy-system has been considered as a realistic case study and modelled with the support of the EnergyPLAN computer tool after an accurate analysis of its complex structure, and after reprocessing available data from various sources in a suitable way for being used as input into the software itself.

At first, constraining effects which appear in case of a high amount of installed intermittent power capacity are analysed in a reference scenario, that is energy system (both demand-side and supply-side) maintained at its current configuration. The numerical results of the actual system modelling are not the primary outcomes of the article. However, it is shown how, if a penetration of intermittent RES can determine problems just at a level of about 10–15 GW, a penetration of Wind power and PV power up to the level of about 30 GW (more than 50% of the available peak power) can produce a reductio of primary fossil fuel of energy consumption if combined with increase of CHP (100% of the present level), and promotion of electric vehicles (up to 20%). Maintaining the present energy end use, a reduction of fossil fuel consumption estimated at 5.7% could be possible. So it is clear from the present study that the way to transform the energy sector from a fossil fuel based one to renewable energy based is not simple.

Nomenclature and Abbreviations

BP_r	Base power reduction factor
C_{ICE}	Internal combustion engine vehicles consumption as km/kWh
C_{elv}	Electric vehicles consumption as km/kWh
CHP_{inc}	Industrial cogeneration increase factor
E_{CHP}	Electricity from CHP plants
E_{ind}	Primary energy consumption in industrial sector
E_t	Industrial thermal energy demand
E_{tr}	Primary energy consumption for internal combustion engines vehicles
E_{TP}	Primary energy consumption in conventional thermal plants
$Elec_{CHP}$	Electricity from CHP plants
$Elec_{RH}$	Electric energy produced by River Hydro Power Plants
$Elec_{IR}$	Electric energy produced by intermittent renewable
$Elec_{PV}$	Electric energy produced by PhotoVoltaic plants
$Elec_{WIND}$	Electric energy produced by Wind power plants
f	Objective function
g	Inequality constraints
h	Equality constraints
p	Parameters
P_D	Power demand (MW)
P_L	Transmission losses in the grid (MW)
P_w	Installed wind power capacity (MW)
P_{grid}	Power required on the grid
s	Kilometres of traffic by internal combustion engine vehicles substituted by electric vehicles
x, y, w	Variables
X, Y, W	Vector of variables
ΔE_{IR}	percentage *production increase* of electricity production from intermittent renewable source

ΔE_{TP}	reduction in fuel consumption for a value of i GW of wind power, due to the reduction of the $j\%$ of Thermoelectric Power
Δ_{elv}	Additional energy saving due to the introduction of electric vehicles
$\Delta E_{elv,tot}$	Overall energy saving due to the introduction of electric vehicles
Δ_{CHP}	Loss of energy saving in cogeneration scenario
ΔE_{elv}	Energy saving due to greater efficiency of electric vehicles compared to internal combustion engines
η_E	Mean efficiency of the plants connected to the electric grid
η_t	Conventional thermal energy production efficiency
$\eta_{E,CHP}$	CHP electrical efficiency
$\eta_{t,CHP}$	CHP thermal efficiency
AEEG	(Italian) authority for electric energy and gas;
CAES	Compressed Air Energy Storage
CHP	Combined Heat and Power (industrial cogeneration)
EEEP	Exportable Excess Electricity Production
EU	European Union
elv	Electric vehicles
IR	Intermittent Renewables
PES	Primary Energy Saving
PV	photovoltaic
RES	Renewable Energy Sources and Renewable Energy Systems
RH	River Hydro Power Plants
T.E.	Thermoelectric (Power Plants)
Toe	Tons of oil equivalent Primary energy (1 Tpe $= 41.86\,$GJoules)

References

R. Banos, F. Manzano-Agugliaro, F. G. Montoya, *et al.*, Optimization methods applied to renewable and sustainable energy: A review, *Renewable and Sustainable Energy Reviews*, **15**, 1753–1766, (2011).

M. Beaudin, H. Zareipour, A. Schellenberglabe and W. Rosehart, Energy storage for mitigating the variability of renewable electricity sources: an updated review. *Energy for Sustainable Development*, **14**, 302–314, (2010).

M. B. Blarke and H. Lund. The effectiveness of storage and relocation options in renewable energy systems. *Renewable Energy*, **33**, 1499–1507, (2008).

A. Campoccia, L. Dusonchet, E. Telaretti and G. Zizzo. Comparative analysis of different supporting measures for the production of electrical energy by solar PV and Wind systems: Four representative European cases, *Solar Energy*, **83**, 287–297, (2009).

D. Connolly, H. Lund, B. V. Mathiesen and M. Leahy, Modelling the existing Irish energy-system to identify future energy costs and the maximum wind penetration feasible. *Energy*, **35**, 2164–2173, (2010a).

D. Connolly, H. Lund, B. V. Mathiesen and M. Leahy, A review of computer tools for analysing the integration of renewable energy into various energy systems, *Applied Energy*, **87**, 1059–1082, (2010b).

Energy Information Agency, International Energy Outlook 2009. OE/EIA-0484 available on the WEB at: www.eia.doe.gov/oiaf/ieo/index.html (accessed July 2010) (2009).

ENEA, Italian Energy and Environment Report, 2007–2008, available on the web at http://www.enea.it/produzione_scientifica/volumi/REA_2007/REA2007_Dati_Prima .html#nazionali (last accessed July 2010) (2009).

A. Franco and P. Salza, Perspectives for the long-term penetration of new renewables in complex energy systems: the Italian scenario, *Environment, Development and Sustainability*, **13**, 309–330, (2011a).

A. Franco and P. Salza, Strategies for optimal penetration of intermittent renewables in complex energy systems, *Renewable Energy*, **36**, (2011b).

V. Fthenakis, J. E. Mason and K. Zweibel, The technical, geographical, and economic feasibility for solar energy to supply the energy needs of the US, *Energy Policy*, **37**, 387–399, (2009).

C. Frangopoulos, M. von Spakovsky and E. Sciubba, A Brief Review of Methods for the Design and Synthesis Optimization of Energy Systems, *Int. J. Applied Thermodynamics*, **5**, 151–160, (2002).

J. B. Greenblatt, S. Succar, D. C. Denkenberger, *et al.*, Baseload wind energy: modeling the competition between gas turbines and compressed air energy storage for supplemental generation, *Energy Policy*, **35**, 1474–1492, (2007).

M. Hoogwijk, D. van Vuuren, B. de Vries and W. Turkenburg, Exploring the impact on cost and electricity production of high penetration levels of intermittent electricity in OECD Europe and the USA, results for wind energy, *Energy*, **32**, 1381–1402, (2007).

A. Lazzaretto, A. Toffolo, M. Morandin and M. R. von Spakovsky, Criteria for the decomposition of energy systems in local/global optimizations. *Energy*, **35**, 1157–1163, (2010).

H. Lund, Large-scale integration of wind power into different energy systems, *Energy*, **30**, 2402–2412, (2005).

H. Lund, Aalborg University. *EnergyPLAN: Advanced Energy Systems Analysis Computer Model*. Aalborg University, (2008). See also: http://energy.plan.aau.dk/manual .php.

H. Lund, *Renewable Energy Systems – The Choice and Modeling of 100% Renewable Solutions*. Elsevier, (2009a).

H. Lund and W. W. Clark, Management of fluctuations in wind power and CHP comparing two possible Danish strategies, *Energy*, **27**, 471–483, (2002).

H. Lund and B. V. Mathiesen, Energy system analysis of 100% renewable energy systems – the case of Denmark in years 2030 and 2050, *Energy*, **34**, 524–531, (2009).

P. D. Lund, Effects of energy policies on industry expansion in renewable Energy. *Renewable Energy*, **34**, 53–64, (2009).

P. S. Moura and A. T. de Almeida, Multi-objective optimization of a mixed renewable system with demand-side management. *Renewable and Sustainable Energy Reviews*, **14**, 1461–1468, (2010).

P. A. Østergaard, Geographic aggregation and wind power output variance in Denmark, *Energy*, **33**, 1453–1460, (2008).

P. A. Østergaard, Reviewing optimisation criteria for energy systems analyses of renewable energy integration. *Energy*, **34**, 1236–1245, (2009).

P. A. Østergaard, Regulation strategies of cogeneration of heat and power (CHP) plants and electricity transit in Denmark. *Energy*, **35**, 2194–2202 (2010).

P. Pinson, H. A. Nielsen, H. Madsen and G. Kariniotakis, Skill forecasting from ensemble predictions of wind power. *Applied Energy*, **86**, 1326–1234, (2009).

G. Salgi and H. Lund, System behaviour of compressed-air energy-storage in Denmark with a high penetration of renewable energy sources. *Applied Energy*, **85**, 182–189, (2008).

I. Stadler, Power grid balancing of energy systems with high renewable energy penetration by demand response, *Utilities Policy*, **16**, 90–98, (2008).

Terna S.p.A – *Italian Electric System*. (a) Statistical Data on electricity Synthesis in Italy 2010: Gross maximum capacity of power plants in major countries of the World at 31 December 2009, (2011a).

Terna S.p.A. – *Italian Electric System*. (b) Loads of Italian Electric Power System, available at http://www.terna.it/, (2011b).

B. Wille-Haussmann, T. Erge and C. Wittwer. Decentralised optimisation of cogeneration in virtual power plants, *Solar Energy*, **84**, 604–611, (2010).

4

The Smart Grid as a Response to Spread the Concept of Distributed Generation

Yi Ding, Jacob Østergaard, Salvador Pineda Morente, and Qiuwei Wu
Centre for Electric Technology, Department of Electrical Engineering, Technical
University of Denmark, Lyngby, Denmark

4.1 Introduction

Global warming and security of energy supply are two major challenges faced by the world. In the energy framework of the future, renewable energy will play an important role in improving the security of our energy supply by drawing upon sustainable natural sources and reducing environmental impact [1, 2]. In the European Union (EU) 2020 agenda, the EU will reach a 20% share of renewable energy in total energy consumption and increase energy efficiency by 20% reducing greenhouse gas emissions for 2020. However, high renewable energy penetration will have a significant impact on power system security and reliability due to fast fluctuation and unpredictable characteristics of renewable energy resources such as wind and solar power [3]. The integration of a large number of renewable energy resources will increase competition for balancing power in the future.

With the open access of the electricity network, an increased number of smaller-scale generating units will be connected into the distribution network. The integration of smaller-scale generating units into the distribution network may provide an important tool for solving challenges of balancing power effectively and environmentally.

Sustainable Development in Chemical Engineering – Innovative Technologies, First Edition.
Edited by Vincenzo Piemonte, Marcello De Falco and Angelo Basile.
© 2013 John Wiley & Sons, Ltd. Published 2013 by John Wiley & Sons, Ltd.

However, there exist several technical and administrative obstacles to fully utilizing the potential of distributed generation (DG) in the existing system structure.

With the development of information and communication technology (ICT), smart grids will form the backbone of the future electricity network. They will also facilitate the integration of DG into the system, which allows more flexible and active power generation.

This chapter discusses the smart grid as response to integrating DG and providing balancing capacity for mitigating the high volatility of renewable energy resources (RES) in the future. The present central power generation system and future generation system with a high penetration of DG and RES are introduced in Section 4.2 and Section 4.3 respectively. Section 4.4 discusses the integration of DG into a smart grid for providing balancing power. The Bornholm power system as a pioneer of smart grids is briefly discussed in Section 4.5.

4.2   Present Electric Power Generation Systems

The electric power industry has been centrally controlled and managed for nearly 100 years. In a relatively large geographical area, most electric power is supplied by centralized power plants and transmitted to customers through transmission and distribution networks controlled by monopoly utilities. Unit commitment and economical generation dispatch are planned and controlled by the system operators.

Generally, there are three major types of power plants used in the vertically integrated power systems [4]:

1. Thermal power plants,
2. Hydro power plants, either run-of-the-river or dam types.
3. Nuclear power plants.

The electric power generation system of China is a typical example of such a structure, which is the second largest system in the world, in terms of both installed generation capacity and electricity generated. Installed power generation capacity exceeds 1056 GW and electricity generation is over 4692.8 terawatt-hours (TWh) [5]. In China, most electric power is generated from coal-fired thermal power plants. In 2011, coal-fired thermal generating capacity accounted for approximately 75% of total generating capacity of power plants and about 82.5% of total generation [6].

Many technical measures are being successfully implemented to raise the efficiency of power generation and reduce pollution. However, thermal power plants bring environmental problems that cannot be ignored. A large proportion of thermal power generation makes the electrical power industry top the list of energy consumption and pollutant emissions in China.

Hydro power plants do not produce pollution and have less environmental impact during their normal operations. However, the capital cost for a new large-scale hydro plant is significant, which will also bring serious environmental damage and socio-economic problems during construction. Moreover, it is increasingly difficult to identify feasible new sites in developed countries [4].

2012

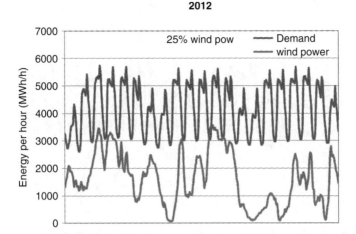

Figure 4.1   *Wind power generation and demand in January 2012 in Denmark*

Currently, around 13–14% of the world's electricity is produced by nuclear power plants [7]. Some developed countries such as France and Japan rely heavily on nuclear power, which accounts for about 50% of their total electricity generation [8]. Though the usage of nuclear power can reduce the consumption of fossil fuel and carbon emissions, the safety and security of nuclear power plants are major concerns now. After the nuclear disaster of Fukushima Daiichi, Japan in 2011, many countries have had to reconsider their nuclear power policies. For example, Germany will permanently shut down its nuclear power plants within the next 10 years, which will replaced by renewable energy resources.

In recent years, renewable energy generation in power generation systems, such as wind energy and solar energy, has undergone rapid development. Denmark is a driver in utilizing renewable energy resources in the European Union (EU). In 2007, 30% of the total electricity consumption came from renewable energy, of which two-thirds was from wind power. The current generation structure of Denmark is a mixture of centralized thermal power plants plus a moderate amount of wind power and other small DG units [9]. Hourly wind power generation and demand in January 2012 in Denmark is shown in Fig. 4.1. Wind generation met 39.04% of the demand in that month as the figure shows. The figure also indicates that the hourly wind generation can occasionally exceed demand, which corresponds to 1.39% of time in January 2012.

4.3 A Future Electrical Power Generation System with a High Penetration of Distributed Generation and Renewable Energy Resources

In the future energy framework, renewable energy resources play an important role in decarbonizing the energy generation, as well as providing a means of improving the energy supply security by drawing upon inexhaustible sources of energy [1, 2].

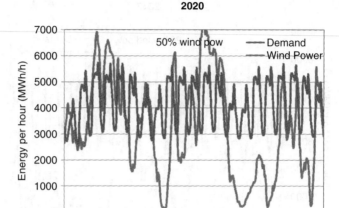

2020

Figure 4.2 *Estimated wind power generation and demand in November in western Denmark with 50% wind power penetration*

The EU Commission proposed roadmaps for developing low carbon technologies for fulfilling the following main targets of utilizing renewable energy by 2020 [2]:

- Up to 20% of electricity in EU will be produced by wind energy resources.
- Up to 15% of electricity in EU will be produced by solar energy resources.
- The electricity grid in Europe will accommodate the integration of 35% renewable electricity in a seamless way, and effectively maintain the balance between generation and demand.
- Cost-competitive and sustainable bio-energy will occupy at least 14% of the EU energy mix.

For achieving these objectives, more efficient and reliable generation technologies concentrating on renewable energy will be implemented. The competitiveness for large scale penetration of wind power and solar power generation will be improved. In Denmark the wind power penetration will reach 50% of electricity consumption by 2025 [10]. Figure 4.2 illustrates the estimated wind power generation and demand with 50% wind power penetration, where the wind generation has been estimated with a simple scaling of the 2012 data, so that wind covers the 50% demand.

However, the high penetration of renewable energy resources will challenge the power system operation. The share of fluctuating and less predictable power generation will increase significantly. Obviously, higher renewable energy penetration will increase the need for balancing power because of the reduction in conventional controllable energy resources and the high volatility of renewable energy resources.

Subtracting wind power from demand leaves a residual demand and an overflow as shown in Fig. 4.3. With the applied simple scaling of the wind power generation, there is 36.9% of time that hourly wind generation will exceed hourly demand.

In the future, most new wind farms will be built as large offshore wind power plants, which may have higher power fluctuation than land based wind power plants. Therefore, the overflow periods will probably be even longer with 50% wind generation [11, 12].

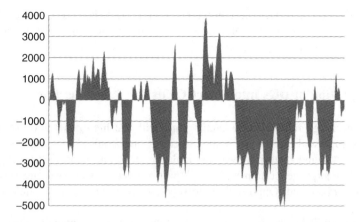

Figure 4.3 *Estimated residual demand and overflow with 50% wind power penetration*

In the future, an increased number of high efficient generators including wind turbines and micro combined heat and power (CHP) will also be connected into the distribution network. Any electrical power generation technology that is integrated into the distribution network can be fitted into the scope of DGs.

The DG is not a new thing and has been used for many years to provide electricity for end users. The critical sectors such as hospitals and military bases use DGs for providing emergency electricity in contingency states. The "conventional" DGs usually passively provide electricity to their own customers, which have limited integration with the system operation. However, technological developments in photovoltaic panels, micro-turbines, wind turbines, and smart controllers have expanded the application range of "modern" DGs, which have become more efficient and economical. Usually "modern" DGs have low or no carbon emission and enhanced controllability for system services [4].

There are various kinds of emerging technologies of DGs with small-scale power generation in the range of 3 kW to 1500 kW [4]: roughly speaking, the power outputs of wind turbines, micro-turbines, residential PEM fuel cells and residential photovoltaic systems can be 1500, 30, 5 and 3 kW, respectively. Comparatively, the power outputs of centralized generation technologies are in the range of 100–10 000 MW.

CHP units are important sources of DGs, which not only generate electricity but also produce heat for customers. In CHP units, turbines are used to drive generators; while the waste heat from the turbines is captured for water heating, space heating, and air conditioning. Therefore the efficiency of total energy utilization can be greatly increased [13]. Another popular source is photovoltaic (PV) systems installed in households and commercial buildings. In the PV system, a number of solar panels are connected in series to form a string, and a number of strings are paralleled and connected to a string inverter, then all the string inverters are connected to the power grid. DGs based on PV technology show many advantages over other generation technologies in terms of low maintenance, low pollution, modularity, and expandability. The high cost of PV systems can be reduced by technology development of PV modules, optimal design, and installment. Therefore, these have drawn great attention and remarkable investment in the past decade. Other sources of DGs include small wind turbines, residential fuel cells and geothermal units.

Incorporation of more DGs may provide a solution to balancing power challenges in the future. However, most the DGs face barriers to provide balancing services in the current market model, which will discussed in the next section.

4.4   Integration of DGs into Smart Grids for Balancing Power

The existing energy infrastructure cannot cope with the high penetration of renewable energy efficiently and effectively. Without serious upgrading of existing electricity network monitoring and metering systems, and operation and control strategies, the use of renewable energy generation will develop at a much slower pace and system security may be jeopardized. Opportunities for energy saving and energy efficiency will be missed [14].

Smart grids will be the backbone of the future electricity network. The vision of smart grid is defined by the European Commission Smart Grids Advisory Council as "an electricity network that grants connection access to all network users and provides efficient and reliable services in order to fulfill customers' needs whilst responding to the changes and challenges ahead [15]."

Smart grids will facilitate the integration of a high penetration of renewable energy and other DGs with a considerable potential benefit in providing a balancing capacity and other ancillary services to the power system. Moreover, smart grids will provide a possible solution for improving the reliability and security of electricity network. The EU Commission is encouraging the implementation of smart grids because they provide an opportunity to boost future competitiveness and worldwide technological leadership from EU technology [14].

Key components of a smart grid include [16]:

- *Smart Meters:* Smart Meters are key enabling components for smart grids and record the electricity consumption in short intervals, for example, 5 min. Smart Meters also enable two-way communication and interaction between the system operator and end users. In turn, these are major drivers behind development of smart controllers for optimal electricity generation or consumption.
- *Intelligent grid management:* Intelligent grid management allows flexible power generation, transmission, and active distribution. The bi-directional power flow between transmission and distribution is also possible, which enables active participation and high integration of DGs in the network.
- *Added-value services and applications:* These services allow DGs and end-customers to economically optimize their electricity production/consumption.
- *System wide oversight and coordination:* A uniform communication infrastructure will enable real-time and system wide communication, which integrates and exchanges information for both producers and consumers in the smart grid. The uniform communication infrastructure will support PMU based Wide Area Measurement Systems (WAMS) for system supervision and control, which will increase the security and reliability of system operation.

In EU countries, national transmission system operators (TSO) have the responsibility of maintaining the balance between generation and demand at all times in electric power systems. The TSO is obliged to purchase and operate different kinds of reserve for maintaining this balance. For example, Energinet.dk is the national TSO for transmission grids in Denmark and has the responsibility of maintaining stable and reliable power system operation [20].

Based on the ENTSO-E terminology, there are three main types of reserves; namely frequency containment reserves (FCR), frequency restoration reserves (FRR) and replacement reserves (RR) [21]: The first type (FCR) is used for quickly restoring system balance and stabilizing system frequency after the occurrence of frequency fluctuation and imbalance events. The activation of FCR is usually automatic and local with the result of a stable frequency deviating from nominal value. The activation time for fully operational FCRs is usually within 30 s. The second type of reserve is implemented for restoring the system frequency to the nominal value; for example, 50 Hz, and reducing the area control error. The FRR is usually provided by generating units with automatic generation control (AGC) and rapid start-up generating units with activation times up to 15 min. Compared with the FCR, FRR is typically activated centrally and can be activated automatically or manually by TSO. The third type of reserve (RR) is deployed to restore FCR and FRR to an appropriate level. The activation time of RR is anything from 15 min up to several hours.

Currently, in Nordic countries the current regulating power (RP) market is the major place for providing RR. The minimum bid for a regulating market is 10 MW. Therefore, an RP market precludes entry for most DGs, whose capacity is usually below 10 MW. DGs could be aggregated to form a virtual power plant (VPP) to participate in the RP market. However, complex bidding procedures, financial obligations for complying with schedules and different cost structures of DGs may result in a significant administrative burden for a VPP. With the increasing penetration of RES, frequency quality has been reduced in recent years. For example, the system operation time outside the 100 MHz frequency band has been increased from 100 min/month to 1100 min/month between 2000 and 2010 in Denmark. This trend is still going on and increases system operation risk and burden of reserves.

For addressing the future needs and challenges of the upcoming power systems, an efficient, market-based tool should be developed. The EcoGrid EU [18, 22] is a large-scale EU-funded project, which first establishes prototype for future European intelligent grids. It will demonstrate a market concept that is designed to incorporate DGs as well as flexible demand into existing electricity markets, balancing tools, and operation procedures. The basic architecture of the Ecogrid EU is based on a new real-time electricity market, which allows small-scale DERs and flexible demand receive and react on variable electricity prices. The new real-time electricity market will remove the obstacles previously faced by DGs, for example, requirements on bidding size, compulsory obligations with schedules, and online monitoring.

The existing market framework must be maintained with the least possible modification but will be expanded for integrating the EcoGrid EU real-time market as shown in Fig. 4.4. The new real-time market extends the current electricity market to a shorter time horizon and to smaller assets, which can be utilized for short-term, intra-hour balancing.

There are two major functions of the Ecogrid EU real-time market. The first major function is to relieve an appropriate level of FCR and FRR, and assist in securing the

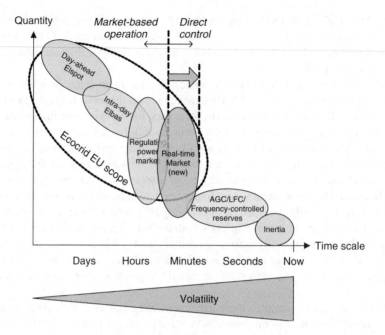

Figure 4.4   *The integration of real-time market into existing markets*

intra-hour balancing. Based on the basic market architecture design, the Ecogrid EU real-time market is an intra-hour market. The second function of the Ecogrid EU market is to handle the intra-hour expected system operational changes such as the intra-hour changes of demand and wind power generation, therefore reducing the burden of the FCR and FRR (use reserves in the opposite direction).

The Ecogrid EU real-time market overlaps/complements the regulating power (RP) market, which mainly addresses larger conventional generating units (several MW). The Ecogrid EU real-time market operates in parallel with the RP market. Consequently, real-time prices are strongly correlated with the price development on the RP market. Figure 4.5 illustrates the correlation of price and quantity in the RP market and the Ecogrid EU real-time market [22]. In the left hand part of the illustration, a certain quantity can be obtained by selecting bids in the RP market, and the price is a result of this. In the right hand part of the illustration, a real-time price is set, which then results in a certain quantity of balancing power in the Ecogrid EU real-time market. The implementation of the activated bids in the RP market and the price signal in the real-time market can obtain the lowest total cost and the required total balancing power, $P = P1 + P2$. The marginal cost of regulating power market equals the real-time price ($C1 = C2$). Because the Ecogrid EU market is another source for providing the balancing services, the total reserve cost and RP price should be reduced.

An overview of the real-time Ecogrid EU market architecture enabling real entry for DG and end-users for providing balancing services is illustrated in Fig. 4.6 [17].

In this architecture, DGs and end-users will receive and react on real-time electricity price, which will update every 5 min to utilize the potential for a dynamic response.

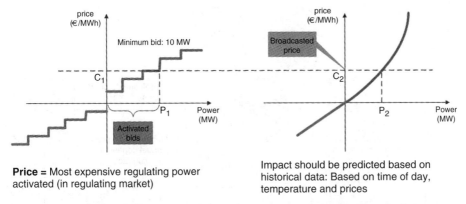

Price = Most expensive regulating power activated (in regulating market)

Impact should be predicted based on historical data: Based on time of day, temperature and prices

Figure 4.5 *Conceptual illustration of the relation between the Nordic regulating power market (left) and the Ecogrid EU real-time market (right).*

The real-time price is set by the real-time market operator, which might be the transmission system operator (TSO), on the basis of the need for up- or down-regulation due to occurring imbalance between production and consumption and/or restrictions in the transmission/distribution system. If the system needs up-regulation, the real-time price will be higher than the day-ahead spot price; while when the system needs down-regulation, the real-time price will be lower than the day-ahead spot price.

The smart controllers connected to DGs and controllable loads are a central part of the concept [17]. They will receive price signals and take control action to optimize electricity production/consumption of their controlled devices based on the settings.

State-of-the-art information and communication technology (ICT) infrastructures in smart grids will enable seamless integration among market players. The price information will be transmitted unidirectionally from price server to smart controllers and this can be through the Internet. The ICT infrastructure should ensure the price information will be received accurately in due time. The standardized ICT interfaces are developed for satisfying requirements of information security and integrity, as well as scalability and adaptability. The following actors can be involved in the communication:

- A real-time market place gateway.
- A prosumer (producer and consumer) gateway at each prosumer node.
- Smart controllers (associated with DER devices).

A simplified architecture of the ICT system is illustrated in Fig. 4.7. The architecture can be extended with a grid manager gateway which can enable communication of dynamic grid tariffs for optimal distribution grid operation.

The real-time marketplace price gateway will send price information to prosumer gateways (e.g., in a household installing DG) when information is ready from the real-time market place. The prosumer gateway is used to receive and distribute the received price signals to one or more smart controllers. The prosumer gateway may host functions for example, storing price information and forecasts and administrating the associated smart controllers. Smart controllers will monitor and optimally control DG based on the real-time price signals, status of the controlled DG device(s) and the associated

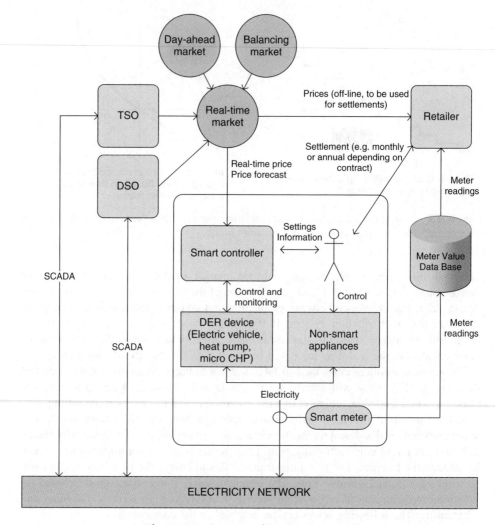

Figure 4.6   *System architecture overview*

environment, and other possible information such as forecasted prices from the real-time marketplace and forecasted weather conditions from other sources and so on. In simple installations the prosumer gateway and a smart controller can be implemented in a single unit, for example a micro CHP unit might be supplied with each gateway and controller.

The basic data structure of the price signal format includes the following information:

- Starting time (date + time in UTC-time).
- Duration (in seconds).
- Price (in DKK/MWh, usually with two digits).
- Signal type (RT = real-time price, FP = forecast price).
- Event trigger (No, start, stop and other commands).
- Checksum.

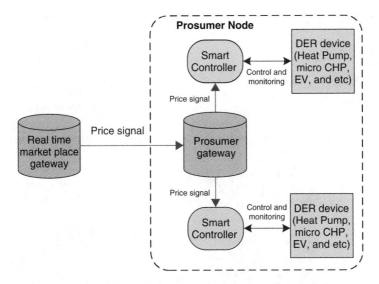

Figure 4.7   *ICT architecture for price signal transmission*

The starting time in the EcoGrid EU implementation will normally be a whole number of 5 min for the real-time price, for example, 9:00, 9:05, 9:10 or 9:01, 9:06, 9:11 and so on. In future implementations using the EcoGrid EU concept the 5 min intervals may be changed. The resolution of the starting time is seconds. The signal type indicates whether the price will be used for settlement (real-time price) or whether it is a forecasted price. Event trigger can be included to enable future direct control schemes for example, a demand response program of contingency events. An event trigger signal set as "start" indicates that the customers or DG should respond to a contingency event. In this case, the customers or DG can be compensated based on the emergency contract. The checksum is also added to the price signal, which is used to check for errors in the price signal.

The smart meters must be able to record electricity production from DGs or consumption from flexible end users with the same time resolution as the price signals, which can be set at 5 min intervals. The smart meters are capable of communicating meter records to the meter-value database, which is a significant amount of information transfer.

4.5   The Bornholm System – A "Fast Track" for Smart Grids

The Danish island of Bornholm is located in the Baltic Sea. The Bornholm power system is a pioneer of future distribution systems and has high penetration of renewable energy. The Bornholm distribution grid is a meshed 60 kV network, which also includes 60/10 kV substations, 10 kV feeders and 10/0.4 kV substations. The Bornholm grid is also connected to the Nordic transmission grid through a 60 kV AC cable.

The Bornholm distribution system is integrated in the electricity market of western Denmark (DK2) and is interconnected with Nordic power system as shown in Fig. 4.8 [18, 19]. The system has a peak load of 55 MW. The capacities of wind power

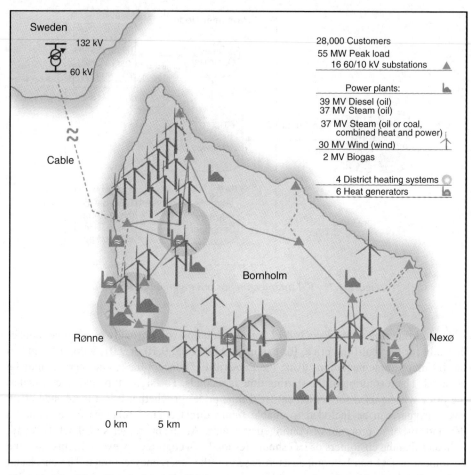

Figure 4.8 *The Bornholm distribution system. Reproduced with permission from J. Østergaard and J. Eli Nielsen, http://www.dtu.dk/sites/powerlabdk/English.aspx [19]*

and CHP units are 30 and 16 MW, respectively. In the future, the amount of renewable energy resources and energy storage devices such as PV, micro-CHP, fuel cell and heat storage will be established for smart grid demonstration purposes.

The public and municipality of Bornholm support and encourage the "Bright Green Island" strategy. The goal of renewable energy resource penetration in the Bornholm power system will reach 100%. The Bornholm system is well suited to testing the feasibility and robustness of EU smart projects.

4.6   Conclusions

Electricity plays an important role in the future energy framework around the world. The future power system will integrate a high penetration of RES, which will significantly challenge the system operation and control. The existing energy infrastructure has to be upgraded and reinforced for handling the fluctuation of RES and demand. With the integration of a large amount of RES in the future, the sufficient balancing power is critical for maintaining system reliability and security. With the open access of the electricity network, an increased number of more efficient and economical DGs, such as micro-CHPs and photovoltaic panels, have been connected into the distribution network. These "modern" DGs have low or no carbon emissions and enhanced controllability for flexible integration with the system operation. They may also provide an important tool for solving the challenges of balancing power effectively and environmentally. However, the current power market design favors conventional production and large generation power plants. Complex bidding procedures, bidding size and compulsory obligations are obstacles to preventing DGs balancing power in the existing power markets. With the development of information and communication technology, the smart grid has become an effective tool for integrating high penetration of RES, DGs and flexible demand. The capability of DGs and flexible demand participating in the power market and contributing to the system operation will be demonstrated in the EcoGrid EU project, which establishes the first prototype of the future European intelligent grids. The EcoGrid EU project will develop a bidless real-time market, where DGs and end-users receive and react on real-time electricity price with 5-min intervals. The basic real-time market concept and ICT architecture have been introduced in this chapter.

References

[1] EU, Communication from the commission to the European Parliament and the Council, Renewable Energy: Progressing towards the 2020 target, Commission Staff Working Document, Brussels, 2011.

[2] EU, Communication from the commission to the European Parliament, the Council, the European Economic and Social Committee and the Committee of the Regions, on Investing in the Development of Low Carbon Technologies (SET-Plan), Commission Staff Working Document, Brussels, 2009.

[3] Y. Ding, P. Wang, L. Goel, *et al.*, "Long Term Reserve Expansion of Power Systems with High Wind Power Penetration using Universal Generating Function Methods", *IEEE Trans. on Power Systems*, Vol. **26**, No. 2, May 2011, pp. 766–774.

[4] H. B. Puttgen, P. R. MacGregor, and F. C. Lambert, "Distributed generation: Semantic hype or the dawn of a new era?", *IEEE Power and Energy Magazine*, Vol. **1**, No. 1, 2003, pp. 22–29.

[5] China Electricity Council, "Statistics Express of China's Electric Power Industry", Available at: http://tj.cec.org.cn/tongji/niandushuju/2012-01-13/78769.html, accessed on January 13, 2012.

[6] China Electricity Council, "China's Electric Power Industry Profiles, January-November 2011", Available at: http://tj.cec.org.cn/tongji/yuedushuju/2011-12-19/75857.html, accessed on December 19, 2011.

[7] World Nuclear Association, "Another drop in nuclear generation", *World Nuclear News*, May 5, 2010.

[8] International Energy Agency, *"Key World Energy Statistics 2007"*, June 2008.

[9] Z. Xu, M. Gordon, M. Lind *et al.*, "Towards a Danish power system with 50% wind: Smart grids activities in Denmark", *IEEE Power & Energy Society General Meeting*, 2009, pp. 1–8.

[10] J. Østergaard, A. Foosnæs, Z. Xu, *et al.*, "Electric Vehicles in Power Systems with 50% Wind Power Penetration: the Danish Case and the EDISON programme", *European Conference Electricity & Mobility*, Würzburg, Germany, 2009.

[11] V. Akhmatov, H. Abildgaard, J. Pedersen, and P. B. Eriksen, "Integration of Offshore Wind Power into the Western Danish Power System," in *Proc. Copenhagen Offshore Wind 2005*, 2005.

[12] P. Sørensen, N. A. Cutululis, A. Vigueras-Rodriguez, *et al.*, "Modelling of power fluctuations from large offshore wind farms." *Wind Energy*, Vol. **11**, 2008, pp. 29–43.

[13] G. M. Masters, *"Renewable and Efficient Electric Power Systems"*, John Wiley & Sons, Inc., Hoboken, 2004.

[14] EU, Communication from the commission to the European Parliament, the Council, the European Economic and Social Committee and the Committee of the Regions, Smart Grids: from innovation to deployment, Brussels, 2011.

[15] Directorate-General for Research Sustainable Energy Systems, *European Smart Grids Technology Platform, Vision and Strategy for Europe's Electricity Networks of the Future*. Available online at http://ec.europa.eu/research/energy/pdf/smartgrids _en.pdf, accessed February 6, 2013. 2006.

[16] SAP AG, Smart Grids for Europe – Benefits, Challenges and best Practices, The SAP Perspective, 2011.

[17] P. Nyeng and J. Østergaard, "Information and Communications Systems for Control-by-Price of Distributed Energy Resources and Flexible Demand", *IEEE Trans. on Smart Grid*, Vol. **2**, No. 2, June 2011, pp. 334–341.

[18] EcoGrid Europe, *A Demonstration Project for the Future*. Available at http:// ec.europa.eu/research/conferences/2009/smart_networks/pdf/behnke.pdf, accessed February 6, 2013. 2009.

[19] J. Østergaard and J. Eli Nielsen, *"The Bornholm Power System –An overview"*, available at http://www.dtu.dk/sites/powerlabdk/English.aspx, accessed February 6, 2013. n.d.

[20] Energinet.dk, *"Ancillary services to be delivered in Denmark, Tender conditions"*, available at http://energinet.dk/, accessed February 6, 2013. January 2011.

[21] ENTSOE, *"UCTE Operational Handbook Policy 1"*, available at http://www.entsoe .eu/index.php?id=57, accessed February 6, 2013. 2012.

[22] J. M. Jørgensen, S. H. Sørensen, K. Behnke, and P. B. Eriksen, "EcoGrid EU-A prototype for European Smart Grids", *IEEE Power and Energy Society General Meeting*, 2011.

5

Process Intensification in the Chemical Industry: A Review

Stefano Curcio

Department of Engineering Modeling, University of Calabria, Rende, Italy

5.1 Introduction

It is difficult to give an appropriate and exact definition of the term "Process Intensification" (PI). Often, process intensification is associated with attractive but vague expressions like "cheaper, smaller, cleaner" or "making more with less". However, all these purely descriptive and qualitative attributes do not contribute to defining the term "process intensification" in a rigorous way. The problem with the PI definition is actually related to the identification of the limit (if any), which allows discrimination between real process intensification and classical process optimization. Generally speaking, PI technologies tend to completely re-design conventional unit operations and, also, make use of novel equipment or processing methods such as multifunctional reactors, micro reactors, enhanced heat exchangers, alternative energy forms, and so on. The recent developments in both climate change and energy supply have confirmed the necessity of faster and wider application of innovative PI-technologies, which can offer significant advantages like decrease of capital and operating expenses, selectivity improvements, lead time reduction and safety improvement.

In recent years, the development and the exploitation of PI methods has seen a world-wide hastening. Starting from the mid-1980s, the interest of both academia and industry in process intensification increased to a significant extent, as testified by the number of articles published in this field. A literature search performed by *Web of Science*, considering

Sustainable Development in Chemical Engineering – Innovative Technologies, First Edition.
Edited by Vincenzo Piemonte, Marcello De Falco and Angelo Basile.
© 2013 John Wiley & Sons, Ltd. Published 2013 by John Wiley & Sons, Ltd.

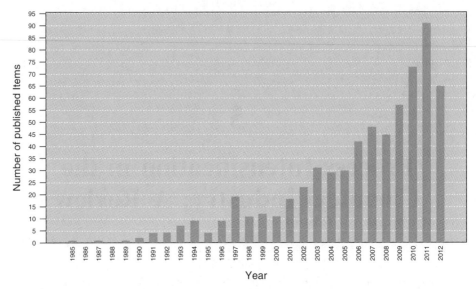

Figure 5.1 *Number of published papers – keywords: "Chemical" AND "Process Intensification". Data collated from www.webofknowledge.com*

the words "chemical" AND "process intensification" returned the result shown in Fig. 5.1: the number of publications in the last few years seems to stabilize in the range 45–90 papers per year, with a peak in 2011. In addition, it is worthwhile remarking that a scientific journal, namely *Chemical Engineering and Processing: Process Intensification*, focuses on PI and on different related fields.

Another fundamental reference in PI field is represented by the European Roadmap for Process Intensification, which presents a valuable review of PI technologies analyzed from an industrial standpoint.

5.2 Different Approaches to Process Intensification

PI methodologies have been extensively applied to improve the performance of several chemical processes. PI has the potential to positively affect a process by increasing its efficiency via a reduction in energy consumption, costs, volumes handled, waste generated, and, by improving the process safety without sacrificing product quality (Reay *et al.*, 2008). Retrofit is a fundamental concept in PI and can be defined as the improvement of an existing plant by substituting or improving some of its constituent unit operations while fitting the rest of the plant and some of the process variables, namely the chemicals processed by the system, the reactions that may be occurring, the product purities that need to be achieved, a sub-set of equipment and operations in the process (Lutzea *et al.*, 2010). Actually, a proper discrimination between drastic improvements and improvements as just a result of qualitative changes has to be done. Microreaction technology, monolithic reactors and reactive separations are classical examples of a drastic enhancement, since they allow, as compared to traditional apparatuses, improved mass and heat transfer rates, thus achieving significant improvements as compared to traditional technologies. Alternative forms of energy supply for chemical reactions such as microwaves and ultrasound, the use of new reaction media such as ionic liquids, micro-emulsions and supercritical phases, or the use of new auxiliary agents such as phase transfer catalysts belong, instead, to the second category, even if they are also capable of improving the existing processes. The principles associated with PI have been classified (Lutzea *et al.*, 2010) considering three different kinds of enhancements, which can be achieved through: (1) the integration of operations; (2) the integration of functions; and (3) the integration of phenomena. In this way, PI is accomplished by adding and enhancing a phenomenon within a function and/or operation for improved equipment design in order to maximize the driving force or to overcome the limitations occurring in the corresponding conventional process. In addition to these three principles, a targeted enhancement of a phenomenon can be considered as well; in this way, PI is achieved through improved use of the respective phenomena/functions to overcome the limitations occurring in a given conventional operation. Van Gerven and Stankiewicz (2009) defined four principles to achieve a completely intensified process: (1) to maximize the effectiveness of intra-and intermolecular events; (2) to optimize the driving forces at every scale and maximize the specific surface area to which these forces apply; (3) to maximize synergistic effects; (4) and to give each molecule the same processing experience. Finally, another possibility for the classification of process intensification measures is to divide them into three superordinate levels (Freund and Sundmacher, 2008). The most detailed level is the so-called *phase level*, at which molecule populations, which build up a thermodynamic phase, are considered. In a whole process, all the thermodynamic phases are embedded into apparatuses, or into individual process spaces. This is the so-called *process unit level*. Usually, the process consists of several such process units. The interconnection between the individual apparatuses and thus the overall process flowsheet can finally be analyzed at the superordinated *plant level*.

5.3   Process Intensification as a Valuable Tool for the Chemical Industry

From an industrial standpoint, PI has been successfully exploited in many different sectors, as discussed in the European Roadmap for Process Intensification.

In petrochemical industry, for instance, energy definitely represents a relevant part of total delivered product costs. Energy efficiency significantly affects the cost competitiveness and sustainability of petrochemical industry, which, in recent years, has already attained considerable improvements in energy efficiency. Given the large size of the petrochemical industry and the highly competitive market in which it operates, it is worthwhile combining reliability and predictability in order to achieve a decrease in energy expenses. Unexpected production interruptions do indeed have serious impacts on operational and logistic costs and may easily erode profits. In this context, PI may play an important role either in reducing energy costs or in lowering capital needs for investments, keeping safety and process reliability at the levels required by both legislation and market. PI does indeed aim at promoting innovation, creating technology breakthroughs for any industrial sector and, therefore, also for petrochemical industry processes (Harmsen, 2010). As far as energy efficiency is concerned, it is expected that the exploitation of process intensification methodologies may determine at least a 20% improvement by 2050 (Sanders *et al.*, 2012). This will be achieved by implementing new technologies, for example, new processes, bio-based and bio-catalytic routes, novel materials, hybrid reactors, novel separation systems, and so on. Like all new technologies, however, PI faces several barriers, which need to be overcome to effect an efficient PI implementation. These barriers can be summarized as: high cost to retrofit PI technologies in current production processes; risks of commercializing breakthrough technology; scale-up of PI; lack of PI knowledge, unfamiliarity with PI technologies and, finally, a long development path (Ren, 2009).

In the pharmaceutical/fine chemicals industry the use of process intensification methods is called to increase the selectivity of reactions (Yadav *et al.*, 2003) and, thus, the material yield, reducing the lead time (i.e., the time between raw materials delivery and the completion of the product) of the entire production process. The achievement of these goals would also determine savings in energy consumption and contribute to related sustainability needs. Currently, the typical selectivity of chemical reactions is about 80%; PI technologies can increase selectivity up to 90%. For a classic multi-step process, this would achieve a rise in material yield from 30–60%, which significantly contributes to the sustainability of the whole process. Moreover, current processes in the fine chemicals industry may have up to 50 process steps, with a lead time of several months. The exploitation of improved design methods based on PI will allow operating integrated and continuous process steps, thus reducing these lead times to a significant extent. It is indeed expected that a multipurpose serial production train, based on PI technologies, can result in 50% reduction of production costs within 10–15 years. However, this vision can be achieved if PI technologies, whose efficiency has been proved on lab-scale plants only, are properly scaled-up and industrialized. However, a successfully implementation of PI technologies in the pharmaceutical/fine chemicals industry is actually dependent on a detailed insight, at molecular level, of both kinetics and thermodynamics of the reactions participating in the process (Gernaey *et al.*, 2012). This knowledge is essential to formulate reliable models, which are called to predict the influence of process and operating variables on system behavior. A dependable model is indeed necessary for the implementation of a proper control system since it allows investigating how the process outputs may change with time under the influence of changes of the external disturbances and manipulated variables. Another important factor that has to be accounted for regards the development of effective on-line analytic techniques aimed at monitoring and validating the "intensified" continuous processes. Also in this case, PI faces several barriers (Lutzea *et al.*, 2010): suppliers for industrial applications of PI technologies are lacking; the design of novel PI apparatuses on pilot-/industrial scale is expensive and technologically uncertain; the capital cost of novel PI modules per production capacity today is often higher than that of traditional reactors. In addition, efficient solutions for downstream processing in combination with PI reactors where the production of multiple products is combined in a single unit, are difficult to implement.

Agro-food industry is characterized by huge volumes or great flow rates of (very) diluted streams. Cost competitiveness of agro-food industry is significantly affected by two factors, namely the energy costs for processing and the wastes treatment cost.

Both these costs can be reduced exploiting process intensification methodologies (Doré *et al.*, 2011) provided that an optimization, performed over the whole agro chain and considering – at least – crop harvesting, transportation, preliminary processing, refining, derivatization process, has been preliminarily implemented. The production chain and processing of agro-food industry is indeed constrained by the limited stability of crops and derived materials. PI can potentially achieve different kinds of optimization. Typical examples are: separation in (or close to) the field of certain crop components determines significant energy and transportation cost savings; preservation of crop components right after harvesting reduces the seasonality effect, leading to much higher capacity utilization of process equipment and related capital costs; valorization of non-foodstuff competitive crop components into the so-called second-generation bio-fuels can be achieved through the exploitation of new crops (waste oils, lignocellulosic materials, whey, etc.) (Sansonetti *et al.*, 2009; De Paola *et al.*, 2009) and novel harvesting technologies. For instance, using a PI standpoint, milk separation into water, proteins and fats can be performed on-site (i.e., at the farm) with low energy-consuming micro separators. In this way, product transportation to and handling at the factory can be limited to the relevant components, proteins and fats, saving energy and reducing CO_2 emissions by avoiding the unnecessary transportation of water. Different factors have to be analyzed before implementing PI technology in the agro-food industry: crops with short storage limits such as milk, fruit juices, sugar beets, and vegetables do indeed determine particular problems; final product quality is strongly affected by the variability of the agro material due to source and seasonality; finally, the significant increase of crops dedicated for bio-ethanol and bio-diesel production is determining a competition for valuable lands utilization, thus determining the necessity of both intensified crops and novel process approaches. Food industry performance is also affected by market conditions, which request accurate cost awareness that directly turn into technology improvements so as to achieve higher yields. Product quality and food safety are definitely essential issues, which need to be accounted for in relation to consumer and legislation requirements. Food companies, therefore, need updated product innovation in order to satisfy consumer trends and changing demand. In addition, energy efficiency increases, plant safety, and flexibility are considered as important factors, which may contribute to the implementation of proper PI methodologies. Different technologies, having a potential high impact on food technology, have been actually identified. In particular, it is expected that PI will lead to an overall energy efficiency increase of 30–40% by using: (1) milder processing techniques which will result also in a better product quality; (2) more selective treatment of raw materials and ingredients; (3) more effective removal of micro traces of selected compounds; (4) increased processing flexibility (Ammar *et al.*, 2012). Moreover, it is expected that a decrease of (bio)-fouling in processing equipment and an increase in cleaning efficiency will determine a 20% capacity growth and a 60% energy saving. An important barrier to PI development in the agro-food industry is the limited availability of true process experts who, generally, concentrate on optimizing the existing plants and pay little attention to new, green, improved processes. This is partially to be ascribed to the fact that novel PI-based technologies are difficult to be incorporated in existing and well-established factories. Food industry is indeed a rather traditional sector not prone to adopting new technologies based, in particular, on PI methods (Fitzpatrick and Ahrné, 2005).

5.4   PI Exploitation in the Chemical Industry

In this section, a description of some process units designed on the basis of PI concepts will be presented, pointing out on their major features, on the advantages determined by the exploitation of these PI units and, in some cases, on the existing barriers that are currently limiting their spread on an industrial scale. Many of these units have already replaced the traditional apparatuses and improved the performance of conventional processes; some others are at a development stage, but it is expected that they may replace the traditional units in the future, hopefully in the next 10–15 years. The following critical review on each of the considered apparatus was readapted from the European Roadmap for Process Intensification where more detailed information can be found.

5.4.1   Structured Packing for Mass Transfer

Structured packing is used in various mass-transfer operations, including distillation, liquid/liquid mixing, absorption, and extraction. It offers high interfacial mass transfer areas and good phase dispersion resulting in high number of transfer units. Many different designs of structured packing are available; most of them are realized in order to form an open honeycomb structure with inclined flow channels and a relatively high exposed surface area (Bessoua *et al.*, 2010).

5.4.2   Static Mixers

Static mixers are pipe inserts which generate radial mixing or, for multiphase systems, produce fine bubbles or droplets (Lobry *et al.*, 2011). Static mixers are particularly useful for the continuous processing of chemicals; their main limitation is sensitivity to clogging. Compared to conventional mixing systems they are characterized by a very high-energy dissipation rate, resulting in very compact and energy efficient units.

5.4.3   Catalytic Foam Reactors

Solid foam catalytic reactors use a solid foam structure as the support for depositing a catalyst. These materials combine high void fraction and high surface area. Catalytic foam reactors (Palma *et al.*, 2012) represent a valid alternative to fixed bed reactors, since they are characterized by lower pressure drop and, hence, lower energy consumption; in addition, in the case of metallic foams, heat transfer is enhanced and more equal temperature profiles are obtained. The major disadvantages are represented by the inherently higher costs of the foam-based catalyst and by the relatively low surface area for deposition of the catalytic material. This technology is still in an early stage of development.

5.4.4   Monolithic Reactors

Monolithic catalysts are made of ceramic materials or corrugated metal sheets and consist of a many narrow parallel channels whose walls are used for catalyst deposition. Monoliths are characterized by narrower residence time distribution, which corresponds, as compared to conventional units, to higher selectivity, higher specific geometric areas, lower mass transfer resistances and, hence, smaller reactor volume, very low pressure drops and therefore, lower energy requirement (Edvinsson and Cybulski, 1994). Monolithic reactors are used in gas-phase cleaning in environmental applications; however, a cost reduction of monoliths and the exploitation of novel materials allowing higher catalyst loading are both required to achieve a wider spread of this technology.

5.4.5   Microchannel Reactors

Microreactors are chemical reactors of extremely small dimensions and usually a sandwich-like structure, consisting of a number of layers constituted by micromachined channels whose diameter may range between 20–400 μm (Fig. 5.2). Microreactors allow very high heat transfer rates (Regatte and Kaisare, 2011), not attainable by other apparatus. Very low reaction-volume-to-surface-area ratios make microreactors attractive for reactions involving poisonous or explosive reactants. Compared to traditional reactors, heat dissipation in microreactors is significantly enhanced by passing the reaction fluid through very small channels, realized in high thermally conductive metal blocks which, depending on the endothermic or exothermic nature of the reaction, are heated or cooled, respectively. The major advantage of microreactors is definitely represented by the excellent control of reaction temperature (Kockmann *et al.*, 2011). The main drawbacks, instead, are high pressure drops, relevant clogging tendency, and high unit cost. Microreactors may be successfully exploited in both fine chemistry and the pharmaceutical industry.

5.4.6   Non-Selective Membrane Reactors

In a reactor equipped with a non-selective membrane, the membrane is used to provide a support for the deposition of a catalyst (Shuit and Ong, 2012; Basile *et al.*, this book).

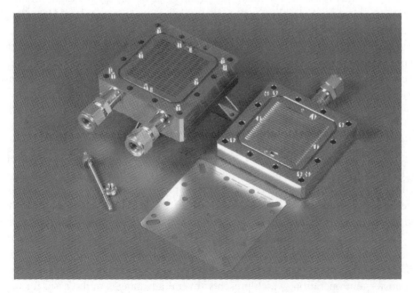

Figure 5.2   *Microchannel reactor. Reproduced from www.vinci-technologies.com*

The reaction, therefore, can take place either inside the pores of the membrane or on its external surface; the membrane is aimed at offering a very effective contact between reactants and catalyst, thus resulting in a short contact time catalytic reactor. An attractive exploitation of non-selective membrane reactor is the controlled dosing of one reactant along the length of the reactor (Rahimpour, 2009), which allows avoiding high local concentrations. The problems that need to be solved for a complete utilization of membrane technology on industrial scale are still represented by the system costs and the membrane stability.

5.4.7   Adsorptive Distillation

Adsorptive distillation is a three-phase mass transfer operation performed in two columns (Mujiburohman *et al.*, 2006); the first one is used to achieve the actual separation, the second one to attain the adsorbent regeneration. The adsorbent is a fine powder, which is fluidized and circulated by an inert carrier. Adsorptive distillation is capable of enhancing separation efficiency, especially in the case of azeotropes or close-boiling components and can be exploited, especially in the fine chemical industry, to perform continuous processes. One significant improvement of this operation is actually represented by the suspension absorptive distillation, where the solid absorptive material acts as an extractant. This technology, however, is still in a very early stage of development and no demonstration of its feasibility or of the actual advantages over traditional alternatives has been yet provided.

5.4.8   Heat-Integrated Distillation

As compared to a classic distillation column, a dividing wall column is capable of delivering pure side fractions, thus reducing the number of necessary distillation columns in a separation sequence. A dividing wall column represents an example of Heat-Integrated Distillation Columns (Kiss and Suszwalak, 2012; Markowski *et al.*, 2007) where the integration is achieved by the combination of the rectifying and the stripping columns in a single arrangement. Dividing wall columns have been effectively used in several chemical processes and can be considered to be a mature technology, which allows, in terms of energy and investment costs, savings even higher than 30%.

5.4.9   Membrane Absorption/Stripping

Membrane absorption occurs when a gaseous component is selectively transported through a membrane and, then, dissolved in an absorbing liquid (Fig. 5.3). One of the most important uses of membrane absorption is represented by CO_2 capture from flue gas (Feron and Jansen, 1997; Belaissaoui *et al.*, 2012). A natural extension of membrane adsorption is represented by either the membrane-based absorption-desorption process, where two liquids are on both sides of the membrane, or the membrane stripping process, in which selected components are removed from the liquid phase through the membrane by a stripping gas. Membrane absorption is actually a bubble-free gas-liquid mass transfer unit operation, which presents several advantages in the case, for instance, of shear-sensitive biological systems.

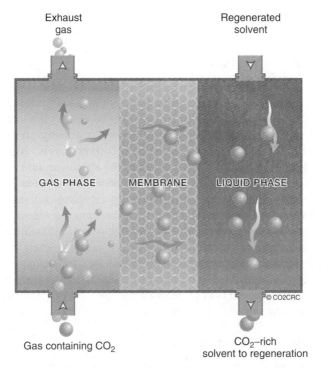

Figure 5.3 *Membrane absorption schematic. Reproduced with permission. Copyright ©
CO2CRC 2011*

5.4.10 Membrane Distillation

Membrane distillation (Calabrò *et al.*, 1994) is a thermally driven unit operation
consisting in transporting a volatile component of a liquid feed stream, as a vapor,
through a porous membrane; the vapor is then condensed on the other side of the
membrane (Fig. 5.4). This technology offers the potential of performing very efficient
separations; however, energy (actually a lower amount as compared to conventional

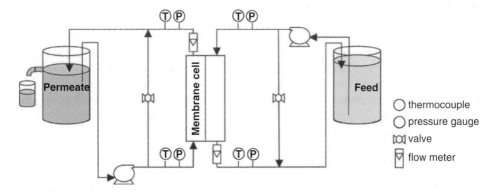

Figure 5.4 *Membrane distillation schematic. Reproduced from www.omicsonline.org*

distillation) is needed for vaporization. Another advantage is represented by the absence of concentration polarization, which, instead, afflicts other membrane operations.

5.4.11   Membrane Crystallization

Membrane crystallization is a novel crystallization technique, used – for instance – for protein crystals growing with enhanced crystallization kinetics. Membrane crystallization allows obtaining better crystal forms under operating conditions that are unsuitable for conventional crystallization processes. This technology is actually given as the combination of crystallization and membrane distillation (Di Profio *et al.*, 2010) and is based on a rather simple principle: the solvent evaporates (by steam/vacuum) at the membrane interface, migrates through the pores of the membrane and eventually condenses on the opposite side of the membrane. The membrane is capable to activate heterogeneous nucleation starting at low super saturation, thus enhancing the kinetics of crystallization, even for large molecules, like proteins (Di Profio *et al.*, 2006). Since all solvent must be evaporated, the process can be rather energy-intensive, even if it operates at a low temperature-difference as compared to the established technology, thus potentially resulting in energy savings. Moreover, due to the lower temperature differences, the crystallization process is carried out at milder conditions and in a more controlled way, which could be of great interest for the pharmaceutical industry.

5.4.12   Distillation-Pervaporation

A pervaporation membrane can be coupled with a conventional distillation column, thus resulting in a hybrid membrane/distillation process (Naidu and Malik, 2011), where the membrane can be placed either on the overhead vapor or on the feed of distillation column. Examples of potential improvements of this technology as compared to conventional processes are breaking an azeotrope, without the addition of a solvent, increasing the capacity of a distillation column and improving quality of the bottom and overhead products. The potential of the technology has been already proved in the case of alcohols dehydration and of the separation of isomeric hydrocarbons.

5.4.13   Membrane Reactors

Membrane reactors represent a very effective system in equilibrium limited reactions, since the products are continuously and selectively removed, thus favoring the forward reaction according to Le Chatelier's principle (Gallucci *et al.*, 2008). Two well-known examples of membrane reactors consist in hydrogen conducting membranes, based on palladium (Basile *et al.*, 2008), and oxygen conducting membranes, based on perovskites (Evdou *et al.*, 2008). A very wide range of principles and characteristics are actually known; this leads to a broad range of operating conditions that can be used to operate a membrane reactor. Figure 5.5 shows a palladium membrane reactor for the direct synthesis of phenols from benzene.

5.4.14   Heat Exchanger Reactors

In a heat exchanger reactor (Anxionnaz *et al.*, 2008, Despènes *et al.*, 2012) the reaction occurs close to a heat exchange surface, which is aimed at removing (or supplying for

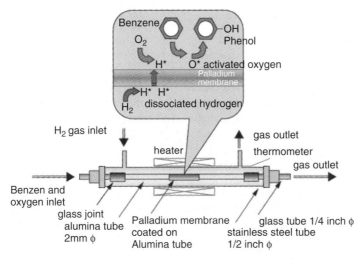

Figure 5.5 *Palladium membrane reactor for the direct synthesis of phenols from benzene in a single-stage reaction. Reproduced from www.aist.go.jp*

an endothermic reaction) the heat generated by the reaction. In this way, an accurate control of the operating temperature is actually attained. Both non-catalytic and catalytic heat-exchanger reactors show interesting promise for (very) fast reactions characterized by high heat of reaction.

5.4.15 Simulated Moving Bed Reactors

The simulated moving-bed reactor combines continuous countercurrent chromatographic separation with chemical reaction (Kundu *et al.*, 2009). This hybrid, not energy-intensive process is competitive with traditional processes in which reaction and separation are performed in different units. Higher conversions and better yields can be attained by separating the reaction product(s) in an equilibrium reaction. This technology is applied in those processes where chromatographic separation is a necessary step, that is, when high quality of separation is required and the product has a high value (fine chemicals and pharmaceuticals).

5.4.16 Gas-Solid-Solid Trickle Flow Reactor

In the Gas-Solid-Solid Trickle Flow Reactor fine adsorbent trickles through a fixed bed of catalyst (Kuczynski *et al.*, 1987); one or more products are selectively removed from the reaction zone. In case of methanol synthesis, for examples, conversions significantly higher than the equilibrium conversions are achieved under the same operating conditions. The major advantages of this technology are energy savings, increased conversion, and high potential for innovations as given by the utilization of multifunctional catalysis or of structured packing (Muzen and Cassanello, 2007). A current barrier is represented by the fact that, generally, solid recycling is less attractive than liquid or gas recirculation.

5.4.17   Reactive Extraction

Reactive extraction encompasses simultaneous reaction and liquid-liquid separation (Dussan *et al.*, 2010). Reactive extraction may be exploited in multi-reaction systems to attain significant improvements in both yield and selectivity of desired products. This leads to a reduction of both recycle streams and wastes. The combination of reaction and liquid-liquid extraction can be also used for the separation of waste byproducts, which are difficult to remove be conventional methods.

5.4.18   Reactive Absorption

Reactive absorption (Noeres *et al.*, 2003) is actually a mature technique used in the production of nitric or sulfuric acid, for the removal, by amine solutions, of the CO_2 contained in the flue gas, for gas desulfurization, in which H_2S is removed and converted to sulfur, and for the separation of light olefins and paraffins (Ortiz *et al.*, 2010). Reactive absorption is the most applied reactive separation technique.

5.4.19   Reactive Distillation

Reactive Distillation is performed in a conventional distillation column where some species react in the presence of a structured catalyst (Noeres *et al.*, 2002); the obtained products are continuously separated by fractionation, thus favoring the forward reaction; in some cases, even total conversion can be reached. The major benefits of this operation are: lower energy requirements, higher yields, good product purity, and lower capital investment. Reactive distillation may also be exploited as an effective separation method in the case of mixtures containing reactive and non-reactive components with near boiling point (Lai *et al.* 2007). Both homogeneous and heterogeneous catalysis can be applied, even if some problems regarding the catalyst development (performance, kinetics, stability, morphology, coating procedures, etc.) are still to be solved.

5.4.20   Membrane-Assisted Reactive Distillation

In some cases characterized by particular limitations, that is, the presence of an azeotrope, reactive distillation cannot fulfill the desired process performance and another unit operation has to be combined. This could be achieved by coupling a reactive distillation unit to a pervaporation module (Buchaly *et al.*, 2007) so as to further purify the distillate. Two commercial applications have been reported so far, that is the production of methyl borate and the production of fatty acid esters (Kiatkittipong *et al.*, 2011).

5.4.21   Hydrodynamic Cavitation Reactors

The energy associated with a liquid in motion can be successfully utilized to promote cavitation, which is aimed at intensifying reactions and other operations, such as homogenization or emulsification. Figure 5.6 shows a typical hydrodynamic cavitation chamber. Hydrodynamic cavitation can be obtained by allowing the liquid pass through a throttling valve, an orifice plate or any other mechanical constriction (Gogate and Kabadi, 2009). If the operating pressure falls below the cavitation pressure, millions of microcavities are obtained. These bubbles then implode when pressure is

Figure 5.6 *Hydrodynamic cavitation chamber. Reproduced from vrtxtech.com*

increased. Hydrodynamic cavitation may improve the performance of several industrial transformations, as compared to conventional processes (Gogate and Pandit, 2005). For example, cavitated corn slurry exhibits higher yields in ethanol production as compared to uncavitated corn slurry processing. A significant improvement is achieved also in the mineralization of bio-refractory compounds (Gogate, 2002), which would otherwise need extremely high temperature and pressure conditions to remove the free radicals generated during processing.

5.4.22 Pulsed Compression Reactor

A novel chemical reactor concept derived from PI approach allows a technological break-through in syngas production (Roestenberg *et al.*, 2011). This novel concept completely modifies the traditional standpoints since, instead of formulating improved catalysts, which may allow decreasing the operating temperature, no catalyst is used and the reaction is performed at very high temperatures, with peak values in the order of 1500–5000 K (Roestenberg *et al.*, 2010). The reactor consists of a double-ended cylinder and a free piston, which separates the cylinder into two compression-reaction chambers. The piston reciprocates compressing the feed gas, until it reacts, in turn in the lower and in the upper chamber. The reciprocation is maintained by the released reaction energy. This technology, even if it is an early stage of development, could provide an excellent potential for energy savings (reaction energy is directly applied); moreover, CO_2 and NO_x emissions are reduced since feed heating is actually not required. Cost effectiveness is also improved by narrower reaction product distribution (less downstream processing) and more compact reactor area design.

5.4.23 Sonochemical Reactors

In a liquid system the exposure to ultrasound determines, in a time period of a few milliseconds, the formation, growth, and subsequent collapse of microbubbles (Parvizian *et al.*, 2011). The micro-implosions are followed by high energy release, which leads to local generation of extremely high temperatures and pressures, which can be up to 5000 K and 2000 bar, respectively (Sutkar and Gogate, 2009). The use of ultrasound significantly enhances the rate of any chemical reaction, thus increasing the product yield. For chemical syntheses, this technology is in the early stages of its development; however, it may have potentially interesting applications in food processing, biotechnology and environmental protection.

5.4.24   Ultrasound-Enhanced Crystallization

Sonocrystallization is a non-invasive method based on the exploitation of ultrasound to control the point of nucleation and the number of nuclei formed in a crystallization process (Nalajala and Moholkar, 2011). This generally enhances the crystal yields, improves the product properties, namely handling and appearance, reduces the crystal's agglomeration and increases the process reproducibility. An additional advantage is given by the formation of ultra-fine, nano-structured materials; moreover, a significant energy and cost saving can be attained since expensive milling or recrystallization steps are avoided. This technology provides very interesting perspectives in both pharmaceutical and food industry, but reasonably will need about 10–15 years of fundamental research to reach full commercial application.

5.4.25   Electric Field-Enhanced Extraction

Electric fields are commercially employed to increase the process rates and to control the droplet size in pain spraying and in surface coating processes. Other well-known applications regard gas cleaning, emulsion breaking, ink-yet spraying, fuel spraying, crude oil desalting and bulk chemical washing. Electric fields can also enhance liquid-liquid or solid-liquid extraction (Grimi *et al.*, 2007, Gachovska *et al.*, 2010) and improvements of about 200%, as compared to conventional processes, were reported. Typical advantages of this technology are energy savings and equipment size reduction; however, even if well developed for some specific applications, no new innovative products are to be expected in the future.

5.4.26   Induction and Ohmic Heating

Inductive heating is a process where electric currents are induced within a food or any other product due to the presence of oscillating electromagnetic fields generated by electric coils (Sastry and Barach, 2000). Inductive heating is different from microwave heating due to the frequency and the nature of the source. In microwaves in fact, the frequency is specifically assigned and a magnetron (not coils) is used to promote heating. Ohmic heating (also known as Joule heating), electro-conductive heating, or direct electrical resistance heating, involves the direct passage of electric current through a material for the purpose of heating it. Ohmic heating necessarily involves electrodes that are put in contact with the product to create a continuous electrical circuit. The most promising application of Ohmic heating is the continuous sterilization of foods (Goullieux and Pain, 2005) by high-temperature-short-time processing, which yields a significant quality improvement. Figure 5.7 shows a typical ohmic heater exploited in the food industry.

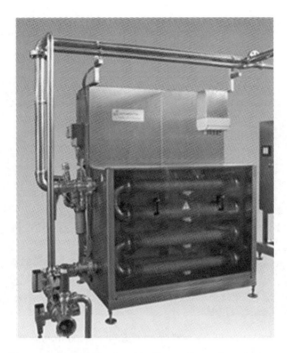

Figure 5.7 *Ohmic heater for the food industry. Reproduced from www.kasag.ch*

5.4.27 Microwave Drying

In general, microwave frequencies range from 0.3–300 GHz, but, in order to avoid any interference with radar and telecommunication applications, the industrial and domestic microwave appliances operate at a standard allocated frequency, most often at 2.45 GHz. Molecules with a permanent dipole moment (e.g., water) can rotate in a fast changing electric field; in addition, in substances where free ions or ionic species are present, the energy is also transferred by the ionic motion in an oscillating microwave field. As a result of both these mechanisms, the substance is heated directly and almost regularly. The heating extent depends on dielectric properties, namely the dielectric constant and the loss tangent, of the substance to be heated. Some materials absorb the microwave energy very easily; others are transparent or impermeable to it; this may allow a selective heating of materials/products (Holtz *et al.*, 2010). Microwave-enhanced drying is used on industrial scale in food, wood (Bartholme *et al.*, 2009), textile and pharmaceutical industries (Mcloughlin *et al.*, 2000). The speed of the MW drying allows preventing unwanted degradation of some thermo-labile components that it is wanted to preserve in the dried material. Microwave heating allows remarkable energy savings compared to conventional processes. Figure 5.8 shows an industrial-scale microwave oven exploited to achieve wood drying.

5.4.28 Microwave-Enhanced Separation and Microwave Reactors

Microwave heating proved to enhance also some extraction operations, particularly the extraction of pharmaceutical ingredients from plant material. A limited number of papers

Figure 5.8   *Microwave oven for wood drying. Reproduced from Jiyuan Electric Co., Ltd.*

regarding the enhancement effects of microwaves on membrane separation and distillation have been published as well. The ability of microwave heating to accelerate chemical reactions and to improve the product yield has been extensively reported (Monsef-Mirzai *et al.*, 1995). For example, in microwave-assisted oxidative coupling of methane on alumina supported La_2O_3/CeO_2 catalyst conversion occurred at temperatures about 250°C lower than those used by the conventional heating. However, no consolidated data exist on the actual energy/cost savings achievable by microwave reactors.

5.4.29   Photochemical Reactors

Photochemical reactors use the energy of light to initiate or catalyze reactions (Gupta *et al.*, 2011). The basic principle is that light quanta are absorbed by chemical species which are excited and become more reactive with regards to other compounds present. The light can originate either from the Sun or from artificial sources, namely medium-pressure mercury or a xenon lamp. In the case of solar photochemical reactors, the energy has to be concentrated so as to attain sufficient efficiency. In non-catalyzed photochemical reactions the light energy is absorbed by a reagent or by a sensitizer, which transfers the electronic energy to the reagent or undergoes a reversible redox reaction with the reagent. Currently, the major applications of photochemical reactors are in production of protective and decorative coatings, inks, packaging, and electronic materials. In other fields, like chemical, pharmaceutical, and food industry the applications are rather rare. The main benefits of this technology regard the use of low temperatures, which means significant energy savings, and very high conversion/yield/selectivity.

5.4.30 Oscillatory Baffled Reactor Technologies

The Oscillatory Baffled Reactor technology generally uses a cylindrical column containing equally spaced orifice baffles and superimposed fluid oscillation (Vilar *et al.*, 2008); this allows plug flow conditions even at low (laminar) flow rates, thus inducing enhanced mass and heat transfers as compared to conventional stirred tank reactors. Vortices are generated when fluid flows through the baffles; the continuous generation and cessation of eddies creates uniform mixing in each baffled cell (Smith and Mackley, 2006). The major advantages of this technology regard a significant energy/utility savings, higher yields and less side product. In addition, capital cost savings are achieved through much more compact designs.

5.4.31 Reverse Flow Reactor Operation

A well-known example of the integration of reaction and heat transfer in a multifunctional single unit is reverse-flow reactors (Simeone *et al.*, 2012) where one or more process variables are intentionally and continuously perturbed (see Fig. 5.9); this dynamic operation results in process improvements, which cannot be achieved by steady state operation. In the case of exothermic processes the periodic flow reversal, typical of such systems, allows for a very good utilization of the heat of reaction, which is maintained within the catalyst bed and, after the reversion of the flow, is used to pre-heat the stream of cold reactants entering the reactor. So far, the reverse-flow reactors have been used in the SO_2 oxidation (Gosiewski, 1993), in the total oxidation of hydrocarbons contained in the off-gases and in the NO_x reduction (Botar-Jid *et al.*, 2007). The major benefits of this technology are represented by energy savings, increased conversion selectivity, and enhanced productivity.

5.4.32 Pulse Combustion Drying

The term pulse combustion originates from the intermittent combustion of the solid, liquid, or gaseous fuel in contrast to the continuous combustion that occurs in conventional burners (Zbiciński *et al.*, 2001). Such periodic combustion generates pressure,

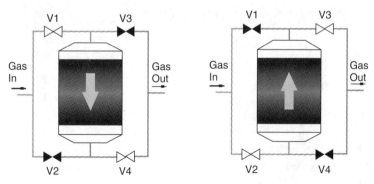

Phase 1. Forward direction of the flow   Phase 2. Reverse direction of the flow

Figure 5.9 *Reverse flow reactor operation. Reproduced from www.matrostech.com*

velocity, and temperature waves which propagate from the combustion chamber to the drying chamber. Due to such wave propagation, the rates of heat and mass transfer and, consequently the drying rate, are significantly improved as compared to conventional processes. The pulse combustion dryer consists of a pulse combustor combined to a spray dryer or a rotary kiln or a fluid-bed dryer (Zbiciński, 2002). Pulse combustion drying represents an energy-efficient and environmentally-friendly technology; as compared to traditional spray dryers, it indeed provides lower energy consumption, lower capital cost, and lower CO_2 emission. In spite of these advantages, the application of pulse combustion is not widespread.

5.4.33   Supercritical Separation

Supercritical fluids have some properties, for example, density, that are quite similar to those of a liquid, and some other properties, such as low viscosity, low surface tension, and high diffusivities for solutes, which makes them similar to a gas (Gere *et al.*, 1997). This is due to the fact that above the supercritical point there is no difference between the phases. Supercritical fluids, due to their unique properties, may significantly improve the performance of some conventional separation processes, like for example extraction. In fact, some compounds which are almost insoluble in a fluid at ambient conditions can become soluble in the fluid at supercritical conditions (see Fig. 5.10). Currently CO_2 is mostly used but other compounds, namely hexane, pentane, and ammonia, can be

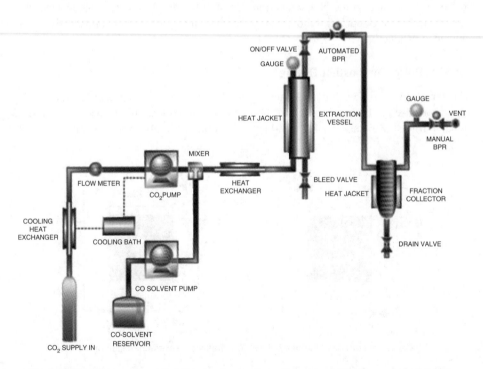

Figure 5.10   *Supercritical fluid extraction system. Reproduced from www.waters.com*

exploited as well. Supercritical separations using CO_2 are well established in extraction of natural materials (Reverchon, 1997). By using CO_2 in supercritical conditions, that is, when the substance is like a liquid, high value compounds can be extracted from mixtures; after pressure reduction, the supercritical fluid becomes a gas and the compounds can be recovered in a pure/not dissolved form. The major benefit is therefore represented by the avoidance of a traditional solvent and its associated waste problems, that is, emissions to the atmosphere or waste treatment.

5.5 Conclusions

Process intensification is a valuable tool for the development of more sophisticated and more efficient processes aimed at sustainable production in industry. Process intensification methodologies have already offered substantial benefits in material and energy efficiency, and the contribution of PI concepts was fundamental in designing novel types of equipment and of processes. Further improvements can be attained if operation and control of a complete process are considered simultaneously and systematically together with different PI design options. In order to find true optimal configuration it is also necessary to exploit a rigorous methodology, which allows determining intensified options by stepwise reduction of the search space through constraints, performance evaluation and objective function. In this context, a set of PI metrics accounting for economic, environmental, safety and process performance has to be developed to compare, on the basis of precise modeling techniques, different possible options, which allow intensifying a process through a standard set of performance measures.

References

Y. Ammar, S. Joyce, R. Norman, *et al.*, Low grade thermal energy sources and uses from the process industry in the UK (2012), *Applied Energy*, **89** (1), 3–20.

Z. Anxionnaz, M. Cabassud, C. Gourdon and P. Tochon, Heat exchanger/reactors (HEX reactors): Concepts, technologies: State-of-the-art (2008), *Chemical Engineering and Processing: Process Intensification*, **47** (12), 2029–2050.

M. Bartholme, G. Avramidis, W. Viöl and A. Kharazipour, Microwave drying of wet processed wood fibre insulating boards (2009), *European Journal of Wood and Wood Products*, **67** (3), 357–360.

A. Basile, F. Gallucci and S. Tosti, Ch. 8: Synthesis, Characterization, and Applications of Palladium Membranes (2008), in: *Inorganic Membranes: Synthesis, Characterization and Applications, in Membrane Science and Technology*, Vol. **13**, 255–323, Elsevier B.V., Amsterdam.

A. Basile, A. Iulianelli and S. Liguori, Process intensification in the chemical and petrochemical industry, Ch. 6, this book.

B. Belaissaoui, D. Willson and E. Favre, Membrane gas separations and post-combustion carbon dioxide capture: Parametric sensitivity and process integration strategies (2012), *Chemical Engineering Journal*, **211–212**, 122–132.

V. Bessoua, D. Rouzineaua, M. Prévosta, *et al.*, Performance characteristics of a new structured packing (2010), *Chemical Engineering Science*, **65** (16), 4855–4865.

C. C. Botar-Jid, P.Ş. Agachi and D. Fissore, Comparison of reverse flow and counter-current reactors in case of selective catalytic reduction of NO_x (2007), *Computer Aided Chemical Engineering*, **24**, 1331–1336.

C. Buchaly, P. Kreis and A. Górak, Hybrid separation processes – Combination of reactive distillation with membrane separation (2007), *Chemical Engineering and Processing: Process Intensification*, **46** (9), 790–799.

V. Calabrò, B. L. Jiao and E. Drioli, Theoretical and experimental study on membrane distillation in the concentration of orange juice (1994), *Industrial & Engineering Chemistry Research*, **33** (7), 1803–1808.

M. G. De Paola, E. Ricca *et al.*, Factor analysis of transesterification reaction of waste oil for biodiesel production (2009), *Bioresource Technology*, **100**, 5126–5131.

L. Despènes, S. Elgue, C. Gourdon and M. Cabassud, Impact of the material on the thermal behaviour of heat exchangers-reactors (2012), *Chemical Engineering and Processing: Process Intensification*, **52**, 102–111.

G. Di Profio, E. Curcio and E. Drioli, Controlling protein crystallization kinetics in membrane crystallizers: effects on morphology and structure (2006), *Desalination*, **200** (1–3), 598–600.

G. Di Profio, E. Curcio and E. Drioli, Membrane Crystallization Technology (2010), in *Comprehensive Membrane Science and Engineering, Volume 4: Membrane Contactors and Integrated Membrane Operations*, ISBN: 978-0-08-093250-7, Elsevier B.V., Amsterdam.

T. Doré, D. Makowski, E. Malézieux, *et al.*, Facing up to the paradigm of ecological intensification in agronomy: Revisiting methods, concepts and knowledge (2011), *European Journal of Agronomy*, **34** (4), 197–210.

K. J. Dussan, C. A. Cardona, O. H. Giraldo, *et al.*, Analysis of a reactive extraction process for biodiesel production using a lipase immobilized on magnetic nanostructures (2010), *Bioresource Technology*, **101** (24), 9542–9549.

R. K. Edvinsson and A. Cybulski, A comparative analysis of the trickle-bed and the monolithic reactor for three-phase hydrogenations (1994), *Chemical Engineering Science*, **49** (24 Part 2), 5653–5666.

A. Evdou, L. Nalbandian and V. T. Zaspalis, Perovskite membrane reactor for continuous and isothermal redox hydrogen production from the dissociation of water (2008), *Journal of Membrane Science*, **325** (2), 704–711.

P. H. M. Feron and A. E. Jansen, The production of carbon dioxide from flue gas by membrane gas absorption (1997), *Energy Conversion and Management*, **38** (Supplement), S93–S98.

J. J. Fitzpatrick and L. Ahrné, Food powder handling and processing: Industry problems, knowledge barriers and research opportunities (2005), *Chemical Engineering and Processing: Process Intensification*, **44** (2), 209–214.

H. Freund and K. Sundmacher, Towards a methodology for the systematic analysis and design of efficient chemical processes Part 1. From unit operations to elementary process functions (2008), *Chemical Engineering and Processing: Process Intensification*, **47**, 2051–2060.

T. Gachovska, D. Cassada, J. Subbiah, *et al.*, Enhanced anthocyanin extraction from red cabbage using pulsed electric field processing (2010), *Journal of Food Science*, **75** (6), 323–329.

F. Gallucci, S. Tosti and A. Basile, Pd–Ag tubular membrane reactors for methane dry reforming: A reactive method for CO2 consumption and H2 production (2008), *Journal of Membrane Science*, **317** (1–2), 96–105.

D. R. Gere, L. G. Randall and D. Callahan, Ch. 11: Supercritical fluid extraction: Principles and applications (1997), in *Techniques and Instrumentation in Analytical Chemistry*, Vol. **18**, 421–484, Elsevier B.V., Amsterdam.

K. V. Gernaey, A. E. Cervera-Padrell and J. M. Woodley, A perspective on PSE in pharmaceutical process development and innovation (2012), *Computers & Chemical Engineering*, **42**, 15–29.

P. R. Gogate, Cavitation: an auxiliary technique in wastewater treatment schemes (2002), *Advances in Environmental Research*, **6** (3), 335–358.

P. R. Gogate and A. M. Kabadi, A review of applications of cavitation in biochemical engineering/biotechnology (2009), *Biochemical Engineering Journal*, **44** (1), 60–72.

P. R. Gogate and A. B. Pandit, A review and assessment of hydrodynamic cavitation as a technology for the future (2005), *Ultrasonics Sonochemistry*, **12** (1–2), 21–27.

K. Gosiewski, Dynamic modelling of industrial SO2 oxidation reactors. Part II. Model of a reverse-flow reactor (1993), *Chemical Engineering and Processing: Process Intensification*, **32** (4), 233–244.

A. Goullieux and J. P. Pain, Ch. 18: Ohmic Heating (2005), in *Emerging Technologies for Food Processing*, 469–505, D.-W. Sun (ed.), Elsevier Ltd., Oxford, ISBN: 9780126767575.

N. Grimi, I. Praporscic, N. Lebovka and E. Vorobiev, Selective extraction from carrot slices by pressing and washing enhanced by pulsed electric fields (2007), *Separation and Purification Technology*, **58** (2), 267–273.

V. K. Gupta, R. Jain, S. Agarwal and M. Shrivastava, Kinetics of photo-catalytic degradation of hazardous dye Tropaeoline 000 using UV/TiO 2 in a UV reactor (2011), *Colloids and Surfaces A: Physicochemical and Engineering Aspects*, **378** (1–3), 22–26.

J. Harmsen, Process intensification in the petrochemicals industry: Drivers and hurdles for commercial implementation (2010), *Chemical Engineering and Processing: Process Intensification*, **49** (1), 70–73.

E. Holtz, L. Ahrné, M. Rittenauer and A. Rasmuson, Influence of dielectric and sorption properties on drying behaviour and energy efficiency during microwave convective drying of selected food and non-food inorganic materials (2010), *Journal of Food Engineering*, **97** (2), 144–153.

W. Kiatkittipong, P. Intaracharoen, N. Laosiripojana, *et al.*, Glycerol ethers synthesis from glycerol etherification with tert -butyl alcohol in reactive distillation (2011), *Computers & Chemical Engineering*, **35** (10), 2034–2043.

A. A. Kiss and D. J. P. C. Suszwalak, Enhanced bioethanol dehydration by extractive and azeotropic distillation in dividing-wall columns (2012), *Separation and Purification Technology*, **86**, 70–78.

N. Kockmann, M. Gottsponer and D. M. Roberge, Scale-up concept of single-channel microreactors from process development to industrial production (2011), *Chemical Engineering Journal*, **167** (2–3), 718–726.

M. Kuczynski, M. H. Oyevaar, R. T. Pieters and K. R. Westerterp, Methanol synthesis in a countercurrent gas-solid-solid trickle flow reactor: An experimental study (1987), *Chemical Engineering Science*, **42** (8), 1887–1898.

P. K. Kundu, Y. Zhang and A. K. Ray, Modeling and simulation of simulated countercurrent moving bed chromatographic reactor for oxidative coupling of methane (2009), *Chemical Engineering Science*, **64** (24), 5143–5152.

I. K. Lai, S. B. Hung, W. J. Hung, *et al.*, Design and control of reactive distillation for ethyl and isopropyl acetates production with azeotropic feeds (2007), *Chemical Engineering Science*, **62** (3), 878–898.

E. Lobry, F. Theron, C. Gourdon, *et al.*, Turbulent liquid–liquid dispersion in SMV static mixer at high dispersed phase concentration (2011), *Chemical Engineering Science*, **66** (23), 5762–5774.

P. Lutzea, R. Gani and J. M. Woodley, Process intensification: A perspective on process synthesis (2010), *Chemical Engineering and Processing: Process Intensification*, **49**, 547–558.

M. Markowski, M. Trafczynski and K. Urbaniec, Energy expenditure in the thermal separation of hydrocarbon mixtures using a sequence of heat-integrated distillation columns (2007), *Applied Thermal Engineering*, **27** (7), 1198–1204.

C. M. McLoughlin, W. A. M. McMinn and T.R.A. Magee, Microwave Drying of Pharmaceutical Powders (2000), *Food and Bioproducts Processing*, **78** (2), 90–96.

P. Monsef-Mirzai, W. R. McWhinnie, M. C. Perry and P. Burchill, Reaction of phosphorylating reagents with substituted phenols and coals, using microwave heating (1995), *Fuel*, **74** (7), 1004–1008.

M. Mujiburohman, W. B. Sediawan and H. Sulistyo, A preliminary study: Distillation of isopropanol–water mixture using fixed adsorptive distillation method (2006), *Separation and Purification Technology*, **48** (1), 85–92.

A. Muzen and M. C. Cassanello, Flow regime transition in a trickle bed with structured packing examined with conductimetric probes (2007), *Chemical Engineering Science*, **62** (5),1494–1503.

Y. Naidu and R. K. Malik, A generalized methodology for optimal configurations of hybrid distillation–pervaporation processes (2011), *Chemical Engineering Research and Design*, **89** (8), 1348–1361.

V. S. Nalajala and V. S. Moholkar, Investigations in the physical mechanism of sonocrystallization (2011), *Ultrasonics Sonochemistry*, **18** (1), 345–355.

C. Noeres, A. Hoffmann and A. Górak, Reactive distillation: Non-ideal flow behaviour of the liquid phase in structured catalytic packings (2002), *Chemical Engineering Science*, **57** (9), 1545–1549.

C. Noeres, E. Y. Kenig and A. Górak, Modelling of reactive separation processes: reactive absorption and reactive distillation (2003), *Chemical Engineering and Processing: Process Intensification*, **42** (3), 157–178.

A. Ortiz, L. M. Galán, D. Gorri, *et al.*, Kinetics of reactive absorption of propylene in RTIL-Ag + media (2010), *Separation and Purification Technology*, **73** (2), 106–113.

V. Palma, A. Ricca and P. Ciambelli, Monolith and foam catalysts performances in ATR of liquid and gaseous fuels (2012), *Chemical Engineering Journal*, **207–208**, 577–586.

F. Parvizian, M. Rahimi and M. Faryadi, Macro- and micromixing in a novel sono-chemical reactor using high frequency ultrasound (2011), *Chemical Engineering and Processing: Process Intensification*, **50** (8), 732–740.

M. R. Rahimpour, Enhancement of hydrogen production in a novel fluidized-bed membrane reactor for naphtha reforming (2009), *International Journal of Hydrogen Energy*, **34** (5), 2235–2251.

D. Reay, C. Ramshaw and A. Harvey, *Process Intensification – Engineering for Efficiency: Sustainability and Flexibility* (2008), Elsevier Ltd.

V. R. Regatte and N. S. Kaisare, Propane combustion in non-adiabatic microreactors: 1. Comparison of channel and posted catalytic inserts (2011), *Chemical Engineering Science*, **66** (6), 1123–1131.

T. Ren, Barriers and drivers for process innovation in the petrochemical industry: A case study (2009), *Journal of Engineering and Technology Management*, **26** (4), 285–304.

E. Reverchon, Supercritical fluid extraction and fractionation of essential oils and related products (1997), *The Journal of Supercritical Fluids*, **10** (1), 1–37.

T. Roestenberg, M. J. Glushenkov, A. E. Kronberg, *et al.*, Heat transfer study of the pulsed compression reactor (2010), *Chemical Engineering Science*, **65** (1), 88–91.

T. Roestenberg, M. J. Glushenkov, A. E. Kronberg, *et al.*, Experimental study and simulation of syngas generation from methane in the Pulsed Compression Reactor (2011), *Fuel*, **90** (5), 1875–1883.

J. P. M. Sanders, J. H. Clark, G. J. Harmsen, *et al.*, Process intensification in the future production of base chemicals from biomass (2012), *Chemical Engineering and Processing: Process Intensification*, **51**, 117–136.

S. Sansonetti, S. Curcio, V. Calabrò and G. Iorio, Bio-ethanol production by fermentation of ricotta cheese whey as an effective alternative non-vegetable source (2009), *Biomass & Bioenergy*, **33**, 1687–1692.

S. Sastry and J. T. Barach, Ohmic and inductive heating (2000), *Journal of Food Science*, **65** (s8), 42–46.

S. H. Shuit, Y. T. Ong, K. T. Lee, *et al.*, Membrane technology as a promising alternative in biodiesel production: A review (2012), *Biotechnology Advances*, **30** (6), 1364–1380.

M. Simeone, L. Salemme and L. Menna, Methane autothermal reforming in a reverse flow reactor on Rh/Al $2O$ 3 catalyst (2012), *International Journal of Hydrogen Energy*, **37** (11), 9049–9057.

K. B. Smith and M. R. Mackley, An experimental investigation into the scale-up of oscillatory flow mixing in baffled tubes (2006), *Chemical Engineering Research and Design*, **84** (11), 1001–1011.

V. S. Sutkar and P. R. Gogate, Design aspects of sonochemical reactors: Techniques for understanding cavitational activity distribution and effect of operating parameters (2009), *Chemical Engineering Journal*, **155** (1–2), 26–36.

T. Van Gerven and A. Stankiewicz, Structure, energy, synergy, time the fundamentals of process intensification (2009), *Ind. Eng. Chem. Res.*, **48**, 2465–2474.

G. Vilar, R. A. Williams, M. Wang and R. J. Tweedie, On line analysis of structure of dispersions in an oscillatory baffled reactor using electrical impedance tomography (2008), *Chemical Engineering Journal*, **141** (1–3), 58–66.

G. D. Yadav, Y. B. Jadhav and S. Sengupta, Selectivity engineered phase transfer catalysis in the synthesis of fine chemicals: reactions of p -chloronitrobenzene with sodium sulphide (2003), *Journal of Molecular Catalysis A: Chemical*, **200** (1–2), 117–129.

I. Zbiciński, Equipment, technology, perspectives and modeling of pulse combustion drying (2002), *Chemical Engineering Journal*, **86** (1–2), 33–46.

I. Zbiciński, C. Strumillo, M. Kwapinska and I. Smucerowicz, Calculations of the pulse combustion drying system (2001), *Energy Conversion and Management*, **42** (15–17), 1909–1918.

6

Process Intensification in the Chemical and Petrochemical Industry

Angelo Basile, Adolfo Iulianelli, and Simona Liguori
Institute of Membrane Technology, Italian National Research Council (ITM-CNR),
University of Calabria, Rende, Italy

6.1 Introduction

The increasing demand for raw materials, energy, and products under constraints imposed by sustainable development is a complex issue that needs concrete solutions. These can be found by the integration and implementation of new economical, industrial, environmental, and social strategies. On this route, many efforts from both chemical engineering and industrial research focus on the development of new equipment and techniques that potentially could transform the concept of chemical plants, leading to compact, safe, energy-efficient and environment-friendly sustainable processes.

In this context, *process intensification* (PI) represents a design approach useful for achieving real benefits in manufacturing and processing, such as the decrease of equipment size, energy consumption, waste production, and capital cost and the enhancement of plant efficiency resulting in cheaper and sustainable technology.

These goals can be achieved by multifunctional reactors, which can be described as reactors combining at least two functions/operations that, conventionally, are performed in separate devices. In particular, multifunctional reactors may integrate operations

Sustainable Development in Chemical Engineering – Innovative Technologies, First Edition.
Edited by Vincenzo Piemonte, Marcello De Falco and Angelo Basile.
© 2013 John Wiley & Sons, Ltd. Published 2013 by John Wiley & Sons, Ltd.

like reaction and heat transfer, reaction and separation and combine reaction and phase transition.

A well-known example of multifunctional reactor is the Membrane Reactor (MR), in which separation and reaction take place in the same tool. This alternative device represents a real model of intensification showing a higher efficiency compared to conventional separation and reaction operations.

The number of possible applications of MRs in chemical and petrochemical industry is wide owing to its benefits. Indeed, by using MRs for carrying out, for example, dehydrogenation, methane steam reforming, oxidative coupling of methane and water gas shift reactions, the performances in terms of conversion, products selectivity and yield of overall process result to be enhanced and, in some case, the thermodynamic restrictions are overcome.

However, the industrial interest towards the membrane technology is due to the fact that the membrane operations have intrinsic characteristics of efficiency and operational simplicity, compatibility between different membrane operations in integrated systems, low energy requirements, stability under operating conditions, environmental compatibility, and scale-up.

For all these reasons, the integrated unit membrane operations are becoming an attractive and economically competitive option with respect to the traditional ones for both chemical and petrochemical industrial processes.

To conclude, an overview is given in this chapter on the importance of PI for realizing real benefits in industrial process, focusing the attention on membrane technology, in particular on MRs.

6.2   Process Intensification

6.2.1   Definition and Principles

The term "process intensification" was originally introduced at ICI (Imperial Chemical Industries) in the UK in the 1970s (Dautzenberg and Mukherjee, 2001). As a general approach, this term was used to describe the strategy for reducing *dramatically* the physical size of a chemical plant achieving, in the same time, a given production objective. During the last two decades, many definitions have been proposed, which, despite to their common point of view on innovation, were often quite different in substance. For instance, for Stitt (2002) and Mae (2007) the apparatus miniaturization with microreactors is the fundamental issue; for Dautzenberg and Mukherjee, (2001) and Huang *et al.* (2007) the PI is based on functional integration with reactive distillation and for Arizmendi-Sanchez and Sharatt (2008) the main PI principles are the combination of different approaches by identifying synergistic integration of process tasks and phenomena. Finally, in 2009 Gerven and Stankiewicz undertook the first attempt to define the fundamentals of PI with the suggestion that PI should follow a function oriented approach. In particular, they distinguish four principles (Table 6.1) taking into account all scales existing in chemical processes, from molecular to meso and macro-scale.

These principles are not completely new to chemical engineering. Nevertheless, in PI they are seen as explicit goals that an intensified process aims to reach. Moreover, the PI interpretation of these principles often goes beyond the limits of the classical chemical

Table 6.1   *Generic principles of PI (Gerven and Stankiewicz, 2009)*

Principle (Goal)	Focus issues related to the principle
Maximize the effectiveness of intra and intermolecular events	Control of spatial orientation and energy in molecular collisions
Give each molecule the same processing experience	Molecules uniformity, residence time and thermal uniformity
Optimize the driving forces at every scale and maximize the specific surface area to which these forces apply	Transport across interfaces
Maximize the synergistic effects from partial processes	Combining functions

engineering approach, as in the first principle, where PI concerns a method of improving the inherent kinetics of chemical reactions, rather than reaching them.

6.2.2   Components

Generally, as shown in Figure 6.1, the whole PI field can be divided into two areas: process-intensifying equipment, such as novel reactors, intensive mixing, heat-transfer, and mass-transfer devices; and process-intensifying methods, such as new or hybrid separations, integration of reaction and separation, heat exchange, or phase transition (in so-called multifunctional reactors), techniques using alternative energy sources (light, ultrasound, etc.), and new process-control methods (Stankiewicz, 2003). Obviously, there could also be some overlap. Indeed, new methods could need equipments already developed and vice versa.

On this route, PI leads to higher process flexibility, improved inherent safety, and energy efficiency, distributed manufacturing capability and ability to use reactants at higher concentrations (Keil, 2007). These goals are achieved, for instance, by multifunctional reactors as reactive distillation or MRs. In particular, MRs are considered the most important class of multifunctional reactors due to their ability to combine distinct tasks as reaction and separation in only one device. On this route, the membrane represents the main key factor for application of membrane-based separations during the reaction in the chemical and petrochemical-related industries.

In particular, the membrane can play various functions in such reactor systems. For instance, it can be used for selective in situ separation of the reaction products realizing, as a benefit, the reaction shift towards a further products formation. It can be also applied for a controlled distributed feed of some reactants in order to increase overall yield or selectivity of a process, as in fixed-bed or fluidized-bed MRs (Tsotsis *et al.*, 1992; Adris and Grace, 1997) or to facilitate mass transfer, as direct bubble-free oxygen supply or dissolution in the liquid phase via hollow-fiber membranes (Ahmed and Semmens, 1992; Shanbhang *et al.*, 1995) The choice of membrane type to be used in MRs depends on various parameters, such as the productivity, separation selectivity, membrane lifetime, mechanical, and chemical integrity at the operating conditions, and particularly, the cost (Julbe *et al.*, 2001).

Currently, membrane technologies are more often used for separation of a wide variety of mixtures in chemical and petrochemical industries competing successfully

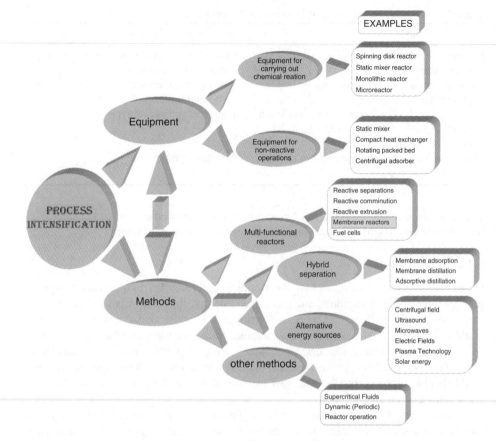

Figure 6.1   *Process Intensification and its components*

with conventional processes (Bessarabov, 1999; Spillman, 1995; Ettouney *et al.*, 1998). From this point of view, MR could be an optimal choice as an alternative solution to the conventional systems in order to carry out the reaction and separation in only one device improving, as a consequence, the overall process performances and using smaller equipment.

6.3   The Membrane Role

Membrane is defined essentially as a barrier, which separates two phases and restricts transport of various chemicals in a selective manner (Ravanchi and Kaghazchi, 2009).

Generally, the membranes can be classified according to their nature, geometry and separation regime (Khulbe *et al.*, 2007). The first classification by their nature distinguishes the membranes into biological and synthetic ones. These two types of membranes differ completely for functionality and structure (Xia *et al.*, 2003). Furthermore, synthetic membranes can be subdivided into organic and inorganic ones. The first kind of membrane usually operates between 100–300 °C (Catalytica, 1998), the second one

Table 6.2 *Benefits and drawbacks of inorganic membranes with respect to the polymeric ones*

Benefits	Drawbacks
Stability at high temperatures	Low capital/operating cost
Resistance to harsh environments (chemical degradation, pH, etc.) and to high pressure drops	Embrittlement phenomenon (in the case of dense Pd membranes)
Inertness to microbiological degradation	Difficulty of achieving high selectivities in large scale microporous membranes
Easy to clean after fouling	Generally low permeability of the highly hydrogen selective (dense) membranes at medium temperatures
Easy catalytic activation	Difficult to seal membrane-to-module at high temperature

above 250 °C (Hseih, 1996). Moreover, the inorganic membranes may be categorized as porous, classified according to their diameter, and no-porous (dense), subdivided into supported and unsupported.

From an industrial processes point of view, it is very important to understand the limits and benefits of both organic and inorganic membranes because the membrane choice in industrial processes depends on its stability, durability, and its chemical and mechanical robustness under specific operating conditions.

Indeed, for example, in the recovery and separation of higher C_2 hydrocarbons from refinery streams or methane in natural-gas processing, the polymeric membranes could be used, whereas to produce high purity hydrogen, it is preferred to use inorganic membranes, in particular dense palladium-based membranes, which are characterized by full selectivity to hydrogen permeation with respect to all other gases.

In Table 6.2, the most important advantages and disadvantages of inorganic membranes with respect to the polymeric ones are reported.

Nowadays in the chemical and petrochemical industry, membrane technology is a well-established unit operation. In particular, in gas separation field the application of membrane separation systems competes with the traditional methods as absorption, adsorption, and cryogenics. Indeed, membranes offer versatility and simplicity in comparison to other methods (Scoth, 1999). An overview of various applications and materials used for gas separation processes can be found in the literature (Spillman, 1989; Walker and Koros, 1991; Kesting and Fritzsche 1993; Cen and Lichtenthaler, 1995; Baker, 2002, 2004).

The industrial interest toward membrane technology is not only limited to the separation of gaseous mixtures, but different membrane operations as reverse osmosis (Lefebvre and Moletta, 2006; Muraleedaaran *et al.*, 2009), pervaporation (Lin *et al.*, 2009), organic solvent nanofiltration (White, 2006; Vandezande *et al.*, 2008), membrane contactors (Yeon *et al.*, 2005) and MRs (Gryaznov and Orekhova, 1998; Sirkar *et al.*, 1999; Shirasaki *et al.*, 2009) have been also taken into account. In particular, in recent years significant progress has been made in developing alternative devices such as MRs. Thus, the chemical and petrochemical industry could take advantage considerably from these innovative technologies in order to satisfy the stringent environmental standards, to control production cost and final product quality.

6.4   Membrane Reactor

MRs have been investigated since the 1970s (Wasewar, 2007). In particular, most of the progress in the MR area has been developed in the last 20 years mainly owing to the development of new membrane materials. The significant progress in this area is reflected in an increasing number of scientific publications, which have grown exponentially over the last few years, as shown in Figure 6.2.

A MR can perform both in flat or tubular geometry (Figure 6.3). In the latter case, multichannel tubular monoliths and a catalyst deposited inside the pores can be used to improve the density of packed bed.

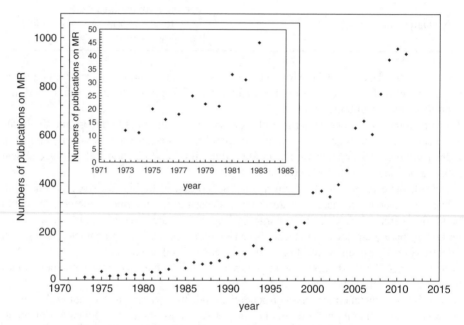

Figure 6.2   *Number of publications versus time (http://scopees.elsevier.com)*

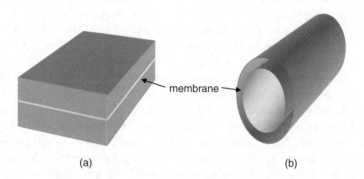

Figure 6.3   *(a) Flat and (b) tubular MR*

Generally, MRs can be also categorized according to the function and position of membrane, as reported in Table 6.3 (Langhendries and Baron, 1998; Wu *et al.*, 2000; Vital *et al.*, 2001; Liu *et al.*, 2002; Tuan *et al.*, 2002).

Hence, the membrane can cover various functions in the reactor system. For instance, a membrane is able to separate products from the reaction mixture or it allows the control

Table 6.3   *Classification of MRs based on the function and position of membrane*

Figure	Membrane Reactor	Main Features
	CMR Catalytic active membrane Perm-selective membrane	The selective product removal improves the conversion. The selective reactant supply increase the selectivity
	CNMR Catalytic active membrane Non perm-selective membrane Interphase contactor	Short contact time
	PBMR Inert membrane Perm-selective membrane	The selective product removal improves the conversion. The selective reactant supply increase the selectivity
	NMR Inert membrane Non perm-selective membrane Reactant distributor	Better controllability Improved selectivity
	PLMR The MR consists of catalyst particles coated with a perm-selective membrane layer Inert membrane	The selective product removal or reactant supply improves the selectivity. Poisoning catalyst is avoided

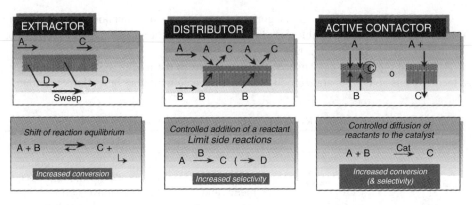

Figure 6.4   *The three main applications of MR*

of the addition of one or more reactants into reaction zone. As a consequence, the MR can be mainly used as an extractor, distributor or active contactor (Figure 6.4a,b,c).

The extractor mode is the most common type in which one or more products, generated during the reaction, are continuously and selectively removed from the reaction zone. In the petrochemical field, this type of MR is usually applied to increase the conversion of numerous reactions limited by thermodynamics, such as light alkanes catalytic dehy-drogenations or water-gas shift, methane steam reforming, by the selective removing of the hydrogen produced (Westermann and Melin, 2009). Although, most of the extractor applications concern H_2 removal, some decomposition reactions in which O_2 is extracted have been also considered (Dixon, 1999).

The distributor mode is used to control the contact within reactants and it is typically applied to consecutive or parallel reaction systems, such as partial oxidation, hydro-carbon oxy-dehydrogenation or methane oxidative coupling. For these applications, the membrane is used for distributive feeding of O_2 along the catalytic bed in order to obtain high reactant conversions and high product selectivities avoiding, at the same time, the back firing effect (Tsai *et al.*, 1995; Coronas and Santamaria, 1999; Saracco *et al.*, 1999; Bredesen *et al.*, 2004; Julbe *et al.*, 2005). Moreover, the feed distribution can represent a promising approach for fast reactions.

In the active contactor mode, the membrane acts as a diffusion barrier and it is cat-alytically active.

In this kind of MR, the reactants can be supplied from the opposite membrane sides (opposing reactant mode) or they can be forced to pass through it (forced flow-mode).

The forced flow contactor mode is usually applied to the total oxidation of volatile organic compounds (Pina *et al.*, 1996; Irusta *et al.*, 1998), whereas, the opposing reactant contactor mode is applied to both equilibrium and irreversible reactions (Sanchez and Tsotsis, 1996).

6.4.1   Membrane Reactor and Process Intensification

Membrane operations offer great potential in the field of PI. Indeed, the advantages of dif-ferent types of membrane operations over conventional processes are summarized next:

- In some membrane gas separation processes, the separation takes place without phase transition. This saves energy during operation (Mulder, 2000);

- In membrane technology, relatively simple and non-harmful materials are used and this is better for the environment (Scoth, 1999);
- Compared to conventional processes, membrane operations are simple, easy to operate and less maintenance is required (Scoth, 1999);
- The recovery of valuable minor components from the main stream can be performed without further energy supply.

It's clear that, the developments of new materials in membrane gas separation technology such as organic polymeric and hybrid organic–inorganic could open new application fields in the chemical and petrochemical industries (Ravanchi and Kaghazchi, 2009).

Concerning MR, the combination of both membrane and reaction characteristics represents a real model of intensification showing a higher efficiency compared to both conventional separation and reaction operations (Drioli and Curcio, 2007).

In particular, the MR is a good example of the applications of the PI principles. Indeed, the continuous product removal from the reaction side through the membrane fits with the second principle of the PI in which it is very important to give each molecule the same processing experience. In particular, the product removal from the system would prevent further formation of unwanted products giving a uniform product and minimum waste.

Moreover, the MR is a clear example of the forth principle of PI "Maximization of the synergistic effects from partial processes".

6.4.2   Membrane Reactor Benefits

MRs are not only a good example of PI but they also present many advantages over conventional reactors. In particular, with respect to the conventional reactors, MRs are able to:

- combine chemical reaction and separation in only one system reducing the capital costs;
- enhance conversion of equilibrium limited reactions, in the case of dense MRs. In fact, by the selective removal of one or more products from the reaction side, the thermodynamic equilibrium restrictions can be overcome, due to the so-called "shift effect";
- achieve higher conversions than conventional reactors, operating at the same MR conditions or, on the contrary, the same conversion, but operating at milder conditions;
- improve the product selectivity.

To confirm this, it is useful to report the work carried out by Keizer *et al.*, (1994), who studied the performances of several MRs, using different kind of membranes and carrying out the cyclohexane dehydrogenation reaction. Moreover, they made a comparison with a conventional reactor working at the same MRs operating conditions. For their scope, they defined the parameter H as the ratio between permeation and reaction rates. Thus, they represented the dependence of the cyclohexane conversion as a function of the parameter H (Figure 6.5). The line with H = 0 corresponds to a conventional reactor, while the other curves correspond to a different type of MRs. In particular, lines 1–2 refer to MRs governed by Knudsen transport mechanisms; lines 3–4 refer to microporous MRs and lines 5–6 refer to dense ones. In the graphs, it is possible distinguish two regions. The first one corresponds to low permeation to reaction rate ratios. In this region, the different MRs show the same behavior. However, the performance of each MR types in terms of conversion is better than the conventional one. At higher H-values, the

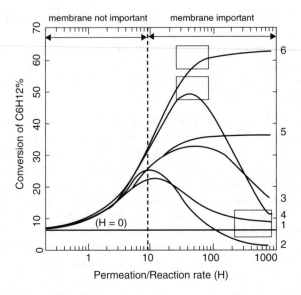

Figure 6.5   *Conversion of cyclohexane versus H (Keizer et al., 1994)*

difference in the MRs' properties are visible. In particular, porous MRs show an optimum permeability/reaction rate region. Above optimum, the reactant loss due to permeation induces a detrimental effect on the conversion. On the contrary, dense MRs maintain the conversion at a high value owing to no reactant loss.

Therefore, the main benefit of using MRs is represented by the combination of reaction and separation, leading to a reduction of capital cost and better reactor performances.

6.5   Applications of Membrane Reactors in the Petrochemical Industry

Refinery operations consist of operations involving multiple processing units for production of different types of fuels and other products. In many steps of complex refinery operations, different types of reactions occur. Many of these reactions are already operated efficiently, mainly, in fixed bed reactors for optimal productivity and selectivity. However, nowadays, there is an evident opportunity for membrane operations in refineries owing to the increasing H_2 demand. Indeed, H_2 is required for many operations in the petrochemical processes such as hydrotreating (where it is used to remove impurities from streams and to hydrogenate aromatics and olefins) and hydrocracking (where it breaks down large hydrocarbons into smaller, higher value molecules) (Armor, 1998). Moreover, regulations on gasoline aromatic composition are limiting reformer operation and removing some of the hydrogen sources, traditionally available to refineries.

All these causes led to a deficit in hydrogen balance in a refinery, resulting in need for investing in additional hydrogen production. This has prompted refiners to examine ways of increasing the efficiency of their hydrogen systems.

An effective solution to resolve this problem could be represented by using MRs. In particular, inorganic MRs could be favored with respect to the organic ones owing to the high temperature and chemically harsh environments of many catalytic industrial processes. Moreover, palladium and its alloys could be considered the dominant materials for preparing this kind of membranes owing to their very high solubility of H_2 and high selectivity towards the H_2 permeation with respect all other gases.

In this section, some promising applications in the petrochemical industry using inorganic MRs such as dehydrogenation, oxidative coupling of methane, water gas shift, and methane steam reforming reactions are discussed.

6.5.1 Dehydrogenation Reactions

One of the most important petrochemical reactions by which hydrogen is produced is dehydrogenation. Some of the dehydrogenation reactions frequently used in the petrochemical industry are shown in Table 6.4.

An example of dehydrogenation of cyclohexane to benzene has been proposed by Itoh (1987), using a Pd/Ag membrane tube packed with 0.5% Pt/Al_2O_3 catalyst. The reaction was carried out at 200 °C and 1 atm, using an argon stream saturated with cyclohexane vapor and the feed rate was kept very low in order to obtain 99% of cyclohexane conversion. Moreover, the sweep gas is used for collecting the H_2 permeating through the Pd-based membrane.

However, carbon-based (Itoh and Haraya, 2000) and alumina-based (Okubo *et al.*, 1991; Tiscarno-Lechuga *et al.*, 1996) MRs have been also used for carrying out this kind of reaction. Recently, Jeong *et al.* (2003) have used a FAU-type zeolite MR packed with a Pt/Al_2O_3 catalyst. The cyclohexane dehydrogenation reaction has been performed at 200 °C and ambient pressure reaching 73% of the hydrocarbon conversion with respect 32% of equilibrium value.

The isobutane dehydrogenation is another dehydrogenation reaction used in petrochemical industry. Kikuchi and co-workers (Kikuchi *et al.*, 1989; Matsuda *et al.*, 1993; Kikuchi, 1995; 1997) have used a Pd alloy coated onto a mesoporous membrane support. In their study, isobutane is passed over a Pt/Al_2O_3 catalyst packed into the membrane. The yield of isobutylene is increased from the equilibrium value of 6–29% at 450 °C using this kind of membrane. Guo *et al.*, (2003) have used a Pd-Ag/ceramic composite MR packed with catalyst pellets of Cr_2O_3-Al_2O_3. Furthermore, they made a comparison with a conventional reactor working at the same MR operating conditions. Thus, at 450 °C, the MR was able to give 51% of isobutene conversion with respect to 16% for the conventional reactor, whereas the equilibrium value was equal to 19%.

On the contrary, Casanave *et al.*, (1999) performed the dehydrogenation reaction in a zeolite-based MR using Pt-Sn/γ-Al_2O_3 as catalyst at 450 °C. The counter and co-current sweep-gas flow configuration was used. As best results, the MR showed 22.4% and 21.7% of isobutylene yield obtained in the counter-current and co-current flow configuration, respectively, with respect to 13% obtained with a conventional reactor working at the same MR operating conditions.

Sznejer and Sheintuch, (2004) carried out the isobutane dehydrogenation on a chrome alumina catalyst by using a MR equipped with a molecular-sieve carbon membrane. Two types of operation mode were studied, using nitrogen as a sweep gas in counter-current flow configuration and using vacuum. At 500 °C, in counter-current flow operation and in the vacuum mode the conversion obtained was of 85% and 40% respectively, compared to 30% of isobutene conversion made in a conventional reactor.

Table 6.4 Several dehydrogenation reactions used in the petrochemical industry

Dehydrogenation Reaction (DR)	Membrane	Catalyst	Operative Conditions	Notes	References
DR of cyclohexane $C_6H_{12} \rightarrow C_6H_6 + 3H_2$	Pd/Al_2O_3	Pt/Al_2O_3 (pellets)	$T = 250\text{–}300\,^\circ C$ $p = 100\text{–}400\,kPa$	$\chi = 97.0\%$ at $300\,^\circ C$ $\chi_{eq} = 18.7\%$ $HR = 93.0\%$	Itoh (1987)
	FAU-type zeolite membrane/Al_2O_3	Pt/Al_2O_3	$T = 150\text{–}250\,^\circ C$ $p = 101\,kPa$	$\chi = 72.1\%$ $\chi_{eq} = 32.2\%$	Jeong (2003)
DR of isobutane $C_4H_{10} \rightarrow C_4H_8 + C_4H_6 + H_2$	Pd/Al_2O_3	$CrO_3\text{-}Al_2O_3$, Pt/Al_2O_3	$T = 350\text{–}450\,^\circ C$	$Y_{C4H8} = 28.5\%$ at $450\,^\circ C$ $Y_{TR} = 6.0\%$ $\chi_{eq} = 10.5\%$	Matsudu (1993)
	zeolite	$Pt\text{-}Sn/\gamma\text{-}Al_2O_3$ (cylindrical pellets)	$T = 450\,^\circ C$	$Y_{C4H8} = 22.4\%$ counter-current $Y_{C4H8} = 21.7\%$ co-current $Y_{TR} = 13.5\%$	Casanave (1999)
	Pd/Ag ceramic composite membrane	$Cr_2O_3\text{-}Al_2O_3$ (pellets)	$T = 350\text{–}500\,^\circ C$ $p = 100\text{–}300\,kPa$	$\chi = 50.5\%$ at $450\,^\circ C$ $\chi_{eq} = 18.8\%$ $\chi_{TR} = 15.5\%$	Guo (2003)
	molecular-sieve carbon membrane	$Cr_2O_3\text{-}Al_2O_3$ (pellets)	$T = 500\,^\circ C$	$\chi = 85\%$ (N_2 sweep gas counter-current) $\chi = 40\%$ (vacuum) $\chi_{TR} = 30\%$	Sznejer (2004)
DR of propane $3C_3H_8 \rightarrow C_3H_6 + H_2$	Pd-Ag (25%) Pd-Ru (2%)	Pt	$T = 550\text{–}570\,^\circ C$ $p = 101\,kPa$	at $550\,^\circ C$: $Y_{C3H6,MR} = 70.0\%$ with Pd-Ru $Y_{C3H6,eq} = 32.0\%$ $Y_{C3H6,MR} = 47.5\%$ with Pd-Ag	Sheintuch (1996)

Reaction	Membrane	Catalyst	Conditions	Results	Reference
	Silica composite membrane	Cr_2O_3-Al_2O_3	$T = 500\,°C$	$S_{C3H6} = 85.0\% \div 95.0\%$ at $WHSV = 0.005 \div 0.1\,h^{-1}$: $\chi = 60.0\% \div 80.0\%$	Weyten (2000)
	Pd-Ag composite	Cr_2O_3-Al_2O_3	$p = 101\,kPa$	at $535\,°C$: $\chi_{MR} = 49.0\%$, $\chi_{CFB} = 37.0\%$, $S_{C3H6,MR} = 70.0\%$, $S_{C3H6,CFB} = 60.0\%$, $Y_{C3H6,MR} = 34.0\%$, $Y_{C3H6,CFB} = 29.0\%$	Schäfer (2003)
	Ceramic membrane	Pt-Sn/Al_2O_3	$T = 550{-}535\,°C$	at $500\,°C$: $S_{C3H6,MR} = 80.0\%$, $S_{C3H6,CFB} = 70.0\%$, $Y_{C3H6,MR} = 22.0\%$, $Y_{C3H6,CFB} = 17.0\%$	
DR of methylcyclohexane $C_7H_{14} \rightarrow C_7H_8 + 3H_2$	Pd-Ag	Pt/Al_2O_3	$T = 320{-}400\,°C$ $p = 1500\,kPa$	at $370\,°C$: $\chi_{MR} = 90.0\%$, $\chi_{eq} = 73.3\%$, $\chi_{TR} = 65.0\%$, Yield toluene $= 78\%$	Ali (1997)
DR of ethylbenzene $C_8H_{10} \rightarrow C_8H_8 + H_2$	Pd	—	$T = 700\,°C$ $p = 3030\,kPa$	With 7 fluized bed in series and 16 membrane tubes: $\chi_{C8H10} = 96.5\%$, $Y_{C8H8} = 92.4\%$	Abdallah (1995)
	Pd-Ag/porous substrate Tube	—	$T = 600\,°C$	Simulation study: $\chi = 62.1\%$ Co-current mode, $\chi = 78.8\%$ Counter-current mode	Gobina (1995a)

(continued overleaf)

Table 6.4 (continued)

Dehydrogenation Reaction (DR)	Membrane	Catalyst	Operative Conditions	Notes	References
	Pd-coated porous stainless steel Tube	–	$T = 620\,°C$ $p = 100–250\,kPa$	Simulation study: $\chi = 88.8\%$ $\chi_{eq} = 71.2\%$ $S_{C8H8} = 95.2\%$ $Y_{C8H8} = 83.8\%$	Hermann (1997)
		$Fe_2O_3\text{-}K_2O$	$T = 600\,°C$ $p = 200\,kPa$	$X_{PBR} = 46.5\%$; $Y_{C8H8} = 33.0\%$ $X_{PBMR} = 52.6\%$; $Y_{C8H8} = 41.3\%$ $X_{FLBR} = 34.2\%$; $Y_{C8H8} = 19.5\%$ $X_{FLBMR} = 67.9\%$; $Y_{C8H8} = 55.6\%$	Elnashaie (2001)
	Pd/porous ceramic	$Fe_2O_3\text{-}K_2O\text{-}CeO_2$		$X_{PBR} = 52.4\%$; $Y_{C8H8} = 28.4\%$ $X_{PBMR} = 58.3\%$; $Y_{C8H8} = 39.2\%$ $X_{FLBR} = 37.7\%$; $Y_{C8H8} = 12.7\%$ $X_{FLBMR} = 73.5\%$; $Y_{C8H8} = 54.0\%$	
		$Fe_2O_3\text{-}K_2O\text{-}CeO_2\text{-}CrO_3$		$X_{PBR} = 51.3\%$; $Y_{C8H8} = 44.3\%$ $X_{PBMR} = 64.8\%$; $Y_{C8H8} = 60.5\%$ $X_{FLBR} = 36.3\%$; $Y_{C8H8} = 23.4\%$ $X_{FLBMR} = 79.1\%$; $Y_{C8H8} = 75.4\%$	

Reaction	Membrane	Catalyst	Conditions	Results	Reference
Oxidative DR of ethane $C_2H_6 + \frac{1}{2}O_2 \rightarrow C_2H_4 + H_2O$	Porous membrane	$MgO/LiO/Sm_2O_3$	$T = 600\,°C$	$\chi_{C2H6} = 95.0\%$ $S_{C2H4} = 53.0\%$ $S_{C2H4,TR} = 8.1\%$ $Y_{C2H4} = 50.5\%$ $Y_{C2H4,TR} = 8.4\%$	Tonkovich (1996)
	Porous membrane	$V_2O_5/\gamma\text{-}Al_2O_3$	$T = 500\text{–}680\,°C$ $p = 101\ kPa$	$Y_{C2H4,FLBMR} = 36.6\%$ $Y_{C2H4,FLBR} = 23.3\%$ $Y_{C2H4,PBMR} = 34.9\%$	Ahchieva (2005)
DR of ethane $C_2H_6 \rightarrow C_2H_4 + H_2$	Pd-Ag /Vycor	Pd/Al_2O_3 (cylindrical pellets)	$T = 387\,°C$ $p = 128.7\ kPa$	Contact time is very important $\chi(\text{measured}) = 18.4\%$ $\chi(\text{predicted}) = 26.1$	Gobina (1995b)
	Pd-Ag	Re/HZSM-5	$T = 500\text{–}585\,°C$ $p = 101\ kPa$	At $545\,°C$ $S_{C2H4} = 70.0 \div 80.0\%$ $\chi_{C2H6} = 9.0 \div 25.0\%$	Wang (2003)
DR of *cis*-3-hexen-1-ol $C_6H_{12}O \rightarrow C_6H_{10}O + C_6H_{12}O + C_6H_{14}O$	Pd-Ag	CuO-ZnO (cylindrical pellet)	$T = 170\text{–}230\,°C$	$S_{1\text{-Hexanol}} = 17.0\%$ $S_{cis\text{-}3\text{-hexenal}} = 9.4 \div 65.0\%$ $S_{Hexanal} = 32.6 \div 80.4\%$	Sato (2007)
DR of 2-butanol $C_4H_{10}O \rightarrow C_4H_8O + H_2$	$Pd\text{-}Ag/Al_2O_3$	Cu/SiO_2 (pellets)	$T = 190\text{–}240\,°C$ $p = 101\ kPa$	$\chi = 93.0\%$ at $240\,°C$ $\chi_{eq} = 80.0\%$ at $240\,°C$ $S_{C4H8O} = 96.0\%$	Keuler (2002)
DR of *cis*-3-hexen-1-ol $C_6H_{12}O \rightarrow C_6H_{10}O + C_6H_{12}O + C_6H_{14}O$	Pd-Ag	CuO-ZnO (cylindrical pellets)	$T = 170\text{–}230\,°C$	$S_{1\text{-Hexanol}} = 17.0\%$ $S_{cis\text{-}3\text{-hexenal}} = 9.4 \div 65.0\%$ $S_{Hexanal} = 32.6 \div 80.4\%$	Sato (2007)

For the dehydrogenation of propane Sheintuch and Dessau (1996) and Weyten *et al.* (2000) used a Pd-based supported MR, whereas Schäfer *et al.* (2003) employed a ceramic MR. Thus, in the case of Pd-based MR, the yield of propylene is increased from the equilibrium value of 32% to 70% at 550 °C (Sheintuch *and* Dessau, 1996), whereas at the same temperature, the ceramic MR was able to give 80% and 22% of propylene selectivity and yield, respectively, compared with 70% and 17% reached in the conventional reactor.

Moreover, Bobrov *et al.*, (2005) investigated the propane dehydrogenation using inorganic membranes with molybdenum (Mo) selective layer. In particular, Mo-ceramic (Mo coated on ceramic support) and Mo-carbon membranes (Mo coated on carbon support) were used. The maximum selectivity to propylene for Mo-ceramic membrane was 63% (at 580 °C and conversion to propane 17%). The maximum selectivity to propylene for Mo-carbon membrane was 84% (at 580 °C and conversion to propane 28%).

Reforming via dehydrogenation of methylcyclohexane is another dehydrogenation reaction. Ali and Baiker (1997) used a Pd-based MR, containing a commercial Pt/Al_2O_3 catalyst to remove H_2 in situ and, consequently, to improve the efficiency of methylcyclohexane dehydrogenation to toluene. With this system at 370 °C and 15 atm, a yield of 78% toluene was obtained.

The ethylbenzene dehydrogenation to styrene is another worthy dehydrogenation reaction used in petrochemical industry. Liu *et al.* (1993) have proposed a two stage packed bed reactor followed by a MR, where the mesoporous membrane is packed with a commercial K promoted iron oxide catalyst used for styrene production. In this system, a 4% yield enhancement to styrene has been observed.

Other dehydrogenation reactions have been considered and studied in MR, as shown in Table 6.4. For instance, Ahchieva *et al.*, (2005) studied the performance of a fluidized bed MR for the catalytic oxidative dehydrogenation of ethane using γ-alumina supported vanadium oxide catalyst. The experimental results were compared to the conventional reactor. The maximum ethylene yield observed was 36.6% for the MR and 23.3% for the conventional one. Moreover, ethylene selectivity is higher in the MR (up to 20%).

In brief, different types of membranes have been used for performing various dehydrogenation reactions and in some cases a comparison with conventional reactor has been also reported. Making a qualitative analysis on the performances realized using different MRs is clear that the MRs achieve higher conversions than conventional reactors. Moreover, Pd-based MRs show better performances in terms of conversion and yield than other inorganic ones.

6.5.2   Oxidative Coupling of Methane

Oxidative coupling of methane (OCM) is a promising process for converting the abundant natural gas into liquid fuels or useful chemicals (Mlezko and Baerns, 1995). There are mainly two different methods that have been studied in converting methane to chemicals and liquid fuels, that is, direct or indirect routs. The direct route consists of carrying out the OCM reaction for producing C_2 hydrocarbons such as ethane and ethylene. The indirect route is the production of syngas (H_2 and CO mixture) by methane steam reforming or partial oxidation reaction of methane followed by a further conversion into higher hydrocarbons by the Fischer–Tropsch process.

Comparatively, the direct route is economically more promising if the C_2 yield can be remarkably improved. Moreover, in order to make OCM a commercial process, the C_2 hydrocarbon yield has to be higher than 30%.

Higher C_2 yield can be realized by either increasing methane conversion, C_2 selectivity, or both. In reality, it is difficult to achieve both high conversion and selectivity in a conventional reactor because the increasing of methane conversion usually leads to lower C_2 selectivity (Jaso *et al.* 2011). Therefore, to achieve an economically attractive C_2 yield (>30–40%), considerable interests have been recently devoted to the development of various MR (Coronas *et al.*, 1994b, 1997; Cheng *and* Shuai, 1995; Tonkovich *et al.*, 1996; Lu *et al.*, 1997, 2000a; Hamel *et al.*, 2003; Kao *et al.*, 2003; Kiatkittipong *et al.*, 2005; Bhatia *et al.*, 2009; Oliver *et al.*, 2009). In particular, the MR acts in a distributor modality allowing the feed of very small amounts of oxygen continuously into the reaction zone, enhancing both conversion and selectivity and, at the same time, limiting the undesired combustion reactions.

Various types of inorganic MRs have been used for performing this kind of reaction. These include the inert porous membrane (Coronas *et al.*, 1994a, 1997; Kao *et al.*, 2003; Lafarga *et al.*, 2005), catalytic membrane (Lu *et al.*, 2000b, Shao *et al.*, 2001), solid-oxide membrane (Guo *et al.*, 1997; Langguth *et al.*, 1997) and dense membrane (Nozaki *et al.*, 1992; Xu and Thomson, 1997; Akin and Lin, 2004; Wang *et al.*, 2005). It has been proved theoretically that by using dense MRs, yields higher than 30% can be achieved (Xu and Thomson, 1997; Akin and Lin, 2004), but practically, only a 28% yield has been obtained (Lu *et al.*, 2000a). Nonetheless, many research efforts have been done in order to enhance the C_2 yield. As an example, Coronas *et al.* (1994a) used a packed bed MR with coating the porous α-alumina membrane by depositing silica sol inside the porous alumina structure and further impregnating with Li_2CO_3 solution, in order to improve the membrane surface area and its acidity. A Li/MgO catalyst was packed within the tubular membrane. At an operating temperature of 750 °C and feed molar ratio $CH_4/O_2 = 4$, 28% of methane conversion, C_2 selectivity 60–65% and C_2 yield of 16.8–18.2% were obtained.

The porous-based MRs have shown low performances in terms of ethylene yield and selectivity, owing to a poor controlling of oxygen permeation.

Recently, some other research efforts have been directed toward searching for ionic conducting perovskite-type oxide membrane (Xu and Thomson, 1997; Lu *et al.*, 2000b) and dense fluorite-structured membranes (Zeng and Lin, 2000; Akin and Lin, 2004), which could enhance the oxygen permeation flux in MR. For instance, Wang *et al.* (2005) achieved 70% of CH_4 conversion, C_2 selectivity and yield of 70% and 15%, respectively, using a reactor of a dense membrane tube made of $Ba_{0.5}Sr_{0.5}Co_{0.8}Fe_{0.2}O_{3-\delta}$, packed with La-Sr/CaO catalyst at 850 °C. Among the perovskyte type ceramic MR, the highest C_2 yield (16.5%) was obtained by Lu and co-workers, (2000a) using a tubular $BaGe_{0.8}Gd_{0.2}O_3$ membrane. Nevertheless, the C_2 yield values remained low.

6.5.3   Methane Steam Reforming

Methane steam reforming (MSR) reaction is the principal method of hydrogen production on a large industrial scale. Indeed, 80–85% of the worldwide hydrogen supply is produced by this kind of reaction in fixed bed reactors (Simpson and Lutz, 2007). MSR is an endothermic reaction that takes place at 800–900 °C and at 15–20 atm in order to achieve high methane conversion (Barelli *et al.*, 2008). Nevertheless, at this elevated temperature the catalyst could undergo deactivation due to carbon formation resulting in blockage of reformer tubes and increasing, as a consequence, the pressure drops (Trimm, 1997).

However, if H_2 is selectively removed from the reaction zone, the reaction can be shifted towards further products formation obtaining efficient conversion of CH_4 to CO_2 and H_2, operating at lower temperatures and avoiding the issues aforementioned. By using a dense Pd-based MR, which allows only hydrogen to permeate through the membrane, it is possible to realize these benefits and, moreover, to produce pure H_2 stream, simplifying, thus, the current operation which includes extensive H_2 purification steps (Gunardson, 1997).

Uemiya *et al.*, (1991a) have shown that SMR reaction can be improved using a Pd alloy membrane coated onto a porous glass tube packed with an alumina supported Ni catalyst. At feed molar ratio H_2O/CH_4 = 3/1, 500 °C and 1 atm, they reached 80% of CH_4 conversion compared to equilibrium value of ~42%.

Kikuchi *et al.*, (2000) demonstrated that a complete methane conversion was achieved at relatively low temperatures (500 °C), employing a membrane constituted of a thick layer of palladium on a porous material surface.

Lin *et al.*, (2003) carried out the SRM reaction in Pd-based MR, packed with a nickel catalyst. The experimental results indicated that a methane conversion exceeding 80% can be achieved at a temperature of 500 °C compared to 850 °C necessary in a conventional reactor. Moreover, CO production (<2%) was considerably lower in MR than the one obtainable in a conventional reactor (>50%), combining also high hydrogen yield and purity.

Tong and Matsumura, (2005) employed the Pd membrane supported on porous stainless steel modified with cerium oxide particles to conduct the MSR reaction. Particles of ceria hydroxide were introduced in order to reduce the surface and prevent the metal inter-diffusion. Moreover, the residual ceria oxide in the pores of the substrate can act as a catalyst (Pd/CeO_2). In the best result, a methane conversion around 97% was achieved at 500 °C.

Chen *et al.*, (2008) used a Pd-based MR packed with a Ni-based catalyst. A 99% of methane conversion, 97% of CO_2 selectivity and 95.0% of hydrogen recovery confirmed that the selective removal of hydrogen from the MR reaction zone allows us to achieve a methane conversion significantly higher than the one obtained in a conventional reactor.

Iulianelli *et al.*, (2010) using a dense self-supported Pd-based MR obtained 50% methane conversion and 70% CO_x-free hydrogen recovery. The lower performance in terms of conversion was probably accounted for the relatively low temperature (450 °C) and for the low Ni phase concentration of the catalyst (0.5%).

In summary, regarding the SMR reaction, it seems to be a great opportunity for application of MRs although big challenges yet remain. Indeed, it is necessary to develop membranes highly selective (preferentially selective to H_2 permeation) with high H_2 flux permeating and resistant to harsh operative conditions. Moreover, it should avoid the possibility of carbon formation on the membrane in order to be competitive with commercial SMR reactors characterized by long term reliability with a five-year catalyst life (Armor, 1999).

Table 6.5 shows an overview of some experimental data from the literature concerning the MSR reaction in Pd-based MRs.

6.5.4 Water Gas Shift

The water gas shift (WGS) reaction is another important process operation within a refinery. In particular, it is used to reduce the CO content in the stream coming from the reformer and to increase, at the same time, the hydrogen production.

The WGS reaction is limited in terms of thermodynamic constrains. Moreover, taking into account the exothermic nature of this reaction, higher CO conversions are favored at lower temperatures and furthermore a large volume of catalyst is necessary owing to slow kinetics. Therefore, a promising approach for this reaction is the use of MRs. Moreover, the industrial practice of using an excess of steam beyond the amount required to prevent coking formation and to favor the reaction could be avoided using the MR.

Indeed, it has been shown that the WGS reaction is favored by the use of a Pd-based MR. Pioneering work on Pd-based MR for the WGS reaction was done by Kikuchi and coworkers (Kikuchi *et al.*, 1989; Uemiya *et al.*, 1991b). The reaction was carried out at 400 °C over a commercial iron-chromium catalyst. In their studies, the authors used a very thin palladium layer ($\sim$10 µm) deposited on a microporous glass tube. As a result, CO conversion in the 80–92% range was realized compared to the equilibrium value of 78%. Moreover, a slight increase in the reaction pressure (0.1–0.5 MPa) improved the conversion from 92–99% owing to the higher driving force for hydrogen permeation.

Basile and coworkers (Basile *et al.*, 1996a, 1996b, 2001, 2008; Criscuoli *et al.*, 2000) have also used Pd-based MRs for carrying out the WGS reaction. For instance, they used a Pd-Ag rolled membrane (thickness: 50 µm) for carrying out the reaction at 1.0 bar, feed molar ratio $H_2O/CO = 1/1$ and nitrogen as sweep gas in the co-current flow configuration (Basile *et al.*, 2001). At 330 °C, the MR exhibited a CO conversion of 93%. In another study, Basile *et al.* (1996a) used a composite membrane obtained by co-condensation technique in which an ultrathin Pd film ($\sim$0.1 µm) was coated on the inner surface of a porous ceramic support (γ-Al_2O_3). In this case, at 320 °C, 1.1 bar and feed molar ratio $H_2O/CO = 2/1$, the 96% of CO conversion versus equilibrium value of 70%, was realized. Moreover, Basile *et al.* (1995) illustrated that a CO conversion of around 98% could be reached by using a composite membrane with a 10 µm Pd film coated on a ceramic support.

WGS reaction has been also carried out in a silica MR. For instance, Brunetti and co-workers (2007) investigated the WGS reaction in a supported silica MR. The macropores of the support were modified by packing with silica xerogel (500 nm) under a pressure of 10 MPa and by coating with an intermediate layer of γ-alumina. Thus, a mesoporous γ-alumina layer was coated on the support disk from a boehmite sol (γ-AlOOH) by a soaking–rolling procedure. The reaction has taken place on a CuO/CeO$_2$ based commercial catalyst at a temperature range of 220–290 °C up to 6.0 bar. At 280 °C and 4.0 bar, they obtained a CO conversion of 95%.

In summary, previous works show that MR use allows achievement of higher CO conversions at higher temperatures and overcomes equilibrium value. Moreover, the results suggested that a high pressure drop across the membrane would act positively

Table 6.5   *Some experimental data from literature concerning the MSR reaction in Pd-based MRs*

Membrane	Catalyst	Operative conditions	Notes	References
Pd supported onto PSS	Ni/Al$_2$O$_3$	T = 500 °C p = 20.0 bar H$_2$O/CH$_3$OH = 3/1	X_{CH4} = 86% HR = 90%	(Lin *et al.*, 2003)
Pd supported onto Al$_2$O$_3$	Ni-La/Mg-Al	T = 550 °C p = 9.0 bar H$_2$O/CH$_3$OH = 3/1	X_{CH4} = 99% HR = 95%	(Chen *et al.*, 2008)
Pd-Ag supported onto PSS	Cu/ZnO	T = 500 °C p = 6.0 bar H$_2$O/CH$_3$OH = 2.9/1	X_{CH4} = 50%	(Jorgensen *et al.*, 1995)
Pd supported on Vycor	–	T = 500 °C p = 9.1 bar –	X_{CH4} = 90%	(Uemiya *et al.*, 1991a)
Pd-Ag supported onto PSS	Ni/Al$_2$O$_3$	T = 500 °C p = 1.36 bar H$_2$O/CH$_3$OH = 3/1	X_{CH4} = 55%	(Shu *et al.*, 1995)
Pd-based	Ni/Al$_2$O$_3$	T = 500 °C p = 1.0 bar H$_2$O/CH$_3$OH = 3/1	X_{CH4} = 100%	(Kikuchi *et al.*, 2000)
Flat Pd-Ag	Ru/Al$_2$O$_3$	T = 300 °C p = 1.0 bar –	X_{CH4} = 16.5%	(Basile *et al.*, 2003)
Dense self-supported Pd-Ag	Ni/Al$_2$O$_3$	T = 450 °C p = 3.0 bar H$_2$O/CH$_3$OH = 2/1	X_{CH4} = 50% HR = 70%	(Iulianelli *et al.*, 2010)
Pd supported onto PSS	Ni/Al$_2$O$_3$	T = 527 °C p = 3.0 bar H$_2$O/CH$_3$OH = 3/1	X_{CH4} = 100%	(Tong and Matsumura, 2005)
Pd supported onto Inconel	–	T = 650 °C p = 4.0 bar H$_2$O/CH$_3$OH = 4/1	X_{CH4} = 97%	(Patil *et al.*, 2007)
Pd-Ag supported onto Inconel	Ni/Al$_2$O$_3$	T = 500 °C p = 2.0 bar H$_2$O/CH$_3$OH = 3/1	X_{CH4} = 80%	(Shu *et al.*, 1994)

on the performance of the MR. As a consequence, the catalyst amount necessary for a given conversion can be significantly reduced. For instance, working at 280 °C and 1.0 bar the catalyst volume necessary to reach the 90% of CO conversion, obtained in the conventional reactor, is reduced by half by using a silica-based MR and to a third by using a Pd-based membrane, working at the same operating conditions (Brunetti *et al.*, 2007).

6.6   Process Intensification in Chemical Industry

The chemical industry covers a wide range of processes incorporating several operating units, which can benefit of PI technology. As already mentioned, the benefits that PI can offer lie primarily in four areas: costs, safety, compactness, and controlled well-defined conditions. The first area covers an important role for chemical companies because PI leads to substantially cheaper processes, in terms of investment costs (smaller equipment), costs of raw materials (higher yields/selectivities), costs of utilities (energy in particular), and costs of waste-stream processing (less waste in general).

However, from an engineering point of view, the main PI benefit is the opportunity to combine different operations in a way that is not possible, or at best difficult, with conventional devices. In particular, the combination of distinct task such as reaction and one or more operations (as heat transfer and/or separation) plays an important role in the industrial processes owing to the reduction of plant size and the increase of overall process efficiency. Moreover, the combined operations have to show a sufficiently large feasible operation window and avoid being too restrictive in terms of process flexibility and control. Table 6.6 reports generic benefits and drawbacks resulting in the combination of reaction/separation units.

Well-known examples of reaction and separation integration are: reactive distillation, reactive extraction, reactive adsorption and membrane reactors (these latter have been widely discussed previously).

6.6.1   Reactive Distillation

Reactive distillation is one of the better known examples of reaction and separation combination used commercially (De Garmo *et al.*, 1992). In this case, the reactor consists of a distillation column filled with catalyst, which can be also in the same phase of the reacting species. The aim of the distillation column is to separate the reaction products by fractionation or to remove impurities or undesired species.

Table 6.6   *Benefits and drawbacks resulting in the combination of operating units*

Benefits	Drawbacks
Enhanced of overall process performances in terms of reaction conversion and selectivity	Relatively new technology
Heat integration benefit	Limited applications
Milder operating conditions	Complex modeling need
New process configuration	Increased operational complexity
Reduced both capital and operating costs	Extensive equipment design effort
Simplified separation units	Increased scale-up risks

The advantages of the reactive distillation, besides the continuous removal of reaction products which allows achievement of higher yields, consist mainly of an energy requirement reduction and lower capital investment (Stadig, 1987).

The main industrial applications of reactive distillation are esterification, etherification and alkylations (Tuchlenski *et al.*, 2001). In particular, the most famous example of esterification (and indeed the most famous example of reactive distillation), is the "Eastman process" for methyl acetate. In this process, the number of main equipment has been reduced from 28 to 3 by using one reactive distillation column and two separation columns (Agreda *et al.*, 1990; Hendershot, 1999).

Other prospective applications of reactive distillation include liquid–liquid hydrolyses, saponification, nitration, oxidation, fermentations, hydrolysis of aqueous methyl formate, and dehydrogenation of cyclohexane to benzene (Stadig, 1987).

Nowadays, the number of processes in which reactive distillation has been implemented on a commercial scale is still quite limited; nevertheless, the potentiality of this technique certainly goes far beyond today's applications.

6.6.2   Reactive Extraction

Reactive extraction is the combination of processes such as reaction and solvent extraction. The main benefit of this integration results in fewer process steps overall, thereby reducing capital cost. Moreover, other possible advantages include the enhancement of both selectivity and efficiency.

The reactive extraction can be considered as an effective way to remove a desired product from the reaction zone preventing any side reactions (Krishna, 2002). For instance, in Krishna's study (2002), reactive extraction was used for carrying out the bromination of dialcohols in the aqueous phase. Particularly, adding an HC to the reaction mixture extracts the monobromide, the formation of the by-product dibrome is avoided.

In the last years, the reactive extraction has been also studied for producing biodiesel trying to avoid the expensive crushing step (Haas *et al.*, 2004; Harvey, 2006). Indeed, currently biodiesel is produced by vegetable crushing and solvent extraction; successively, in the presence of a liquid alkaline catalyst, the produced oil reacts with methanol (occasionally ethanol) to produce the ester (the biodiesel). The idea in the use of reactive extraction is that the seeds should be directly macerated and put in contact with an alcohol containing the catalyst. In this way, the alcohol would cover the role of both solvent and reactant.

6.6.3   Reactive Adsorption

Adsorption is a separation technique that can be used in parallel with a reaction for enhancing the performance of the process. Reactive adsorption combines the separation role of, for instance, a solid adsorbent with the reaction taking place on a different surface. The ability to remove the adsorbent or to desorb one or more reaction products can be convenient, although adsorbents do not respond rapidly.

An interesting type of an adsorptive reactor is the so-called Gas-Solid-Solid Trickle Flow Reactor, in which fine adsorbent trickles through the catalyst fixed bed selectively removing in situ one or more products from the reaction zone. In case of the methanol synthesis this led to conversions significantly exceeding the equilibrium conversions under given conditions (Stankiewicz, 2003).

However, the industrial scale applications of adsorptive reactors are not yet reported. Indeed, efforts involving new materials development for catalysts/adsorbents and matching of process conditions (same temperature) for both reaction and adsorption to realize both high yields and selectivities have to be still achieved.

6.6.4 Hybrid Separation

In the chemical industry, besides to the use of multifunctional reactors, the hybrid separation process also plays an important role. Indeed, over the years now the hybrid separation processes have attracted considerable attention owing to the synergistic interaction of two unit operations that contrasts the conventional configuration of different apparatuses. This process represents a good example of the applications of the PI principles. Indeed, many of the developments in this area involve integration of membranes with another separation technique. Membrane distillation is probably the best known of hybrid separation (Godino *et al.*, 1996; Lawson and Lloyd, 1997). The technique is widely considered as an alternative to reverse osmosis and evaporation. In comparison with conventional distillation, the membrane distillation possesses typical basic advantages of membrane separation, that is, simple up-scaling, possibility for high membrane surface/volume ratios, possibility to treat flows with heat sensitive components and/or a high suspended particle content at atmospheric pressure and a temperature below the boiling point of the supply.

Other examples of hybrid separation are membrane absorption and stripping, in which the membrane serves as a permeable barrier between the gas and liquid phases. By using hollow-fiber membrane modules, large mass-transfer areas can be realized resulting in compact equipment. Moreover, absorption membranes offer operation independent of gas and liquid flow rates, without entrainment, flooding or foaming (Jansen *et al.*, 1995; Poddar *et al.*, 1996).

Among hybrid separation processes not involving membranes, adsorptive distillation (Yu *et al.*, 1996) offers interesting advantages over conventional methods. In this technique, a selective adsorbent is added to a distillation mixture. This increases separation ability and can present an attractive option in the separation of azeotropes or close boiling components. Adsorptive distillation can be used, for instance, for the removal of trace impurities in the manufacturing of fine chemicals. Despite their advantages, hybrid separation processes are rarely introduced in industry due to a general skepticism towards new process types, a lack of reliable design methodologies and a lack of process details arising from increased complexity of combined unit operations.

6.7 Future Trends

The key to PI lies in novel designs that substantially enhance mass transfer rates for equipment miniaturization and that combine distinct tasks such as reaction and separation improving overall performances.

In this chapter, the state of the art of the various reactions of different petrochemical feedstocks performed in MRs has been proposed, paying special attention to the effect of Pd-based membrane applications. Indeed, dense Pd-based MRs seem to be the dominant applications in this field, particularly owing to the complete selectivity towards hydrogen permeation.

The scientific literature on this matter is exceptionally rich, although, from our best knowledge, only a very small number of them are focused on the analysis of the costs. For this reason, there are some discrepancies concerning the industrial feasibility of Pd and Pd-alloy MRs as a more competitive alternative to the conventional ones. In particular, Lu *et al.* (2007) and Armor (1998) have pointed out that the cost of Pd and the limited membrane life are the most relevant commercial limitations for using dense Pd and Pd alloy membranes. On the contrary, Criscuoli *et al.* (2001) reported that, even though both MR capital and operating costs are higher than for conventional ones, it is possible to determine a range in which they could be cost effective by reducing the Pd thickness. Indeed, MRs could represent a possible alternative to conventional reactors employing a Pd thickness below 20 μm. From the viewpoint of cost reduction and in order to use low cost MRs, there is also a strong need to develop no Pd-based membranes. Therefore, further efforts should be done for preparing defect-free inorganic membranes able to work for long periods and under hard operating conditions.

Currently, no large scale industrial applications have been reported so far. The primary reason is the relatively high price of membrane units, although other factors, such as low permeability as well as low mechanical and thermal resistance, also play an important role. As a consequence, further developments in the field of material engineering surely will change this scenario.

6.8   Conclusion

A PI definition often used is the following: the PI is strategy for realizing dramatic reductions in the size of the plant at a given production volume. Other researchers define PI in different ways. However, two words are common to PI definitions. One of these words is "innovative": PI is characterized by the innovation. The other common word is "substantial" that clearly defines the target of PI.

The opportunities that PI offers to chemical and petrochemical companies on an industrial scale lie mainly in four areas: costs, safety, compactness, and controlled well-defined conditions. Indeed, PI leads to considerably cheaper processes, investment costs (smaller equipment, reduced piping, etc.), costs of raw materials (higher yields/selectivities), costs of utilities (energy in particular), and costs of waste-stream processing (less waste in general). It is obvious that smaller is safer. However, PI offers not only smaller equipment but also better possibilities for keeping processes under control, for instance, by extremely efficient heat removal from exothermic reactions.

The majority of these goals are collected in the MRs in which distinct task as reaction and separation are combined in only one device. It seems attractive to consider MR in industrial applications as in chemical and petrochemical industry.

The application of different MRs to reactions of dehydrogenation, oxidative coupling of methane, steam reforming of methane and water gas shift were investigated and valuable experimental results were also obtained.

In particular in this chapter, MR's ability to achieve higher reactans conversion than those obtained in a convention reactor for different reactions has been illustrated, as well as the further advantage of producing a highly pure hydrogen stream or improving the products' selectivity and yield.

In summary, a chance for realistic application of MRs in the petrochemical industry will be possible if membranes are defect-free, mechanically and thermally resistant, low cost will be achieved. Moreover, the scale-up of MRs for reforming reactions is one of the most important issues and more experimental analyses on the lifetime of MRs should be realized to validate them as a possible alternative to the conventional systems at larger scales.

Nomenclature

CFB: Circulating Fluidized Bed
FLBMR: Fluidized Bed Membrane Reactor
FLBR: Fluidized Bed Reactor
H: parameter (ratio between permeation and reaction rates)
HR: Hydrogen Recovery
MR: Membrane Reactor
OCM: Oxidative Coupling of Methane
PBMR: Packed Bed Membrane Reactor
PBR: Packed Bed Reactor
PI: Process Intensification
S = Selectivity
SMR: Steam Methane Reforming
TR: Traditional Reactor
WGS: Water Gas Shift
Y = Yield
χ = Conversion
χeq = Equilibrium Conversion

References

B.K. Abdallah and S.S.E.H. Elnashaie (1995). "Fluidized bed reactors without and with selective membranes for the catalytic dehydrogenation of ethylbenzene to styrene", *J Membrane Sci*, **101**, 31–42.

A.E.M. Adris and J.R. Grace (1997). "Characteristics of fluidized-bed membrane reactors: scale-up and practical issues", *Ind Eng Chem Res*, **36**, 4549–4556.

V.H. Agreda, L.R. Partin and W.H. Heise (1990). "High-purity methyl acetate via reactive distillation", *Chem Eng Prog*, **86**, 40–46.

D. Ahchieva, M. Peglow, S. Heinrich, *et al.* (2005). "Oxidative dehydrogenation of ethane in a fluidized bed membrane reactor", *Appl Catal A: Gen*, **296**, 176–185.

T. Ahmed and M.J. Semmens (1992). "Use of sealed end hollow fibers for bubble-less membrane aeration: experimental studies", *J Membrane Sci*, **69**, 1–10.

F.T. Akin and Y.S. Lin (2004). "Oxygen permeation through oxygen ionic or mixed-conducting ceramic membranes with chemical reactions", *J Membrane Sci*, **231**, 133–146.

J.K. Ali and A. Baiker (1997). "Dehydrogenation of methylcyclohexane to toluene in a pilot-scale membrane reactor", *Appl Catal A: Gen*, **155**, 41–47.

J.A. Arizmendi-Sanchez and P.N. Sharatt (2008). "Phenomena-based modularization of chemical process models to approach intensive options", *Chem Eng J*, **135**, 83–94.

J. Armor (1998). "Applications of catalytic inorganic membrane reactors to refinery products", *J Membrane Sci*, **147**, 217–233.

J. Armor (1999). "The multiple roles for catalysis in the production of H_2", *Appl Catal A: Gen*, **176**, 159–176.

R.W. Baker (2002). "Future directions of membrane gas separation technology", *Ind Eng Chem Res*, **41**, 1393–1411.

R.W. Baker (2004). *Membrane Technology and Applications*, 2nd edn. John Wiley & Sons, Ltd., Chichester.

L. Barelli, G. Bidini, F. Gallorini and S. Servili (2008). "Hydrogen production through sorption-enhanced steam methane reforming and membrane technology: A review", *Energy*, **33**, 554–570.

A. Basile, E. Drioli, F. Santella, *et al.* (1996a). "A study on catalytic membrane reactors for the water-gas shift reaction," *Gas Sep Pur*, **10**, 53–61.

A. Basile, A. Criscuoli, F. Santella and E. Drioli (1996b). "Membrane reactor for water gas shift reaction," *Gas Sep Pur*, **10**, 243–254.

A. Basile, V. Violante, F. Santella and E. Drioli (1995). "Membrane integrated system in the fusion reactor fuel cycle", *Cat Today*, **25**, 321–326.

A. Basile, G. Chiappetta, S. Tosti and V. Violante (2001). "Experimental and simulation of both Pd and Pd/Ag for a water-gas shift membrane reactor," *Gas Sep Pur*, **25**, 549–571.

A. Basile, L. Paturzo and A. Vazzana (2003). "Membrane reactor for the production of hydrogen and higher hydrocarbons from methane over Ru/Al_2O_3 catalyst", *Chem Eng J*, **93**, 31–39.

A. Basile (2008). "Hydrogen production using Pd-based membrane reactors for fuel cells", *Top Catal*, **51**, 107–122.

D. Bessarabov (1999). "Membrane gas-separation technology in the petrochemical industry", *Membrane Technol*, **1999**, 9–13.

S. Bhatia, C.Y. Thien and A.R. Mohamed (2009). "Oxidative coupling of methane (OCM) in a catalytic membrane reactor and comparison of its performance with other catalytic reactors", *Chem Eng J*, **148**, 525–532.

V.S. Bobrov, N.G. Digurov and V.V. Skudin (2005). "Propane dehydrogenation using catalytic membrane", *J Membrane Sci*, **253**, 233–242.

R. Bredesen, K. Jordal and O. Bolland (2004). "High-temperature membranes in power generation with CO_2 capture", *Chem Eng Process*, **43**, 1129–1158.

A. Brunetti, G. Barbieri, E. Drioli, *et al.* (2007). "A porous stainless steel supported silica membrane for WGS reaction in a catalytic membrane reactor", *Chem Eng Sci*, **62**, 5621–5626.

D. Casanave, P. Ciavarella, K. Fiaty and J.A. Dalmon (1999). "Zeolite MR for isobutane dehydrogenation: Experimental results and theoretical modeling", *Chem Eng Sci*, **54**, 2807–2815.

Catalytica® (1998). Catalytic membrane reactors: concepts and applications, *Catalytica Study N.*, 4187 MR.

Y. Cen and R.N. Lichtenthaler (1995). Vapor permeation, in R.D. Noble and S.A. Stern (eds.), *Membrane Separation Technology: Principles and Applications*, Elsevier, Amsterdam.

Y. Chen, Y. Wang, H. Xu and G. Xiong (2008). "Efficient production of hydrogen from natural gas steam reforming in palladium membrane reactor", *Appl Catal B: Env*, **80**, 283–294.

S. Cheng and X. Shuai (1995). "Simulation of a catalytic membrane reactor for oxidative coupling of methane", *AIChE J*, **41**, 1598–1601.

J. Coronas, M. Mendez and J. Santamaria (1994a). "Methane oxidative coupling using porous ceramic membrane reactors. II. Reaction studies", *Chem Eng Sci*, **49**, 2015–2025.

J. Coronas, M. Menendez and J. Santamaria (1994b). "Development of ceramic membrane reactors with a non-uniform permeation pattern. Application to methane oxidative coupling", *Chem Eng Sci*, **49**, 4749–4757.

J. Coronas, A. Gonzalo, D. Lafarga and M. Menendez (1997). "Effect of the membrane activity on the performance of a catalytic membrane reactor", *AIChE J*, **43**, 3095–3104.

J. Coronas and J. Santamaria (1999). "Catalytic reactors based on porous ceramic membranes", *Cat Today*, **51**, 377–389.

A. Criscuoli, A. Basile and E. Drioli (2000). "An analysis of the performance of membrane reactors for the water-gas shift reaction using gas feed mixtures," *Cat Today*, **56**, 53–64.

A. Criscuoli, A. Basile, E. Drioli and O. Loiacono (2001). "An economic feasibility study for water gas shift in membrane reactor" *J Membrane Sci*, **181**, 21–27.

F.M. Dautzenberg and M. Mukherjee (2001). "Process intensification using multifunctional reactors", *Chem Eng Sci*, **56**, 251–267.

J.L. De Garmo, V.N. Parulekar, and V. Pinjala (1992). "Consider reactive distillation", *Chem Eng Prog*, **88**, 43–50.

A.G. Dixon (1999). "Innovations in catalytic inorganic membrane reactors", *Specialist Period Reports: Catal*, **14**, 40–92.

E. Drioli and E. Curcio (2007). "Perspective membrane engineering for process intensification: a perspective", *J Chem Tech Biotech*, **82**, 223–227.

S.S.E.H. Elnashaie, B.K. Abdallah, S.S. Elshishini, *et al.* (2001). "On the link between intrinsic catalytic reactions kinetics and the development of catalytic processes. Catalytic dehydrogenation of ethylbenzene to styrene", *Cat Today*, **64**, 151–162.

H.M. Ettouney, H.T. El-Dessouky and W.A. Waar (1998). "Separation characteristic of air by polysulfone hollow fiber membranes in series", *J Membrane Sci*, **148**, 105–117.

T. Gerven and A. Stankiewicz (2009). "Structure, energy, synergy, time: the fundamentals of process intensification", *Ind Eng Chem Res*, **48**, 2465–2475.

E. Gobina, K. Hou, and R. Hughes (1995a). "Mathematical analysis of ethylbenzene dehydrogenation: comparison of microporous and dense membrane systems", *J Membrane Sci*, **105**, 163–176.

E. Gobina, K. Hou, and R. Hughes (1995b). "Ethane dehydrogenation in a catalytic membrane reactor coupled with a reactive sweep gas", *Chem Eng Sci*, **50**, 2311–2319.

P. Godino, L. Peña and J.I. Mengual (1996). "Membrane distillation: theory and experiments", *J Membrane Sci*, **121**, 89–93.

V.M. Gryaznov and N.V. Orekhova (1998), in A. Cybulski and J.A. Moulijin (eds) *Structured Catalysts and Reactors*, Marcel Dekker Inc., New York, pp. 435–461.

H. Gunardson (1997). *Industrial gases in Petrochemical Processing*, Marcel Dekker, New York.

X.M. Guo, K. Hidajat and C.B. Ching (1997). "Oxidative coupling of methane in a solid oxide membrane reactor", *Ind Eng Chem Res*, **36**, 3576–3582.

Y. Guo, G. Lu, Y. Wang and R. Wang (2003). "Preparation and characterization of Pd–Ag/ceramic composite membrane and application to enhancement of catalytic dehydrogenation of isobutene", *Sep Pur Technol*, **32**, 271–279.

M.J. Haas, K.M. Scott, W.N. Marmer and T.A. Foglia (2004). "In situ alkaline transesterification: an effective method for the production of fatty acid esters from vegetable oils", *J Amer Oil Chem Soc*, **81**, 83–89.

C. Hamel, S. Thomas, K. Schädlich and A. Siedel-Morgenstern (2003). "Dosing concepts to improve the performance of parallel–series reactions", *Chem Eng Sci*, **58**, 4483–4492.

A.P. Harvey (2006). *Presentation at the PIN Meeting*, Newcastle upon Tyne. (www.pinetwork.org)

D.C. Hendershot (1999). "Designing safety into a chemical process". *Proceedings of the 5th Asia Pacific Responsible Case Conference and Chemical Safety Workshop, Shanghai*, November.

C. Hermann, P. Quicker, and Dittmeyer (1997). "Mathematical simulation of catalytic dehydrogenation of ethylbenzene to styrene in a composite palladium membrane reactor", *J Membrane Sci*, **105**, 161–172.

H.P. Hseih (1996) in *Inorganic Membranes for Separation and Reaction*, Elsevier, Amsterdam. ISBN 0-444-81677-1.

K. Huang, S.J. Wang, L. Shan, *et al.* (2007). "Seeking synergistic effect – a key principle in process intensification". *Sep Pur Technol*, **57**, 111–120.

S. Irusta, M.P. Pina, M. Menendez, J. Santamaria (1998). "Development and application of perovskite-based catalytic membrane reactor", *Cat Lett*, **54**, 69–78.

N. Itoh (1987). "A membrane reactor using palladium", *AIChE J*, **33** 1576–1578.

N. Itoh and K. Haraya (2000). "A carbon membrane reactor", *Cat Today*, **56**, 103–111.

A. Iulianelli, G. Manzolini, M. De Falco, *et al.* (2010). "H_2 production by low pressure methane steam reforming in a Pd-Ag membrane reactor over a Ni-based catalyst: Experimental and modeling", *Int J Hydr En*, **35**, 11514–11524.

A.E. Jansen, R. Klaassen and P.H.M. Feron (1995). "Membrane gas absorption – a new tool in sustainable technology development", *Proceedings, 1st Int Conf Proc Intensif for the Chem Ind*, 18, BHR Group, London, 145–153.

S. Jaso, H. Arellano-Garcia and G. Wozny (2011). "Oxidative coupling of methane in a fluidized bed reactor: Influence of feeding policy, hydrodynamics, and reactor geometry", *Chem Eng J*, **171**, 255–271.

B.H. Jeong, K.I. Sotowa and K. Kusakabe (2003). "Catalytic dehydrogenation of cyclohexane in an FAU-type zeolite membrane reactor", *J Membrane Sci*, **224**, 151–158.

S. Jorgensen, P.E.H Nielsen and P. Lehrmann (1995). "Steam reforming of methane in a membrane reactor", *Cat Today*, **25**, 303–307.

A. Julbe, D. Farrusseng and C. Guizard (2001). "Porous ceramic membranes for catalytic reactors – overview and new ideas", *J Membrane Sci*, **181**, 3–20.

A. Julbe, D. Farrusseng and C. Guizard (2005). "Limitations and potentials of oxygen transport dense and porous ceramic membranes for oxidation reactions", *Cat Today*, **104**, 102–113.

Y.K. Kao, L. Lei and Y.S. Lin (2003). "Optimum operation of oxidative coupling of methane in porous ceramic membrane reactor", *Cat Today*, **82**, 255–273.

F. Keil (2007). *Modeling of Process Intensification*, John Wiley & Sons, Inc., Hoboken Keizer K, Zaspalis VT, De Lange RSA, *et al.* (1994), "Membrane reactors for partial oxidation and dehydrogenation reactions", in J.G. Crespo and K.W. Boddeker (eds), *Membrane Processes in Separation and Purification*, Elsevier, Amsterdam, 415–429.

R.E. Kesting and A.K. Fritzsche (1993). *Polymeric Gas Separation Membranes*, John Wiley & Sons, Ltd, Chichester.

J.N. Keuler and L. Lorenzen (2002). "The dehydrogenation of 2-butanol in a Pd–Ag membrane reactor", *J Membrane Sci*, **202**, 17–26.

K.C. Khulbe, C.Y. Feng and T. Matsuura (2007). *Synthetic Polymeric Membranes, Characterization by Atomic Force Microscopy*, Springer-Verlag, Heidelberg.

W. Kiatkittipong, T. Tagawa, S. Goto, *et al.* (2005). "Comparative study of oxidative coupling of methane modeling in various types of reactor", *Chem Eng J*, **115**, 63–71.

E. Kikuchi, S. Uemiya, N. Sate, *et al.* (1989). "Membrane reactor using microporous glass-supported thin film of palladium. Application to the water gas shift reaction", *Chem Lett*, **3**, 489–492.

E. Kikuchi (1995). "Palladium-ceramic membranes for selective hydrogen permeation and their application to membrane reactor", *Cat Today*, **25**, 333–337.

E. Kikuchi (1997). "Hydrogen–permselective membrane reactor", *CaTTech*, **1**, 67–74.

E. Kikuchi, Y. Nemoto, M. Kajiwara, *et al.* (2000). "Steam reforming of methane in membrane reactors: comparison of electroless-plating and CVD membranes and catalyst packing modes", *Cat Today*, **56**, 75–81.

R. Krishna (2002). "Reactive separations: more ways to skin a cat", *Chem Eng Sci*, **57**, 1491–1504.

D. Lafarga, J. Santamaria and M. Menendez (2005). "Methane oxidative coupling using porous ceramic membrane reactors-I. reactor development", *Chem Eng Sci*, **49**, 2005–2013.

J. Langguth, R. Dittmeyer, H. Hofmann and G. Tomandl (1997). "Studies on oxidative coupling of methane using high-temperature proton-conducting membranes", *Appl Catal A: Gen*, **158**, 287–305.

G. Langhendries and G.V. Baron (1998). "Mass transfer in composite polymer-zeolite catalytic membranes", *J Membrane Sci*, **141**, 265–275.

K.W. Lawson and D.R. Lloyd (1997). "Membrane distillation: a review", *J Membrane Sci*, **124**, 1–25.

O. Lefebvre and R. Moletta (2006). "Treatment of organic pollution in industrial saline wastewater: A literature review", *Water Res*, **40**, 3671–3682.

L. Lin, Y. Zhang and Y. Cong (2009). "Recent advances in sulfur removal from gasoline by pervaporation", *Fuel*, **88**, 1799–1809.

Y.M. Lin, S.L. Liu, C.H. Chuang and Y.T. Chu (2003). "Effect of incipient removal of hydrogen through palladium membrane on the conversion of methane steam reforming. Experimental and modelling", *Cat Today*, **82**, 127–139.

B.S. Liu, L.Z. Gao and C.T. Au (2002). "Preparation, characterization and application of a catalytic NaA membrane for CH_4/CO_2 reforming to syngas", *Appl Catal A: Gen*, **235**, 193–206.

P. Liu, G. Gallaher and T. Gerdes (1993). "Experimental evaluation of dehydrogenation using catalytic membrane processes", *Sep Sci Tech*, **28**, 309–326.

Y. Lu, A.G. Dixon, W.R. Moser and Y.H. Ma (1997). "Analysis and optimization of cross-flow reactors with staged feed policies isothermal operation with parallel–series, irreversible reaction systems", *Chem Eng Sci*, **52**, 1349–1363.

Y. Lu, A.G. Dixon, W.R. Moser and Y.H. Ma (2000a). "Oxidative coupling of methane in a modified γ-alumina membrane reactor", *Chem Eng Sci*, **55**, 4901–4912.

Y. Lu, A.G. Dixon, W.R. Moser, *et al.* (2000b). "Oxygen-permeable dense membrane reactor for the oxidative coupling of methane", *J Membrane Sci*, **170**, 27–34.

G.Q. Lu, J.C. Diniz de Costa, M. Duke, *et al.* (2007). "Inorganic membranes for hydrogen production and purification: A critical review and perspective" *J Coll Interf Sci*, **314**, 589–603.

K. Mae (2007). "Advanced chemical processing using microspace", *Chem Eng Sci*, **62**, 4842–4851.

T. Matsuda, I. Koike, N. Kubo, E. Kikuchi (1993). "Dehydrogenation of isobutene to isobutene in a palladium membrane reactor", *Appl Catal A: Gen*, **96**, 3–13.

L. lezko and M. Baerns (1995). "Catalytic oxidative coupling of methane – reaction engineering aspects and process schemes", *Fuel Proc. Technol.*, **42**, 217–248.

M. Mulder (2000). *Basic Principles of Membrane Technology*, Kluwer Academic Publishing, The Netherlands.

S. Muraleedaaran, X. Li, L. Li and R. Lee (2009). "Is Reverse Osmosis Effective for Produced Water Purification? Viability and Economic Analysis", in *SPE Western Regional Meeting*, San Jose, California (USA).

T. Nozaki, O. Yamazaki, K. Omata and K. Fujimoto (1992). "Selective oxidative coupling of methane with membrane reactor", *Chem Eng Sci*, **47**, 2945–2950.

T. Okubo, K. Haruta, K. Kusakabe, *et al.* (1991). "Equilibrium shift of dehydrogenation at short space–time with hollow fiber ceramic membrane", *Ind Eng Chem Res*, **30**, 614–616.

L. Oliver, S. Haag, H. Pennemann, *et al.* (2009). "Oxidative coupling of methane using catalyst modified dense perovskite membrane reactors", *Cat Today*, **142**, 34–41.

C.S. Patil, M.V.S Annaland and J.A.M. Kuipers (2007). "Fluidized bed membrane reactor for ultrapure hydrogen production via methane steam reforming: Experimental demonstration and model validation", *Chem Eng Sci*, **62**, 2989–3007.

M.P. Pina, M. Menendez and J. Santamaria (1996). "The Knudsen diffusion catalytic membrane reactor: an efficient contactor for the combustion of volatile organic compounds", *Appl Catal B: Env*, **11**, 19–27.

T.K. Poddar, S. Majumdar and K.K. Sirkar (1996). "Removal of vocs from air by membrane-based absorption and stripping", *J Membrane Sci*, **120**, 221–237.

M. Ravanchi and T. Kaghazchi (2009). "Application of membrane separation processes in petrochemical industry: a review", *Desalination*, **235**, 199–244.

J. Sanchez and T.T. Tsotsis (1996). "Current developments and future research in catalytic membrane reactors", in A.J. Burggraaf and L. Cot (eds), *Fundamentals of*

Inorganic Membrane Science and Technology, Membrane Science and Technology Series, Vol. **4**, Elsevier, Amsterdam, (Chapter 11).

G. Saracco, H.W.J.P. Neomagus, G.F. Versteeg and W.P.M. van Swaaij (1999). "High-temperature membrane reactors: potential and problems", *Chem Eng Sci*, **54**, 1997–2017.

T. Sato, H. Yokoyama, H. Miki, and N. Itoh (2007). "Selective dehydrogenation of unsaturated alcohols and hydrogen separation with a palladium membrane reactor", *J Membrane Sci*, **289**, 97–105.

R. Schäfer, M. Noack, P. Kölsch, *et al.* (2003). "Comparison of different catalysts in the membrane supported dehydrogenation of propane", *Cat Today*, **82**, 15–23.

K. Scoth (1999), *Handbook of Industrial Membranes*, 2nd edn, Elsevier, Amsterdam.

P.V. Shanbhang, A.K. Guha and K.K. Sirkar (1995). "Single-phase membrane ozonation of hazardous organic compounds in aqueous streams," *J Haz Mat*, **41**, 95–104.

Z. Shao, H. Dong, G. Xiong, *et al.* (2001). "Performance of a mixed-conducting ceramic membrane reactor with high oxygen permeability for methane conversion", *J Membrane Sci*, **183**, 181–192.

M. Sheintuch and R.M. Dessau (1996). "Observations, modelling and optimization of yield, selectivity and activity during dehydrogenation of isoutane and propane in a Pd membrane" *Chem Eng Sci*, **51**, 535–547.

Y. Shirasaki, T. Tsuneki, Y. Ota, *et al.* (2009). "Development of membrane reformer system for highly efficient hydrogen production from natural gas", *Int J Hydr En*, **34**, 4482–4487.

J. Shu, B.P.A. Grandjean and S. Kaliaguine (1994). "Methane steam reforming in asymmetric Pd- and Pd-Ag/porous SS membrane reactors", *Appl Catal A: Gen*, **119**, 305–325.

J. Shu, B.P.A. Grandjean and S. Kaliaguine (1995). "Asymmetric Pd-Ag/stainless steel catalytic membranes for methane steam reforming", *Cat Today*, **25**, 327–332.

A.P. Simpson and A.E. Lutz (2007). "Energy analysis of hydrogen production via steam methane reforming", *Int J Hydr En*, **32**, 4811–4820.

K.K. Sirkar, P.V. Shanbhag and A.S. Kovvali (1999). "Membrane in a reactor: a functional perspective", *Ind Eng Chem Res*, **38**, 3715–3737.

R.W. Spillman (1989). "Economics of gas separation membranes", *Chem Eng Progr*, **85**, 41–62.

R.W. Spillman (1995). "Economics of gas separation membrane processes", in R.D. Noble and S.A. Stern (eds), *Membrane Separation Technology: Principles and Applications*, Elsevier, Amsterdam.

W.P. Stadig (1987). "Catalytic distillation: combining chemical reaction with product separation" *Chem Proc*, **50**, 27–32.

A. Stankiewicz (2003). "Reactive separations for process intensification: an industrial perspective", *Chem Eng Prog*, **42**, 137–144.

E.H. Stitt (2002). "Alternative multiphase reactors for fine chemicals. A world beyond stirred tanks?", *Chem Eng J*, **90**, 47–60.

G. Sznejer and M. Sheintuch (2004). "Application of a carbon membrane reactor for dehydrogenation reactions", *Chem Eng Sci*, **59**, 2013–2021.

F. Tiscarno-Lechuga, C.G. Hill Jr, and M.A. Anderson (1996). "Effect of dilution in the experimental dehydrogenation of cyclohexane in hybrid membrane reactors", *J Membrane Sci*, **118**, 85–92.

J. Tong and Y. Matsumura (2005). "Effect of catalytic activity on methane steam reforming in hydrogen-permeable membrane reactor", *Appl Catal A: Gen*, **286**, 226–231.

A.L.Y Tonkovich, D.M. Jimenez, J.L. Zilka and G.L. Roberts (1996). "Inorganic membrane reactors for the oxidative coupling of methane", *Chem Eng Sci*, **51**, 3051–3056.

D.L. Trimm (1997). "Coke formation and minimization during steam reforming reactions", *Cat Today*, **37**, 233–238.

C.Y. Tsai, Y.H. Ma, W.R. Moser and A.G. Dixon (1995). "Modeling and simulation of a non-isothermal catalytic membranes reactor", *Chem Eng Commun*, **134**, 107–132.

T.T. Tsotsis, A.M. Champagnie and S.P. Vasileiadis, *et al.* (1992). "Packed bed catalytic membrane reactors", *Chem Eng Sci*, **47**, 2903–2908.

V.A. Tuan, S. Li, J.L. Falconer and R.D. Noble (2002). "In situ crystallization of beta zeolite membranes and their permeation and separation properties", *Chem. Mater*, **14**, 489–492.

A. Tuchlenski, A. Beckmann, D. Reusch, *et al.* (2001). "Reactive distillation–industrial applications, process design and scale-up", *Chem Eng Sci*, **56**, 387–394.

S. Uemiya, N. Sato, H. Ando, *et al.* (1991a). "Steam reforming of methane in a hydrogen-permeable membrane reactor", *Appl Catal*, **67**, 223–230.

S. Uemiya, N. Sato, H. Ando and E. Kikuchi (1991b). "The water-gas shift reaction assisted by a palladium membrane reactor", *Ind Eng Chem Res*, **30**, 585–589.

P. Vandezande, LEM Gevers and IFJ Vankelecom (2008). "Solvent resistant nanofiltration: separating on a molecular level", *Chem Soc Rev*, **37**, 365–405.

J. Vital, A.M. Ramos, I.F. Silva and J.E. Castanheiro (2001). "The effect of α-terpineol on the hydration of α-pinene over zeolites dispersed in polymeric membranes", *Cat Today*, **67**, 217–223.

D.R.B. Walker and W.J. Koros (1991). "Transport characterization of a polypyrrolone for gas separations", *J Membrane Sci*, **55**, 99–117.

L. Wang, K. Murata, and M. Inaba (2003). "Production of pure hydrogen and more valuable hydrocarbons from ethane on a novel highly active catalyst system with a Pd-based membrane reactor", *Cat Today*, **82**, 99–104.

H. Wang, Y. Cong and W. Yang (2005). "Oxidative coupling of methane in $Ba_{0.5}Sr_{0.5}Co_{0.8}Fe_{0.2}O_{3-\delta}$ tubular membrane reactors", *Cat Today*, **104**, 160–167.

K.L. Wasewar (2007). "Modeling of pervaporation reactor for benzyl alcohol acetylation", *Int J Chem React Eng*, 5, DOI 10.2202/1542-6580.1362.

T. Westermann and T. Melin (2009). "Flow-through catalytic membrane reactors-Principles and applications", *Chem Eng Proc*, **48**, 17–28.

H. Weyten, J. Luyten, K. Keizer, *et al.* (2000). "Membrane performance: the key issues for dehydrogenation reactions in a catalytic membrane reactor", *Cat Today*, **56**, 3–11.

L.S. White (2006). "Development of large-scale applications in organic solvent nanofiltration and pervaporation for chemical and refining processes", *J Membrane Sci*, **286**, 26–35.

S. Wu, J.E. Gallot, M. Bousmina, *et al.* (2000). "Zeolite containing catalytic membranes as interphase contactors", *Cat Today*, **56**, 113–129.

Y. Xia, Y. Lu, K. Kamata, *et al.* (2003). "Macroporous materials containing three-dimensionally periodic structures", in P. Yang (ed.) *Chemistry of Nanostructured Materials*, World Scientific Publishing, 69–100.

S.J. Xu and W.J. Thomson (1997). "Perovskite-type oxide membranes for the oxidative coupling of methane", *AIChE J*, **43**, 2731–2740.

S.H. Yeon, K.S. Lee, B. Sea, *et al.* (2005). "Application of pilot-scale membrane contactor hybrid system for removal of carbon dioxide from flue gas", *J Membrane Sci*, **257**, 156–160.

K.T. Yu, M. Zhou and C.J. Xu (1996). "A novel separation process: distillation accompanied by adsorption", *Proceedings, 5th World Congress of Chem Eng*, San Diego, 347–352.

Y. Zeng and Y.S. Lin (2000). "Oxygen permeation and oxidative coupling of methane in yttria doped bismuth oxide membrane reactor", *J Catal*, **193**, 58–64.

7

Production of Bio-Based Fuels: Bioethanol and Biodiesel

Sudip Chakraborty[1,2], Ranjana Das Mondal[1], Debolina Mukherjee[3], and Chiranjib Bhattacharjee[1]

[1]*Department of Chemical Engineering, Jadavpur University, West Bengal, India*
[2]*Department of Chemical Engineering and Materials, CNR-ITM, University of Calabria, Rende, Italy*
[3]*Department of Geology, University of Calabria, Rende, Italy*

7.1 Introduction

7.1.1 Importance of Biofuel as a Renewable Energy Source

Achieving a sustainable energy future depends increasingly on renewable energy sources. At the present time renewable energy represents only about 14% of the total world energy supply, of which the largest fraction is traditional biomass. Renewable energy, which experienced no downturn in 2009, continued to grow strongly in all end-use sectors – power, heat, and transport – and supplied an estimated 16% of global final energy consumption. Renewable energy accounted for approximately half of the estimated 194 GW of new electricity capacity added globally during the year. Renewables delivered close to 20% of global electricity supply in 2010, and by early 2011 they comprised one quarter of the global power capacity from all sources. In the US, renewable energy accounted for about 10.9% of domestic primary energy production. Renewables accounted for about 26% of China's total installed electric capacity, 18% of generation, and more than 9% of final energy consumption in 2010. Germany met 11% of its total final energy consumption with renewable sources, which accounted for 9.8% of

Sustainable Development in Chemical Engineering – Innovative Technologies, First Edition.
Edited by Vincenzo Piemonte, Marcello De Falco and Angelo Basile.
© 2013 John Wiley & Sons, Ltd. Published 2013 by John Wiley & Sons, Ltd.

heat production (mostly from biomass), and 5.8% of transport fuel consumption. Several other countries met higher shares of their electricity demand with wind power in 2010 including Denmark (22%), Portugal (21%), Spain (15.4%), and Ireland (10.1%). Non-traditional biomass currently provides about 2% of the total energy consumption in US Fossil fuels, which supply 80% of the world's energy, are projected to be depleted within one or two generations at the present rate of consumption and carry environmental and security concerns (Goldemberg, 2007). It is estimated that forest and agricultural land alone – the two largest potential biomass sources – could produce enough biofuels to meet more than one-third of the current demand for transportation fuels (Perlack *et al.*, 2005). Though biomass is often waste and has fewer environmental concerns than fossil fuel, it is still limited for large scale application due to costly processing routes: this is expected to be fully realized in the near future by developing large scale bioenergy industries. Rapid growth in the production and consumption of biofuels for powering combustion engines for transportation has been witnessed in recent years. Liquid biofuel currently in commercial use comprises primarily ethanol derived fuels based on cereals (corn), sugar crops (sugar cane and sugar beet) and biodiesel based on vegetable oil esters. Ethanol and biodiesel have already expanded the existing market and other renewable fuels have begun to emerge as potential viable alternatives; in particular, bio-methane, bio-hydrogen, bio-butanol (Scheffran, 2010). In 2007, the global production of biofuels amounted to 62 billion liters per year corresponding to 1.8% of total transport fuel consumption in energy terms (OEDC, 2008). Fuel ethanol accounts for most of the world's biofuels, with a production of 49.6 gigaliters in 2007. Almost half of the bio-ethanol is produced in US, 38% in Brazil, 4.3% in the European Union and 3.7% in China. The transportation market is almost completely dependent on energy from fossil fuels such as gasoline, diesel and kerosene, which makes that sector of the economy vulnerable to fluctuations in petroleum price and supply, particularly for the developing countries, whose oil supply depends largely on the import. Recent trends suggest that large scale biofuel production in the US by 2012 would reduce crude oil imports by 2 billion barrels per year (RFA, 2006). Thus, the use of fuels based on plant biomass would offer a significant opportunity to diversify energy sources in the transportation sector, building on the existing infrastructure of gas stations and automobile technology. Advanced bio-energy could possibly help to satisfy the growing energy demands of developing countries. Currently, about 2.4 billion people depend on traditional energy uses of biomass which are often inefficient, unhealthy and non-sustainable (Ezzati and Kammen, 2001). Due to the high productivity of energy crops in tropical and subtropical regions locally produced advanced bio-energy like bio-ethanol for sugar cane or bio-diesel from palm oil could potentially provide income and employment in rural areas and in turn facilitate sustainable development in these regions (Hazell and von Braun, 2006). A study conducted by the United Nations in 2008 on the potential of 'green' jobs found that renewable energy generates more jobs than employment in the fossil fuel industry. An economic study of the existing bioethanol plant (Iowa State University, 2006) estimates that a 50 million gallon ethanol plant with 75% of local ownership would create 220 new jobs created per million liters of annual production, and for biodiesel it would result in 1.16 jobs created per million liters of annual production.

Concerns about global warming have also spurred on searches for low carbon energy alternatives to fossil fuels, as their global scale implementation would help accomplish targets for greenhouse gas (GHG) emission reduction as set in the Kyoto Protocol and follow on agreements (Worldwatch Institute, 2007). Biofuels are carbon neutral because the carbohydrate and starch used to manufacture them originates from atmospheric carbon fixed by photosynthesis. Ethanol from sugar, over its whole bio-industrial cycle, may reduce GHG emission by 80% or more relative to emission from fossil fuel. Biofuel from wheat, sugar beet and vegetable oil provides a GHG saving of 30–60%, ethanol from corn allows a saving of just below 30%. These differences are attributed to the sugar content and the use of fossil fuels in production (OECD, 2008). By early 2011, at least 119 countries had some type of policy target or renewable support policy at the national level, up from 55 countries in early 2005. India is fifth worldwide in total existing wind power capacity and is rapidly expanding many forms of rural renewables such as biogas and solar photovoltaic power. In this chapter we review the potential sources of biofuel, related technologies, all probable prospects, economic aspects and environmental aspects and future prospects of biofuel.

7.2   Production of Bioethanol

With efforts to reduce the global reliance on fossil fuels and lower greenhouse gas emissions, an increasing search for renewably sourced materials, which can be used as feedstock for biofuel production, is ongoing in the past few decades. At the present, ethanol is the most common alternate fuel and is already produced on a fair scale, representing a sustainable substitute for gasoline in passenger cars. In 2010, global production of fuel ethanol reached an estimated 86 billion liters, an increase of 17% over 2009 (Figure 7.1, Table 7.1). The US and Brazil accounted for 88% of ethanol production in 2010, with the US alone producing 57% of the world's total. The United States accounted for most of the increase, producing 8.4 billion more liters in 2010 than in 2009, for a total of 49 billion liters. Well over 90% of US gasoline was blended with

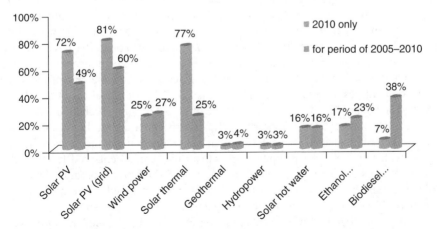

Figure 7.1 *Worldwide annual growth rate of renewable energy capacity and biofuel production for 2005–2010 (Adapted from* Renewables 2011 Global Status Report, *2011)*

Table 7.1   *Biofuel production (in billion liters) for the top 15 countries, 2010 (Adapted from 'Renewables 2011 Global Status Report')*

Ranking	Country	Fuel ethanol	Biodiesel	Total
1	United States	49	1.2	50.20
2	Brazil	28	2.3	30.30
3	Germany	1.5	2.9	4.40
4	France	1.1	2	3.10
5	China	2.1	0.2	2.30
6	Argentina	0.1	2.1	2.20
7	Spain	0.6	1.1	1.70
8	Canada	1.4	0.2	1.60
9	Thailand	0.4	0.6	1.00
10	Italy	0.1	0.8	0.90
11	Indonesia	0.1	0.7	0.80
12	Belgium	0.3	0.4	0.70
13	Poland	0.2	0.5	0.70
14	UK	0.3	0.4	0.70
15	Colombia	0.4	0.3	0.70
–	**World Total**	86	19	105.00

ethanol. Approximately 3% of the world's grain supply was used to produce this ethanol, with almost 32.5 million tons of animal feed as a co-product.

7.2.1   Bioethanol from Biomass: Production, Processes, and Limitations

Growing worldwide concerns about dependence on imported oil, the depletion of fossil fuels, and environmental pollution brought about new policies in biofuel production. Increasing fuel prices and emission of GHG necessitates the development of renewable energy sources, in particular, liquid biofuels for transportation (Huesemann, 2006). Extensive research and developments of so called 'next generation biofuels' ethanol from lingocellulosic biomass is underway and hoped to lead a dramatic expansion in the production of transportation biofuels, without competition for arable land that would impact upon food and feed supplies (Huesemann *et al.*, 2010). Although the basic process of ethanol production has not changed greatly for several decades, significant improvements were made throughout the 1980s and 1990s in the design and engineering of fuel ethanol production facilities. Liquid biofuels provided about 2.7% of global road transport fuels in 2010. The US and Brazil accounted for 88% of global ethanol production; after several years as a net importer, the US overtook Brazil to become the world's leading ethanol exporter. The EU remained the center of biodiesel production, but due to increased competition with relatively cheap imports, growth in the region continued to slow. Brazil has expanded its production of sugar cane ethanol with *Proalcool* program (Nass *et al.*, 2007), whereas the US launched a fuel ethanol program based on corn (Bothast and Schlicher, 2005), when commonly used oxygenate methyl tertiary-butyl ether (MTBE) was banned due to ground water contamination. In 2005, The US endorsed the 'Renewable Fuel Standard' (Energy Policy Act of 2005) that require the use of 7.5 billion gallons (BG) of biofuel for transportation by 2012. Furthermore, comprehensive legislation made in US Law, 2007 mandated a phasing in of all renewable fuels, including conventional (corn starch) ethanol, advanced biofuel (ethanol and biodiesel from biomass), cellulosic biofuel and renewable diesel, from a total of 9 BG (billion gallons) in 2008 to 36 BG by 2022 (Energy Independence and Security Act of 2007). The cost per gallon of ethanol produced from sugar crops is competitive with the cost of gasoline ($2–3), indeed in Brazil the sugar cane crop has overtaken gasoline as the main transportation fuel (FAPRI, 2008). Although technical challenges remain, the need to identify sustainable transportation of fuel sources will dictate the development of advanced bio-refineries producing ethanol from cellulosic feed stocks at a competitive price.

7.2.2 Substrate

7.2.2.1 Bioethanol from Starchy Mass by Fermentation

Bioethanol can be produced from nearly any readily available crop. Starch is the fundamental form of energy storage in many species of plants and constitutes an essential element in the diet of human beings. It is particularly abundant in organisms such as cereals and tubers and is the fundamental raw material in a large number of industrial processes such as the production of adhesives, paints, biodegradable plastics, biofuels, and so on. Some compositional data for raw materials is shown in Table 7.2. Conventionally, sugarcane based bioethanol has been the major automobile fuel in Brazil for almost three decades due to the abundance and renewable nature of sugarcane. Similarly corn based ethanol production in US has expanded from 175 million gallons in 1980 to 4.8 billion gallons in 2006. The corn to ethanol production industry has grown 350% from 2000 to 2008 (RFA, 2008) due to strong and sustainable demand for bioethanol as fuel. Parallel improvements in fermentation technology and process design for ethanol have been reported for a decade. After hiking the price of corn to make the bioethanol industry profitable, improvements in fermentation technology and enzyme system were made. Industrial fermentation technologies are broadly dependent on continuous operation with immobilized cell systems for conversion of sugar and starch based materials to ethanol. These days, ethanol production processes also employ 'growth arrested' cells at very high cell densities, in conjunction with cell recycling membrane system (Inui *et al.*, 2004b). Compared to existing fermentation processes, growth arrested processes have several advantages including high productivity (Inui *et al.*, 2004a) and improved tolerance to fermentation inhibitors (Sakai *et al.*, 2007).

Table 7.2 Some general composition data for corn, barley, rye, wheat, and oats

Component	% Content of dry matter				
	Corn	Barley	Rye	Wheat	Oats
Protein	9.0–12.0	10.0–11.0	10.0–15.0	12.0–14.0	13.0–16.0
Fat	4.5	2.5–3	2.0–3.0	3.0	6.0–7.0
Starch	65–72	52–64	55–65	67–70	54–64
Ash	1.5	2.3	2.0	2.0	2.0
Total dietary fiber	13–15	14–24	15–17	13–10	11–13
Without soluble fiber	–	8.0–10.0	3.0–4.0	1.0–2.0	3.0–5.0

Batch process   To date, simple batch fermentation technology is used in most of the beverage industries using carbohydrate feed stocks. The substrate and the cell are introduced into a bioreactor together with nutrients. Depending on the characteristics of the carbohydrate feed stock, the conversion efficiency attained by yeast has been reported in the range of 90–95% of the theoretical value, with a final ethanol concentration of 10–16% (w/v) (Lin and Tanaka, 2006). Batch processing is inherently simple, with investment cost, easy operation, low requirement for complete sterilization use of less qualified labor, lower risk of financial loss easy management of feed stock and reduced risk of contamination. However, drawbacks are; significant idle time, need for frequent sterilization and existence of a lag phase at the start of new batches. Despite the disadvantages, batch processing is preferred when required to prepare small amounts of product or the product cost is high. Nevertheless one of the most successful batch systems was the 'Melle–Bionot' process which reduces the fermentation time and increases yield by recycling the yeast while simultaneously employing several fermenters operated at staggered intervals. This approach is limited in getting ethanol from lignocellulosic biomass due to presence of fermentation inhibitors such as sugar degradation products, organic acids, polyphenolics, and other compounds found in plant materials, which are inhibitors of fermenting organisms (Bothast *et al.*, 1999). The reaction scheme and the process design for the production of bioethanol is described in Figure 7.2.

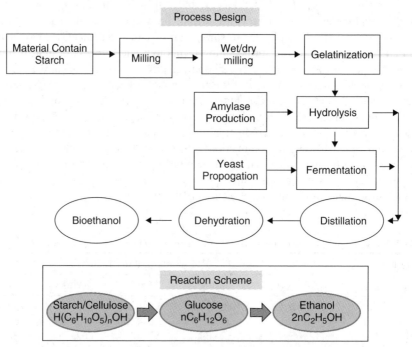

Figure 7.2 *Process design and reaction scheme for the production of bioethanol from starch based materials*

Fed batch process   The fed batch process, which is a combination of batch and continuous process, is the most popular process in bioethanol industries. Fed batch processes are able to solve the problem of substrate inhibition by maintaining low substrate concentrations in the bioreactor. Use of such process enables the problem of mutation and plasmid instability, which is frequently encountered in continuous culture, to be avoided. This process does not suffer from any cell washout problems resulting good yield, high flexibility and well defined cultivation times. Fed batch processes are often employed when there is significant risk of the microorganism mutating or becoming infected, or when batch process would result in too low productivity. For lignocellulosic feed stocks this process is also well established with major advantages in conducting an in situ detoxification of the hydrolysates by a direct action of the fermenting microorganisms (Nilsson *et al.*, 2001; 2002; Taherzadeh *et al.*, 2008; Chaudhary and Qazi, 2011).

Continuous process   Continuous ethanol production considerably reduces industrial inputs such as operation time or equipment downtime. The process is operated in such a way that the composition of the solution in the bioreactor remains constant. In the bio-reactor cells are continuously grown while older cells are continuously washed out so that a steady state is achieved when the growth and washout rates are equal. Despite its advantages, this process has stringent requirements with regard to the raw material quality, as it is extremely difficult to adapt the operating conditions of such systems to large process variations. There is also a high risk that the microorganisms used may undergo mutation over the long cultivation period. Hence, continuous fermentation processes are used only for products that demand both high production rates for economic viability and employ only those microorganisms that have a high stability and resist mutation. Despite the potential productivity advantages offered by this technology, it is not widely adapted for lignocellulosic biomass hydrolysis, which suffers from very low dilution rates that lead to low conversion rates of the inhibitors and reduce the growth rate of the biomass with increased biomass washout. In order to carry out the lignocellulosic conversion process in the continuous mode, cell growth should be maintained at the same rate as dilution to avoid cell washout.

Immobilized cell systems   A new approach in bioethanol industries is the use of immobilized cells to improve the yield value and enzyme life span. The use of immobilized yeast has attracted considerable attention during the past few years due to advantages over the processes with free yeast. Immobilization of the cells can eliminate inhibition caused by high concentration of substrate and product, and also enhance ethanol yield and productivity (Najafpour *et al.*, 2004). Immobilized cell systems have advantages over suspension cultures and permit high cell concentration, not susceptible to cell washout at high dilution rate, recovery of expensive cell systems and recycling, which ultimately leads to high volumetric productivities, minimized shear damage, good nutrient product gradient and good pH gradient. Nonetheless, the key factor to a successful immobilized cell process is the genetic stability of the immobilized cells. In general, four different types of immobilization techniques have been recognized based on the physical mechanism of cell localization and the nature of the support mechanisms (Karel *et al.*, 1985): (2) attachment to the surface; (2) entrapment within the porous matrix; (3) containment behind a barrier; and (4) self-aggregation. The simultaneous enzymatic saccharification and fermentation (SSF) of corn meal using immobilized cells of *Saccharomyces cerevisiae var. ellipsoideus* yeast has been done in a batch system where, the yeast cells were immobilized in Ca-alginate by the electrostatic droplet generation method (Nikolić *et al.*, 2009). Ivanova *et al.* (2011) have also conducted a study on *Saccharomyces cerevisiae* cells which were entrapped in matrix of alginate and magnetic nanoparticles and covalently immobilized on magnetite-containing chitosan and cellulose-coated magnetic nanoparticles.

7.2.2.2   Bioethanol from Lignocellulosic Biomass

The need to use lignocellulosic materials as alternative renewable feed stocks for ethanol production was recognized during the 1970s (Wang *et al.*, 1997), and extensive research in the area of fermentation has been carried out over the subsequent four decades (Flickinger, 1980; Han *et al.*, 1981; Lawford and Rousseau, 1993; Sun and Cheng, 2002; Saha, 2003; Sánchez and Cardona, 2008). Lignocellulosic raw materials primarily consist of lignin, cellulose, and hemicellulose (Figure 7.3). Lignin is heterogeneous polymer of substituted aromatic building blocks, cellulose is a linear polymer of glucose and hemicellulose is a branched polymer which, in addition to glucose, may also contain the hexose sugars mannose and galactose, and the pentose sugars xylose and arabinose (Lynd *et al.*, 2005). The use of waste agricultural biomass as raw material has already been described elsewhere (Chakraborty *et al.*, 2012) as a source of bioethanol and biodiesel production. Pentose sugars are primarily present hemicelluloses derived from hard wood, agricultural residue and grasses, and can make up more than 35% of the total dry matter (Hayn *et al.*, 1993). It has been reported that the complete conversion of pentose sugar to ethanol would reduce the production cost of bioethanol by 22% (Sassner *et al.*, 2008). As a result, current biomass to ethanol technology has been successful in pilot

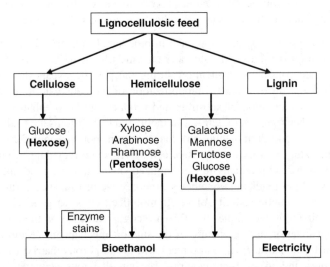

Figure 7.3   *Conversion of lignocellulosic feedstock into ethanol (Adapted from Neves et al., 2007)*

plant production; however, the process of achieving the conversion of lignocellulosic biomass to ethanol is more complex than the conventional starch to ethanol process. The complexities of the lignocellulosic biomass economic value chain (biomass production, varying composition and recalcitrance to degradation) and the overall biomass to biofuel commercial production remains a costly process (Liu, 2010). Considerable investigations into constructing and developing microorganisms that efficiently ferment both hexose and pentose to ethanol have also been made (Dien *et al.*, 2003; Hahn-Hägerdal *et al.*, 2007a,b). Major barriers to the cost effective production of ethanol from lignocellulosic biomass feedstocks include both logistical and technical constraints, such as (1) the efficient transportation of bulky biomass feedstocks to conversion sites; (2) efficient degradation of biomass materials into hydrolysates containing fermentable sugars; (3) efficient fermentation of mixed sugars present in biomass hydrolysates into ethanol; (5) cost effective downstream processing for recovery of ethanol and byproducts; (5) costly processing and use of hydrolytic enzymes for breaking down heterogeneous and recalcitrant lignocellulosic biomass into simple fermentable sugars; and (6) a lack of efficient biocatalysts and optimal processes for converting mixed sugars derived from lignocellulosic mass into ethanol and valuable byproducts (Liu, 2010).

The conversion of lignocellulosic biomass to fermentable sugar involves a series of operation steps (Figure 7.4). The first step is pretreatment of the biomass to make the carbohydrate available for hydrolysis by an acid or enzyme followed by saccharification and fermentation. During this pretreatment and hydrolysis step, multiple molecular events take place such as: liberation of acetic acid from hemicellulose; phenolic compounds from lignin fraction; furans and weak acids from solubilized hexose and pentose sugar (Almedia *et al.*, 2007). Unlike starch, biomass hydrolysates usually contain mixed sugars, sugar degradation products, organic acids, polyphenolics and other compounds found in plant materials which are inhibitors of fermenting organisms (Bothast *et al.*, 1999). Fermenting microorganisms require the hydrolysate to be detoxified prior to fermentation (Hahn-Hägerdal *et al.*, 2007a), which adds to the production cost of ethanol. Microbial sensitivity to the toxicity of ethanol can affect fermentation productivity, the major impacts of which are lower fermentation rates, reduced ethanol yields and a decreased microbial lifetime (Stanley *et al.*, 1997). These issues spur on more research into highly ethanol sensitive and novel microorganisms that can improve the yield of ethanol and make the process more economically feasible.

A possible way of optimizing enzymatic breakdown of lignocellulosic material is the use of efficient microorganism systems for digesting plant biomass, that is, to use an advanced enzyme system. A new enzyme system includes: (1) glycosyl hydrolyses (GHs), and (2) cellulosomes. GHs are modular enzymes that include both cellulose and

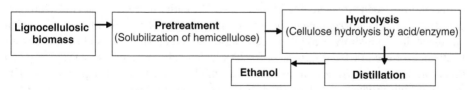

Figure 7.4 *Schematic representation of conversion of lignocellulosic biomass to ethanol by simultaneous saccharification and fermentation*

noncellulosic structural polysaccharides. The modules consist of different combinations of catalytic domains (Bayer *et al.*, 1998; 2000). The cellulose cleaves the β-1, 4-glucosidic bonds of cellulose, resulting in the production of cellobiose. Noncellulosic structural polysaccharidases are a diverse group of enzymes which can cleave many different types of bonds (Bayer *et al.*, 2000). Plant cell wall degradation by bacteria and fungi is coordinated by a multitude of enzymes, which are assembled into a complex molecular matrix referred to as the cellulosome (Bayer *et al.*, 1998; 2000; 2004), that is, a multi-enzyme complex specializing in cellulose degradation. Cellulose degradation by anaerobic bacteria is achieved by the relatively small quantities of enzymes they produce, compared to aerobic microorganisms.

The ability to increase the ethanol tolerance of microorganisms can potentially improve the productivity of lignocellulosic based processes by decreasing fermentation times, with potentially considerable gains in ethanol yields when solid loadings in fermenters are increased. Increasing ethanol tolerance can also provide cross tolerance to the inhibitors present in lignocellulosic hydrolysate (Barber *et al.*, 2002). Research in this area has generated a lot of information on recombinant yeasts as potential elements which minimize byproduct formation. Metabolic engineering strategies have also been employed on commonly available enzymes to reduce byproduct formation (Panagiotou *et al.*, 2006). Gram negative bacteria *Zymomonas mobilis* and *Saccharomyces cerevisie* unable to ferment xylose and arabinose have been engineered for co-fermentation of those sugars to ethanol (Zhang *et al.*, 1995; Deanda *et al.*, 1996). Pilot scale processes have also been developed (Bio-Hol process and Glucotech process) (Lawford, 1987; Doelle *et al.*, 1991), using *Z. mobilis* for ethanol production. The gram negative bacterium *E. coli* was also found to be a potential for ethanol formation due to its natural capacity for hydrolyzing pentose. In *E. coli*, sugars enter the carbon metabolic path way via the pentose phosphate pathway. Some gram positive bacteria also possess the capacity to produce bioethanol and the ability to ferment multiple sugars with low pH tolerances. Some stains are also reported to be capable of growing in a relatively large temperature range with a high tolerance to environmental stress (Bothast *et al.*, 1999). In this family, lactic acid bacteria (LAB) are ethanol tolerant (10–16%) (Gold *et al.*, 1996). Certain species *L. brevis* and *L. buchneri* are known to grow in the presence of inhibitors (Sakamoto and Konings, 2003). *L. buchneri* produces $12\,g\;l^{-1}$ of ethanol from $125\,g\;l^{-1}$ of mixed sugar, in addition to lactate, acetate, and cell mass accumulation.

Consolidated Bioprocessing (CBP), is also a new approach for the lignocellulosic ethanol industry which includes consolidated cellulose production, cellulose hydrolysis and sugar fermentation steps in a single unit operation (Lynd *et al.*, 2005). CPB technology requires lower capital, materials and utilities costs compared to conventional processes. CBP costs about 25% of simultaneous saccharification and co-fermentation with cellulose production and 50% of the cost for the entire process (Lynd *et al.*, 2005). A unique bacterium *Clostridium phytofermentans* has been isolated from forest soil, which has the ability to hydrolyse cellulose, cellobiose and xylan, and also pentose and hexose to ethanol for use in CBP (Warnick *et al.*, 2002). Generally recognized as a safe (GRAS) organism, the *S. cerevisiae* stain has also been modified for CBP for high ethanol tolerance and high productivity. This strain has been found to directly ferment cellulose to ethanol and also ferments hexose and pentose sugar to ethanol.

7.2.2.3   Bioethanol from Microalgae and Seaweeds

Seaweeds and microalgae have cultivation lot of evidence as source of commercial products (McHugh, 2003; Pulz and Gross, 2004). These photosynthetic organisms have also been subjected to extensive investigations relating to their potential as 'green sources' of liquid transportation fuels since the 1970s. Seaweed research has received less attention in US, but was the subject of a major effort by DOE from 1978–1983 (Bird and Benson, 1987), and is now the focus of considerable interest in Europe, Japan and Korea.

There are significant advantages of using algae as source of fuel over other higher plants due to greater productivity in water, avoiding water and nutrient limitations, continuous cultivation without any effect of climatic change and very few nonproductive parts (root and stem). Algal cultivation does not require arable land, and they can grow in shallow ponds, using saline or brackish water. Microalgae systems also require an enriched source of CO_2, such as flue gases from fossil fuel fired power plants and can be collocated with such sources. These positive aspects favor large scale production of algal biomass avoiding any competition with food and feed crops that has bothered the world wide supply–demand balance, economic, social and even ethical problems (Rosegrant *et al.*, 2006). Still, the production of fuel from algal biomass is limited due to its high cost and the undeveloped nature of production routes but is hoped to be used as a potential alternative in near future. Microalgae can potentially generate a wide range of biofuels (ethanol, hydrogen, methane, oil, electricity or power) (Tsukahara *et al.*, 2005). Even more simply, harvested algal biomass (typically 80–90% moisture content) could be dried and combusted to generate electricity but this process is limited due to high drying costs and the elevated generation of NO_x during the high temperature thermal destruction of nitrogen fertilizer content of biomass (Matsumoto *et al.*, 1995; Kadam, 2002). Production of synthesis gas and oil at a laboratory level has also been an optional route for biofuel utilization by thermo chemical conversion of microalgal biomass through gasification or pyrolysis (Minowa *et al.*, 1995; Minowa and Sawayama, 1999; Sawayama *et al.*, 1999; Peng, 2000, 2001a,b; Miao, 2004; Miao and Wu, 2004). Scaling up of this process is industrially challenging due to the high temperature (>400 °C) and pressure (18 MPa) which suggests a rather modest net energy output ratio (EOR) (Huesemann *et al.*, 2010). Entire algal biomass can also be converted to biogas by anaerobic digestion. Another potential approach is to extract the vegetable oil from algal biomass and conversion to biodiesel by transesterification or conversion to diesel like or 'green diesel' by hydrocracking. Microalgae *Botrycoccus braunii*, was reported to produce hydrocarbon that might be refined like petroleum, some others are having stored carbohydrate that can be converted by bacterial fermentation to bioethanol and biobutanol or other liquid fuel (Huesemann *et al.*, 2010).

Two different approaches to getting bioethanol from microalgae have been practiced to date. The first option is the fermentation of stored carbohydrates in algal biomass, such as starch from green algae and glycogen in cyanobacteria. Several authors have demonstrated yeast mediated fermentation (Hirano *et al.*, 1997; Shirai *et al.*, 1998; Matsumoto *et al.*, 2003), but these processes are less attractive due to low yields. The second option is the self-fermentation of carbohydrate storage products by endogenous algal enzyme induced in the absence of oxygen (for *Chlamydomonas*), but this is also limited by insignificant yield value (Akano *et al.*, 1996; Hon-Nami, 1996; Hirayama *et al.*, 1997).

Deng and Coleman (1999) have reported the genetic modification of a cyanobactrium to improve the production of ethanol but result showed a very low yield of ethanol.

The economic contribution of seaweeds as source of renewable energy source is trifling though their potential as feed stocks of biofuel has been recognized for several decades (Bird and Benson, 1987; Horn *et al.*, 2000). In comparison to microalgae, seaweed is less appealing as an energy resource but for some coastal European nations, such as the UK (House of Commons, 2006); seaweeds have been considered for reduction of climatic CO_2 levels. Unlike algal biomass, seaweeds do not contain extractable oil but their high carbohydrate content suggests that they might be utilized as a feed stock for ethanol, butanol and other fermentation products. Extracts from *Laminaria hyperborea* can be fermented to ethanol by conversion of mannitol and laminaran and the best yield of ethanol was reported to date was 0.43 g/gallon in batch culture using bacterium *Zymobactor palmae* and yeast *Picia angophorae*. Seaweed *Sargassum horneri* could be a source of raw material for manufacturing of liquid biofuel feed stock using a bacterial consortium that has been identified as optimal for this bioconversion. Japan has implemented the 'Ocean Sunrise Project' to produce 5 billion liters of bioethanol using *S. horneri* (Aizawa *et al.*, 2007). Danish scientists are also assessing the prospect of converting seaweeds to ethanol from green algae *Ulva lactuca*, which are abundant on Danish shoreline. *Ulva lactuca* was reported to contain 60% of carbohydrate comparable to that of wheat and maize making it a potential bioethanol feed stock. At present, economic feasibility of the processes of getting bioethanol from microalgae and seaweed are unknown. The production potential of both types of biomass is far from being realized and scientific and technological challenges remain unaddressed. Though the biofuel productions from algae and seaweeds are not likely to be major energy resources they must not be ignored as no single biofuel can bring solutions to the energy crisis. Table 7.3 shows some of the companies involved in algae biofuel production. Such approaches need more research into technological developments in order to emerge as competitive against other biomass resources.

7.2.3   Future Prospects for Bioethanol

Bioethanol is not as efficient as petroleum; its energy content is 70% of that of petrol. In order for bioethanol to become more sustainable and replace petrol, the production process has to be more efficient, that is, reduce the cost of conversion , and increase the yield and the diversity of the crop used. Investigations are also in progress to utilize lignin and cellulose into sugars for fermentation. In this respect microbiology, biotechnology, and genetic engineering are helpful to meet the future developments of the bioethanol technology. Though agricultural supplies are major issues for food and fuel economies, these are likely to be insufficient in the near future, presenting great challenges for food processors and biofuels producers in the twenty-first century. The conversion of cellulosic material into ethanol is a relatively old process compared to sugar or starch crops, leading to the need to develop fermentation processes that can convert energy crops such as grasses, and agricultural by-products such as straw and corn stover, into bioethanol allowing high conversion rates of both hexoses and the difficult-to-ferment pentoses into ethanol at high yields. Therefore, the search for a technological breakthrough is on the rise, aiming to develop technologies for effectively converting agricultural and forestry

Table 7.3 *Companies dealing with biofuel production from algal biomass*

Company	Location	Business	Finance support
Aquaflow Bionomic	Marborough, New Zealand	Biodiesel production with sewage treatment	Public held
Aurora Biofuels	California, USA	GM algae for biodiesel	Gabriel Venture Partners, Noventi; Oak Investment Partners; Business angels.
AXI, LLC	Boston, USA	Biodiesel	Allied Minds (seed venture)
Bionavitas	Snoqualmie, Wash., USA	Develop technology for algae bioreactors	N/A
Blue Marble Energy	Seattle, USA	Biofuel production with water treatment	N/A
Bodega Algae	MIT, Massachusetts, USA	Develop low-cost and efficient algae bioreactors	N/A
Cellena	Hawaii	Biodiesel	HR Biopetrolem; Shell
Eni	Italy	Biodiesel production	Public held
GreenFuel Technologies	Cambridge, MA, USA	Biodiesel	Polaris Ventures; Draper Fisher Jurvetson; Imperium Renewables; Cedar Grove Investment; Brighton Jones Wealth Management; BAs
Inventure Chemical	Seattle, USA	Algae-jet fuel	
LiveFuels	Menlo Park, CA, USA	Open-pond algae bioreactors for biodiesel refinery	Quercus Trust; Sandia National Labs
Mighty Algae Biofuels	California, USA	Closed bioreactors for biofuel biodiesel	N/A
Oil Fox SA	Argentina		German Capital
OriginOil	California, USA	Algae biodiesel technology	Public held
ALG Western Oil	South Africa	Algae biodiesel technology	Odyssey Oil & Gas
Petro Sun*	Rio Honda, Texas, USA	Open pond for algae biofuel production	Public held
Sapphire Energy	San Diego, USA	Biodiesel	ARCH Venture Partner; Venrock; The Wellcome Trust
Seambiotic	Israel	Produce algae for applications	Work with Inventure; Chemical, Israeli Electric Co.
Solazyme	San Francisco, CA, USA	Bioreactor	N/A
Solena	Washington, USA	Gasify algae	N/A
Solix Biofuels	Fort Collins, CO, USA	Closed-tank bioreactors for biodiesel production	Colorado State University

lignocellulosic residues to fermentable sugars. Actions towards the development in technology by ethanol producers, vendors of process technology, government and academic laboratories have been initiated. The continuous need for technological development, rapid increase of the biofuel industry urges governments to create and implement new standards for the transport sector based on liquid biofuel from renewable resources, in order to reduce fossil fuel dependency, as well as to lower the contribution of petroleum derivatives to climate change and air pollution (Neves *et al.*, 2007). Energy savings as a result of the integration of a bioethanol plant into a power station are an important element of DONG Energy's bioethanol concept. Bioethanol production is traditionally energy-intensive, but DONG Energy's concept has the bioethanol plant located at a power station, so that the waste heat from power generation can be utilized for processing the straw in the production of bioethanol (DONG Energy, 2003). The Renewable Energy Policy act for the twenty-first century (REN21) has been developed to promote appropriate policies that increase the wise use of the renewable energies in developing and industrialized economies. It also connects international institutions, governments, nongovernmental organizations, industry associations and other partnership and initiatives, strengthening their influence for the rapid expansion of renewable energy worldwide (Renewables 2011 Global Status Report).

7.3   Biodiesel and Renewable Diesels from Biomass

Global biodiesel production increased 7.5% in 2010, to nearly 19 billion liters, giving a five-year average (late 2005 through 2010) of 38% growth. Biodiesel production is far less concentrated than ethanol, with the top 10 countries accounting for just under 75% of total production in 2010. The European Union remained the center of global biodiesel production, at more than 10 billion liters and representing nearly 53% of total output in 2010. Germany remains the world's top biodiesel producer at 2.9 billion liters in 2010, followed by Brazil, Argentina, France, and the US. Consumption in Germany has declined significantly since the elimination of Germany's biodiesel tax credit. The greatest drop in demand has been in pure vegetable oil and B100 (100% unblended biodiesel). In contrast, the use of blended biodiesel has increased during this period due to the national blending quota, and total production rose in 2010. The greatest production increase was seen in Brazil (up 46% to 2.3 billion liters) and in Argentina. Almost 12% of biodiesel production occurred in Asia (up from 10% in 2009), with most of this from palm oil in Indonesia and Thailand (Renewables 2011 Global Status Report).

The term biodiesel is widely used around the world to describe alkyl esters produced by the transesterification of vegetable oil and animal fats with simple alcohol (Figure 7.5). This definition has been used by the American Society of Testing Materials (ASTM) Specification D 6751, the European Specification EN 14214 and many other international specifications (ASTM, 2008; CEN, 2008) for a few years. Another term is also used but not as well defined, is renewable diesel, which is generally used to designate fuels other than alkyl esters that are produced from vegetable oils or animal fats. These days, no specification is accessible regarding definition of this term. This is a class of fuel constituted by fuels that are produced at the petroleum refinery from vegetable oils and animal fats. The refinery removes the oxygen and adds

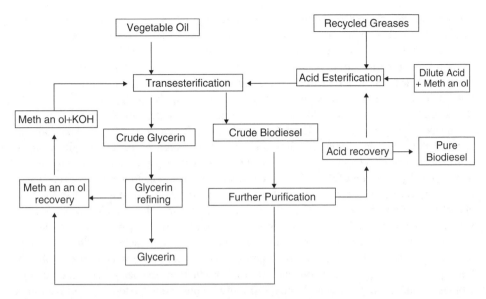

Figure 7.5   *Steps of the preparation of pure biodiesel from vegetable oil and recycled greases*

hydrogen, thus producing fuel that is rich in normal alkanes (Furimsky, 2000). The fuel may either be composed with a conventional petroleum stream (ConocoPhilips, 2006; 2007) or may be processed and maintained as an identifiable stream (Rantanen *et al.*, 2005) Normal alkanes are present in diesel and when blended with renewable diesel it become difficult to distinguish. These also provide good 'Cetane numbers' and are compatible with existing engines and fuel distribution infrastructure. Oxygen is removed so energy content is comparatively higher than original vegetable oil or oil esters, but since it is of a lighter density, so energy content is slightly lower than conventional diesel fuel. Having this insignificant difference the mileage difference remains unnoticeable by consumers. Hydrogenated oils have several advantages over biodiesel, including good performance at low temperatures, no storage stability problems and no susceptibility to microbial attack (Evans, 2008). Research into renewable diesel suggests that n-alkane based renewable fuels have very poor lubricant properties and poor cold properties, but external addition of the additives has been shown to improve the performance. Results of all recent studies suggest the efficiency of renewable diesel as potential alternative of conventional diesel fuels. Monnier *et al.* (1998) has suggested a new approach of getting renewable diesel like fuel by 'hydro-treating' the vegetable oil, animal fat wood oil and tall oil. This diesel was reported to be constituted of n-alkanes with Cetane numbers of approximately 100 and known as 'SuperCetane'.

'Green diesel' is another class of conventional diesel additives from biomass based diesel fuel, although this generally refers to alkane rich fuels produced from carbohydrates, following processes described by Dumesic and Huber (Huber *et al.*, 2005; 2007; Carlson *et al.*, 2008; West *et al.*, 2008). These are also reported as n-alkane based fuels but with simple sugar as a feed stock rather than oils and fats.

Biodiesel from algal biomass is also a new approach to sustainable liquid transportation fuel development (Sheehan *et al.*, 1998). Many microalgae in particular green algae and diatoms have been reported to accumulate significant quantity of the neutral oil (triglyceride oil), which can be extracted from biomass and converted to biodiesel or green diesel as substitutes for petroleum-derived transportation fuels. The crucial factor in controlling the quantity of accumulated lipid in the biomass is limited cellular growth. Lipid synthesis was reported to be triggered under conditions when cellular growth is limited by nutrient deficiency, but metabolic energy supply via photosynthesis is not (Roessler, 1990). Nutrient deficiencies include nitrogen deficiency for green algae and silicon deficiency for diatoms (Piorreck *et al.*, 1984; Suen *et al.*, 1987; Harwood and Jones, 1989; Roessler, 1990; McGinnis *et al.*, 1997; Takaji *et al.*, 2000). Green algae *Botrycoccus braunii* can produce up to more than 50% of its dry weight as pure hydrocarbons, typically between C20–C40, so it can also serve as a potential source of renewable diesel but, very low growth rates makes it difficult for mass culture and opens a new challenge to fuel researchers. Thus the primary challenges of producing economically viable microbial oil is to simultaneously get high cellular growth and high oil productivities. Some other challenges still remaining in making these approaches feasible on a large scale, are (1) the high cost of on shore production of microalgae which ranges 10–100 fold higher than required for conventional fuel production; (2) design and operation of biomass production system and biomass to biofuel conversion system; (3) cultivation at a scale that would make a significant impact on global energy economy. This renewable energy resource requires intense culturing to make the process economically feasible.

7.3.1   Potential of Vegetable Oil as a Diesel Fuel Substitute

Vegetable oils provide a fuel source that utilizes well known technologies and integrates easily into the existing diesel fuel infrastructure. They provide the additional benefits of decreased carbon dioxide emission. Use of pure vegetable oil or blends with diesel produces equivalent power in spite of the low energy content of the fuel. Higher fuel density increases fuel energy delivery so energy can be equal or greater than that of conventional diesel fuel. Nitrogen oxides emission in exhaust gas also found to reduce in considerable extent foe vegetable oil blended diesel but particulate emission has been found to increase slightly at high load. Deposition on the injection nozzle reduces the performance of the engine over prolonged operation time but can be regenerated after proper cleaning to the original one (Humke and Barsic, 1981). In a pre-chamber type diesel engine, similar performance has been achieved with crude-degummed vegetable oil but excessive piston ring and linear wear were observed at high load (Suda, 1984), an increased propensity to deposit buildup on the nozzle tips and on the piston and liner were also noted by Baranescu and Lusco (2004). Emulsions of vegetable oil with low viscosity of liquids were investigated as an alternative to reducing the overall viscosity of the fuel. Emulsions with liquids such as alcohols and water which have high enthalpy of vaporization, offer the benefit of lower temperatures during combustion and reduced emission of NO_x (Sii *et al.*, 1995; Van Gerpen *et al.*, 2007). The emulsions of palm oil methyl esters, diesel fuel and water are found satisfactory compared with ordinary diesel fuel. In spite of the success of using emulsion the problem still arises due to phase separation of water and alcohol at low temperatures. The problems of excessive deposits, wear on the ring and liner motivated the researchers to modify the triglyceride oil as efficient diesel replacement. Early research on biodiesel by Hawkins and Fuls

Table 7.4 *Exhaust emission for B20 and B100 (Adapted from Sharp et al., 2000)*

Test engine	Test Fuel	Transient emission g/(hp-h)					
		HC	CO	NO$_x$	Particulate emission		
					Total PM	VOF	Soot
Cummins N14	B100	−95.6	−45.3	+13.1	−28.3	+42.8	−60.9
Cummins N14	B20	−17.4	−14.7	+4.2	−3.80	+31.4	−20.3
Detroit Diesel 50	B100	−83.3	−38.3	+11.3	−49.0	+10.3	−71.4
Detroit Diesel 50	B20	0	−7.4	+3.60	−13.7	+13.8	−23.8
Cummins B5.9	B100	−74.2	−38.0	+4.30	−36.7	+14.6	−62.9
Cummins B5.9	B20	−32.3	−21.5	+1.90	−14.8	+8.30	−20.0

B100-100 volume percent of biodiesel; B20-20 volume percent of biodiesel with No. 2 diesel fuel; HC- Unburned hydrocarbon

(1982), concluded that esters prepared from sunflower oil could constitute an appropriate substitute fuel for direct injection in diesel engine.

In 1982, Pischinger and Falcon also confirmed the result and reported that soybean oil methyl esters having high compatibility with existing diesel engine. In 1995, Schumacher concluded that oil esters can provide large incentives for substituting biodiesel blends for diesel fuel in urban buses with reduced emission of CO, unburned hydrocarbon and particulate emission. In 2007, Krahl *et al.*, compared the emission resulting from the use of rapeseed oil, methyl ester of rapeseed oil and conventional diesel fuel, confirming the results of Sharp *et al.* (2000) (see Table 7.4) on exhaust properties but, surprisingly, higher mutagenecity (30-fold higher) for crude rapeseed oil as compared to biodiesel and diesel fuel.

7.3.2 Vegetable Oil Ester Based Biodiesel

Vegetable oil and animal fats are composed of triglycerides, with a glycerol backbone to which fatty acids are attached. Chemical modification of triglyceride oil involves treatment of oil with excess of alcohol (preferably methanol) and alkaline catalyst (NaOH, KOH, NaOCH$_3$) to produce glycerol and alkyl ester, known as transesterification (Van Gerpen *et al.*, 2007). Studies have also done to develop solid catalysts (Van Gerpen *et al.*, 2006, 2007; Xie, 2007). Commonly edible oils (soy, rapeseed, mustard, flax, sunflower, palm oil) and fats are normally used for transesterification but after the hiking of price for the vegetable oils (edible type) researchers have now concentrated to the non-edible type of oils like jatropha, mahua, karanja. Types of primary oils, in particular, a fatty acid composition of TGs, used for transesterification control the properties of the final processed biodiesel. Saturated fatty acid rich triglycerides when modified to biodiesel tend to solidify at low temperature and exhibited poor 'cold flow' properties. The amount of saturated alkyl esters (extent of unsaturation) is the parameter that has most significant impact on the Cetane number, oxidative stability and cold flow properties of the final product. Saturated alkyl esters have a high Cetane number (Knoth, 2005), and are not susceptible to oxidative attack that is most common for unsaturated types. The oxidation reaction is crucial in minimizing it as it can result in the breaking of the fatty acid

Table 7.5   *Properties of common biodiesels*

Raw materials	Cetane number	Viscosity (cSt)	Cloud point (°C)	Pour point (°C)	Reference
Methyl ester of Soybean	49.6	4.18	−1.1	−3.9	Yahya and Marley, 1994
Methyl ester of tallow	61.8	4.99	15.6	12.8	Yahya and Marley, 1994
Methyl ester of canola	57.9	4.75	−1.0	−6.0	Peterson, 1994
Methyl ester of high erucic rapeseed	61.8	5.65	0.0	−15.0	Peterson, 1994

chains into shorter chain acids, and can also cause cross-linking and polymerization that generates harmful deposits. Methyl esters have the highest cloud point temperatures, with esters of higher alcohols slightly lower and when branched cahin alcohols are used, cloud point temperatures are much lower (Wang *et al.*, 2005). Mustard oil and a variety of rapeseed oils contain significant amount of glucosinolates that are known to cause the 'hot' flavor of the condiment mustard. The oils commonly preferable are rapeseed of 'double zero' quality that have low glucosinolate and low erucic acid, and are similar to canola oil.

A comparison was also made by Knoth (2005) on the Cetane number of the commonly available biodiesel to assess the effect of the fatty acid composition, which revealed that tallow, more saturated feedstock with 40% saturated fats, has a Cetane number 61.8, while unsaturated vegetable oil esters have a Cetane number between 49.6–56 (Table 7.5). The cloud point for tallow biodiesel is 15.5°C, which is above the values of other unsaturated rich esters.

7.3.3   Several Approaches to Biodiesel Synthesis

Many different researchers have made a preliminary study of the two approaches to economically produce biodiesel from waste cooking oil (WCO) and flaked cottonseed. One was the use of ultrasound-assisted synthesis of biodiesel from WCO. Ultrasonication provides the mechanical energy for mixing and the required activation energy for initiating the transesterification reaction. Ultrasonication increases the chemical reaction speed and yield of transesterification of vegetable oils into bio-diesel. Not only that, it also consumes less energy than the conventional mechanical stirring method. The other was the application of in situ transesterification from flaked cottonseed. Research is also in progress to develop the solid catalysts for transesterification of the triglycerides (TG) to biodiesel (Refaat, 2011). Heterogeneous transesterification is a green process which requires neither catalyst recovery nor aqueous treatment and the yields of methyl esters is very high. The most efficient transesterification catalysts are those with a high specific surface area, strong base strength and high concentration of base sites. The Ca series catalysts have higher catalytic activity for transesterification reactions. To catalyze transesterification of soybean oil to biodiesel with methanol, calcium methoxide has been studied since it has strong basicity and high catalytic activity as a heterogeneous solid base catalyst (Liu *et al.*, 2008).

7.3.4   Sustainability of Biofuel Use

The sustainability of biofuels is subject to different countries or places and the kind of species used to produce the biofuels. However, there are some issues that universally plague the biofuels industry and need to be solved.

7.3.4.1   Food versus Fuel

A large scale mechanized production of fuel crops is neither advisable nor practical in the case of developing countries where food shortage is an issue. This becomes more important when high quality arable land is predominantly occupied by fuel crops and is not available for food cultivation. In order to make biofuel sustainable the biofuel crops should be grown where they will not encroach on food crops (Singh *et al.*, 2011). Using micro-algae for producing biofuels is a way around this dilemma but until now, there has been no commercially viable technique.

7.3.4.2   Water Usage

Studies have also shown that the production of biofuel crops can have significant impact on water demand. Water is required for many activities during production for mixing, washing and evaporative cooling. The biggest water requirement, however, arises from irrigation. While plants like *Jatropha* can do well in semi-arid areas, they may require some irrigation. The fundamental question is, whether it is wise to use irrigated land to produce biofuel crops (e.g., sugarcane).

7.3.4.3   Environmental Issues

The main reason behind the thrust in researching biofuels is that they are assumed to be greenhouse gas (GHG) neutral. However, there has been evidence that under certain circumstances a particular biofuel may emit more GHG gas than it saves (Pearce and Aldhous, 2007). For example, natural forest cover is removed to acquire land for biofuel crops, thus when these crops are burnt in the form of biofuel they effectively are releasing GHG in the atmosphere which won't be absorbed without the natural sponges that were present earlier. Many other issues are location specific, which may include gender issues, impact on poverty alleviation, and governmental policies. All these can be addressed with different strategies; however, the strategies will be as location specific as the problems arise.

7.3.5   Future Prospects

Biodiesel production practice is well established and no specific problem is observed when this renewable fuel is used to power conventional diesel engines. Although biodiesel reduces the emission of CO, HC and PM, there is still concern for a slight increase in the emission of NO_x. The effect of biodiesel on emission is likely to become most promising when current emission regulations are fully implemented in the major cities. The exhaust after-treatment devices will reduce emission to such low levels, regardless of fuel, that differences between fuels may not be significant or promote a particular fuel as a selling point. Approximately 80% of the cost of producing biodiesel is the cost of the feed stock oil (Van Gerpen *et al.*, 2006), which requires further studies in this field to find more revenue and improve the competitiveness of biodiesel with petroleum diesel fuel unless the cost of the feed stock can be reduced. These days, research is underway to find new cheap sources of oils (Altiparmak *et al.*, 2007; Sahoo *et al.*, 2007). However, we need more studies into mass production, optimizing the requirement of fertilizers and pesticides, and environmental impact of the utilization of byproducts to try and make this overall technology more profitable. Biodiesel has the potential to contribute to a significant portion of global diesel demand and the extent of its impact will depend on the adequate supply of low cost feed stocks and overall process developments.

7.4   Perspective

Recently, there has been a trend for biofuel resource transition, moving from first generation biofuel derived from food crops such as corn, sugarcane, and oilseed, towards the next generation – cellulosic biofuel, made from lignocellulosic feed stocks such as corn stove, grasses, wood chips, and waste biomass. The growth of the ethanol industry depends on the success of next generation (2G) technologies. The viability of the first generation of biofuel production is questionable because of the conflict with food supply. Algal biofuel is a viable alternative as renewable and carbon neutral biofuel are necessary for environmental and economic sustainability. Algae offer economic potential as the more sustainable energy supply, especially if coupled with pollution treatment. It was stated in the recent report on the *Economics of Climate Change* that markets for low carbon energy products are likely to be worth at least $500 billion per year by 2050 (Stern, 2007). Substantial progress has been made in facilitating algae biomass production and enhancing production efficiencies through engineering of biophotoreactor systems. These technologies will enhance the cost-effectiveness of the algae biofuel strategy. However, the technology for better biofuel is still far away, although with the government subsidies support change could be more swiftly effected. Today, governments, industry and experts promote biotechnology as a solution to many problems and the emergence of bioethanol and biodiesel as viable 'green' energy sources are promoted by some government policies and investors and biotechnology has been emerging as a solution to the problem to feed developing countries (Zhang and Cooke, 2008).

List of Acronyms

GW	Gigawatt
GHG	greenhouse gas
SSF	saccharification and fermentation
GHs	Glycosyl hydrolyses
LAB	lactic acid bacteria
CBP	Consolidated Bioprocessing
GRAS	Generally Recognized As Safe

EOR   Energy output ratio

CO   *carbon monoxide*

NO_x *nitrogen oxides*

PM   *particulate matter*

VOF   *volatile organic fraction*

WCO   waste cooking oil

TG   triglycerides

References

M. Aizawa, K. Asaoka, M. Atsumi and T. Sakou, (2007), Seaweed bioethanol production in Japan-The Ocean Sunrise Project, in Oceans, Vancouver, Canada.

T. Akano, Y. Muura, K. Fakatsu, *et al.*, (1996), Hydrogen production by photosynthetic microorganisms, *Appl. Biochem. Biotechnol.*, **8**(2), 57–58, 677–688.

J.R. Almedia, T. Modig, A. Petersson, *et al.*, (2007), Increased tolerance and conversion of inhibitors in lignocellosic hydrolysates by *Saccharomyces cerevisiae*, *J. Chem. Technol. Biotechnol.*, **82**, 340–349.

D. Altiparmak, A. Keskin, A. Koca and M. Guru, (2007), Alternative fuel properties of tall oil fatty acid methyl ester-diesel fuel blends, *Biores. Technol.*, **98**, 241–246.

ASTM (2008) D6751-12 *Standard Specification for Biodiesel Fuel Blend Stock (B100) for Middle Distillate Fuels.* Available at: http://enterprise.astm.org/filtrexx40.cgi?+REDLINE_PAGES/D6751.htm (accessed February 22, 2013).

R.A. Baranescu and L.L. Lusco, (2004), Performance, durability and low temperature evaluation of sunflower oil as a diesel fuel extender, *Proceedings of the International Conference on Plant and Vegetable Oil as Fuels*, ASAE, Fargo, ND.

A.R. Barber, M. Henningsson and N.B. Pamment, (2002), Acceleration of high gravity yeast fermentations by acetaldehyde addition, *Biotechnol. Lett.*, **24**, 891–895.

E.A. Bayer, Y. Soham and R. Lamed, (2000), Cellulose decomposing bacteria and their enzyme systems, in: *The Prokaryotes: An Evolving Electronic Resource for the Microbiological Community*. Spinger-Verlag: New York, pp. 1–62.

E.A. Bayer, J.P. Belaich, Y. Soham and R. Lamed, (2004), The cellulosomes: multi enzyme machines for degradation of plant cell wall polysaccharides. *Ann. Rev. Microbiol.*, **58**, 521–554.

E.A. Bayer, H. Chanzy, R. Lamed and Y. Soham, (1998), Cellulose, cellulases, cellulosome, *Curr. Opi. Struc. Biol.*, **8**, 548–557.

T.K. Bird and J. Benson, (1987), *Seaweed Cultivation for Renewable Resources*, p.369, Elsevier Science Ltd, Amsterdam.

R.J. Bothast and M.A. Schlicher, (2005). Biotechnological process for conversion of corn into ethanol, *Appl. Microbiol. Biotechnol.*, **67**, 19–25.

R.J. Bothast, N.N. Nichols and B.S. Dien, (1999) Fermentation with new recombinant organisms, *Biotechnol. Prog.*, **15**, 867–875.

T.R. Carlson, T.P. Vispute and G.H. Huber, (2008), Green gasoline by catalytic fast pyrolysis of solid biomass derived compounds, *ChemSusChem*, **1**, 93–400.

CEN (2008) *Automotive Fuels – Fatty Acid Methyl Esters (FAME) For Diesel Engines – Requirements and Test Methods.* Available at: http://www.novaol.it/novaol/export/sites/default/allegati/EN14214.pdf (accessed February 22, 2013).

S. Chakraborty, V. Aggarwal, D. Mukherjee and K. Andras, (2012), Biomass to biofuel: a review on production technology. *Asia-Pacific Journal of Chemical Engineering*, 7, S254–S262.

N. Chaudhary and I. Qazi Javed, (2011), Lignocellulose for ethanol production: A review of issues relating to bagasse as a source material, *Afr. J. Biotechnol.*, **10**, 1270–1274.

ConocoPhilips, (2006), News Release: *ConocoPhilips begins production of renewable diesel fuel at Whitegate refinery in Cork Ireland*, http://www.conocophilips.com.

ConocoPhilips, (2007), News Release: *ConocoPhilips and Tyson food announce strategic Alliance to produce next generation renewable diesel fuel*, http://www.conocophilips.com.

K. Deanda, M. Zhang, C. Eddy and S. Picataggio, (1996), Development of a rabinose-fermenting *Zymomonas mobilis* stain by metabolic pathway engineering, *Appl. Environ. Microbiol.*, **62**, 4465–4470.

M.D. Deng and J.R. Coleman, (1999), Ethanol synthesis by genetic Engineering in *Cyanobacteria*, *Appl. Environ. Microbiol.*, **65**, 523–528.

B.S. Dien, M.A. Cotta and T.W. Jeffries, (2003), Bacteria engineered for fuel ethanol production: Current Status, *Appl. Microbiol. Biotechnol.*, **63**, 258–266.

H.W. Doelle, L.D. Kennedy and M.B. Doelle, (1991), Scale up of ethanol production from sugarcane using *Zymomonas mobilis. Biotechnol. Lett.*, **13**, 131–136.

DONG Energy, (2003), *Bioethanol: the CO_2 neutral fuel of the future*, Available from: http://www.dongenergy.com (accessed February 22, 2013).

Energy Independence and Security act of (2007), H.R. 6, 110th US Congress, Washington, DC.

Energy Policy Act of (2005), Public Law 109–158, 109th US Congress, Washington, DC.

G. Evans, (2008), *Liquid Transport Biofuels – Technology Status Report*, National Non-Food Crops Centre.

M. Ezzati and D.M. Kammen, (2001), Quantifying the effects of exposure to indoor air pollution from biomass combustion on acute respiratory infections in developing countries, *Environmental Health Perspectives*, **109**(5) 418–488.

M.C. Flickinger, (1980), Current biological research on conversion of cellulosic carbohydrates into liquid fuels: how far have we come? *Biotechnol. Bioeng.*, **22**, 24–48.

E. Furimsky, (2000), Catalytic hydrodeoxygenation, *Appl. Catal. A: General*, **199**, 147–190.

R.S. Gold, M.M. Meagher, S. Tong, *et al.*, (1996), Cloning and expression of the *Zymomonas mobilis* 'production of ethanol' genes in *Lactobacillus casei, Curr. Microbiol.*, **33**, 256–260.

J. Goldemberg, (2007), Ethanol for a sustainable energy future, *Science*, **315**, 808–810.

B. Hahn-Hägerdal, K. Karhumaa, C. Fonseca, *et al.*, (2007a), Towards industrial pentose-fermenting yeast stain, *Appl. Microbiol. Biotechnol.*, **74**, 937–953.

B. Hahn-Hägerdal, K. Karhumaa, M. Jeppsson and M.F. Gorwa-Grauslund, (2007b), Metabolic engineering for pentose utilization in *Saccharomyces cerevisiae*, *Adv. Biochem. Eng. Biotechnol.*, **108**, 147–177.

Y.W. Han, J. Timpa and A. Ceigler, (1981), Gamma-ray-induced degradation of lignocellulosic materials, *Biotechnol. Bioeng.*, **XXIII**, 2525–2535.

J.L. Harwood and A.L. Jones, (1989), Lipid metabolism in algae, *Adv. Bot. Res*, **16**, 1–53.

C.S. Hawkins and J. Fuls, (1982), Comparative combustion studies on various plant oil esters and the Long Term Effects Of An Ethyl Ester On A Compression Ignition Engine, *Proceedings Of The International Conference on Plants and Vegetable Oils as Fuels*, ASAE, Fargo, ND.

M. Hayn, W. Steiner, R. Klinger, *et al.*, (1993), Basic research and pilot plant studies on the enzymatic conversion of lignocellulosics, in *Bioconversion of Forest and Agricultural Plant Residues*, Saddler J.N. (ed.), CAB International, Wallingford, UK.

P. Hazell and J. von Braun, (2006), Biofuels: a win-win approach that can serve the poor. *International Food Policy Research Institute*, IFPRI Forum.

A. Hirano, R. Ueda, S. Hiryama and Y. Ogushi, (1997), CO_2 fixation and ethanol production with microbial photosynthesis and intracellular anaerobic fermentation, *Energy*, **22**(2,3), 137–142.

S. Hirayama, R. Ueda, Y. Ogushi, *et al.* (1997), Ethanol production from Carbon dioxide by fermentive microalgae, *Fourth International Conference on Carbon Dioxide Utilization*, Kyoto, Japan.

K. Hon-Nami, (1996), Unique feature of Hydrogen recovery in endogenous starch to alcohol fermentation of the marine microalga *Chlamaidomonas perigranulata*, *Appl. Biochem. Biotechnol.*, **131**(1–3), 808–828.

S.J. Horn, I.M. Aasen and K. Ostgaard, (2000), Ethanol production from seaweed extract, *J. Ind. Microbiol. Biotechnol.*, **25**, 249–254.

House of Commons, Environment, Food and Rural Affairs Committee (2006), *Renewable Energy : The Potential of Marine Biomass*. Available at: www.publications.parliament .uk/pa/cm200506/cmselect/cmenvfru/965/6042606.htm (accessed February 20, 2013).

G.W. Huber, J.N. Chheda, C.J. Barret and J.A. Dumensic, (2005), Production of liquid alkanes by aqueous phase processing of biomass derived carbohydrates, *Science*, **308**, 1446–1450.

G.W. Huber, P.O. Connor and A. Corma, (2007), Processing biomass in conventional oil refineries: production of high quality diesel by hydrotreating vegetable oils in heavy vacuum oil mixtures, *Appl. Catal. A.: General*, **329**, 120–129.

M. Huesemann, G. Roesjadi, J. Benemann and F.B. Metting, (2010), Biofuels from Microalgae and Seaweeds, in, *Biomass to Biofuels*, A.A Vertès, N. Qureshi, H.P Blaschek and H. Yukawa (eds), pp. 165–184, John Wiley & Sons, Ltd, Chichester.

M.H. Huesemann, (2006), Can advances in science and technology prevent global warming? A critical review of limitations and challenges. *Mitigation and Adaptation Strategies for Global Change*, **11**, 539–577.

L. Humke and N.J. Barsic, (1981), Performance and emission characteristics of a naturally aspirated diesel engine with vegetable oil fuels. *SAE Paper* **810955**.

M. Inui, S. Murakami, S. Okino, *et al.*, (2004a), Metabolic analysis of *Corneybacterium glutamicum* during lactate and succinate productions under oxygen deprivation conditions. *J. Mol. Microbiol. Biotechnol.*, **7**, 182–196.

M. Inui, H. Kawaguchi, S. Murakami, *et al.*, (2004b), Metabolic Engineering of *Corneybacterium glutamicum* for fuel ethanol production under oxygen- deprivation conditions, *J. Mol. Microbiol. Biotechnol.*, **8**, 243–254.

Iowa State University, (2006), *Determining the regional economic values of ethanol production in Iowa considering different levels of local investment.*

V. Ivanova, P. Petrova and J. Hristov, (2011), Application in the Ethanol fermentation of immobilized yeast cells in matrix of alginate/magnetic nanoparticles, on chitosan-magnetite microparticles and cellulose-coated magnetic nanoparticles, *Int. Rev. Chem. Eng.*, **3**, 289–299.

K.L. Kadam, (2002), Environmental implications of power generation via coal – microalgae cofiring, *Energy*, **27**, 905–922.

S.F. Karel, S.B. Libicki and C.R. Robertson, (1985), The immobilization of whole cells: Engineering principles, *Chem. Eng. Sci.*, **40**, 1321–1354.

G. Knoth, (2005), Cetane numbers – heat of Combustion – why vegetable oils and their derivatives are suitable a diesel fuel, in *The Biodiesel Handbook*, G. Knoth, J. van Gerpen and J. Krahl (eds) AOCS Press, Champaign, IL.

J. Krahl, A. Munack, Y. Raschel, *et al.*, (2007), Comparison of emissions and mutagenecity from biodiesel, vegetable oil, GTL and diesel fuel. SAE 2007-01-4042.

H.G. Lawford and, J.D. Rousseau, (1993), Production of ethanol from pulp mill hardwood and softwood spent sulfite liquors by genetically engineered *Escherichia coli. Appl. Biochem. Biotechnol.*, **39–40**: 667–685.

H.J. Lawford, (1987), Ethanol production by high performance bacterial fermentation. George Weston Ltd. (Toronto, CA), *US Patent 4647534.*

Y. Lin and S. Tanaka, (2006), Ethanol fermentation from biomass resources: current state and prospects, *Appl. Microbiol. Biotechnol.*, **69**, 627–642.

S. Liu, (2010), Conversion of biomass to ethanol by other organisms, in *Biomass to Biofuels*, A.A Vertès, N. Qureshi, H.P Blaschek and H. Yukawa (eds), pp.293–310, John Wiley & Sons, Ltd, Chichester.

X. Liu, X. Piao, Y. Wang, *et al.*, (2008), Calcium methoxide as a solid base catalyst for the transesterification of soybean oil to biodiesel with methanol. *Fuel*, **87** (7), 1076–1082.

L.R. Lynd, W.H. van Zyl, J.E. McBride and M. Laser, (2005), Consolidated bioprocessing of cellulosic biomass: an update, *Curr. Opin. Biotechnol.*, **16**, 577–583.

H. Matsumoto, N. Shioji, A. Hamasaki, *et al.*, (1995), Carbon dioxide fixation by microalgae photosynthesis using actual flue gas discharged from a boiler, *Appl. Biochem. Biotechnol.*, **51–52**, 661–692.

H. Matsumoto, H. Yokouchi, N. Sujuki, *et al.*, (2003), Saccharification of marine microalgae using marine bacteria of ethanol production, *Appl. Biochem. Biotechnol.*, **105–108**, 147–154.

K.M. McGinnis, T.A. Dempster and M.R. Sommerfeld, (1997), Characterization of the growth and lipid content of the diatom *Chaetoceros muelleri*, *J. Appl. Phycol.*, **9**, 19–24.

D.J. McHugh, (2003), *A Guide to Seaweed Industry*, FAO Fisheries Technical Paper 441. Food and Agricultural Organization of the United Nations, Rome, Italy.

X. Miao and Q. Wu, (2004), High yield bio-oil production from fast pyrolysis by metabolic controlling of *Chlorella protothecoides*, *J. Biotechnol.*, **110**, 85–93.

X. Miao, Q. Wu and C. Yang, (2004), Fast pyrolysis of microalgae to produce renewable fuel, *J. Anal. Appl. Pyrol.*, **71**, 855–863.

T. Minowa and S. Sawayama, (1999), A novel microalgal system for energy production with nitrogen cycling, *Fuel*, **78**, 1213–1215.

T. Minowa, S.K. Yokoyama, M. Kishimoto and T. Okakura, (1995), Oil production from algal cells of *Dunaleilla tertiolecta* by direct thermochemical liquefaction, *Fuel*, **12**, 1735–1738.

J. Monnier, G. Tourigny, D.W. Soveran, *et al.*, (1998), Conversion of biomass feed stock to diesel fuel additives, *U.S. Patent* **5**,705,722.

G. Najafpour, H. Younesi, K. Syahidah and K. Ismail, (2004), Ethanol fermentation in an immobilized cell reactor using *Saccharomyces cerevisiae*. *Biores Technol.*, **92**, 251–260.

L.L. Nass, P.A. Arraes Pereira and D. Ellis, (2007), Biofuel in Brazil: An Overview. *Crop Sci.*, **47**, 2228–2237.

M.A. Neves das, T. Kimura, N. Shimizu and M. Nakajima, (2007), *State of the Art and Future Trends of Bioethanol Production Dynamic Biochemistry, Process Biotechnology and Molecular Biology*, Global Science Books.

S. Nikolić, L. Mojović, M. Rakin and D. Pejin, (2009), Bioethanol production from corn meal by simultaneous enzymatic saccharification and fermentation with immobilized cells of *Saccharomyces cerevisiae var. ellipsoideus*, *Fuel*, **88**, 1602–1607.

A. Nilsson, M.J. Taherjadeh and G. Liden, (2001), Use of dynamic step response for control of fed batch conversion of lignocellulosic hydrolysate to ethanol, *J. Biotechnol.*, **89**, 41–53.

A. Nilsson, M.J. Taherjadeh and G. Liden, (2002), On-line estimation of sugar concentration for control of fed batch fermentation of lignocellulosic hydrolysates by *Saccharomyces cerevisiae*, *Biopross. Biosyst. Eng.*, **25**, 183–191.

OECD, (2008), *Economic Assessment of biofuel support policies*, Paris: OECD Directorate for Trade and Agriculture.

Ó.J. Sánchez and C.A. Cardona, (2008), Trends in biotechnological production of fuel ethanol from different feedstocks, *Biores. Technol.*, **99**, 5270–5295.

G. Panagiotou, P. Christakopoulos, T. Grotkjaer and L. Olsson, (2006), Engineering of the redox imbalance of *Fusarium oxysporum* enables anaerobic growth on xylose, *Metab. Eng.*, **8**, 474–482.

F. Pearce and P. Aldhous, (2007), Biofuels may not be answer to climate change. *New Scientist*, **196**, 6–7.

W. Peng, Q. Wu and P. Tu, (2001a), Pyrolytic characteristics of heterotrophic *Chlorella protothecoides* for renewable biofuel production, *J. Appl. Phycol*, **13**, 5–12.

W. Peng, Q. Wu, P. Tu and N. Zhao, (2001b), Pyrolytic characteristics of microalgae as renewable energy source determined by thermogravimetric analysis, *Biores. Technol.*, **80**, 1–7.

W. Peng, Q.L.L Wu and P. Tu, (2000), Effects of temperature and holding time on production of renewable fuels from pyrolysis of *Chlorella protothecoides*, *J. Appl. Phycol*, **12**, 147–152.

R.D. Perlack, L.L. Wright, A.F. Turhollow, *et al.*, (2005), Biomass as feedstock for a bioenergy and byproducts industry: the technical feasibility of a billion – ton annual supply, *A joint study sponsored by U.S. Department of Energy and U.S. Department of Agriculture.*

C.L. Peterson, D.L. Reece, B.J. Hammond and J. Thompson, (1994), Processing, characterization and performance of eight fuels from lipids, *ASAE Paper* **946531**.

M. Piorreck, K.H. Baasch and P. Pohl, (1984), Biomass production, total protein, chlorophylls, lipids and fatty acids of fresh water green and blue algae under different nitrogen regimes, *Phytochem.*, **23**, 207–216.

G.H. Pischinger and A.M. Falcon, (1982), Methylesters of plant oil as diesel fuels, either straight or in blends, *Proceedings of the International Conference on Plants and Vegetable Oils as Fuels*, ASAE, Fargo, ND.

O. Pulz and W. Gross, (2004), Valuable products from biotechnology of microalgae, *Appl. Microbiol. Biotecnol.*, **65**, 635–648.

L. Rantanen, R. Linnaila, P. Aakko, *et al.* (2005), Biodiesel of the second generation, *SAE Paper* **01–3771**.

A.A. Refaat, (2011), Biodiesel production using solid metal oxide catalyst, *Int. J. Environ. Sci. Tech.*, **8**, 203–221.

REN21, (2011), *Renewables 2011 Global Status Report*. Online, available at http://www.un-energy.org/sites/default/files/share/une/ren21_gsr2011.pdf (accessed February 22, 2013).

Renewable Fuels Association (RFA) (2006), Just the Facts: Ethanol Markets, http://www.ethanolrfa.org/pages/ethanol-facts

RFA, (2008), *Renewable Fuel Outlook 2008*. Available at http://www.ethanolrfa.org. (accessed February 11, 2013).

P.G. Roessler, (1990), Environmental control of glycerolipid metabolism in microalgae: commercial implications and future research directions, *J. Phycol.*, **26**, 393–399.

M.W. Rosegrant, S. Msangi, T. Sulser and R. Valmonte-Santose, (2006), Biofuels and the global food balance, in *Bioenergy and Agriculture: Promises and Challenges*, International Food and Policy Research Institute, 2020 Focus 14 Report. Available at: http://www.ifpri.org/sites/default/files/pubs/2020/focus/focus14/focus14_03.pdf (accessed February 22, 2013).

C. Saha, (2003), Hemicellulose bioconversion, *J. Ind Microbiol. Biotechnol.*, **30**, 279–291 (2003).

P.K. Sahoo, L.M. Das, M.K.G. Babu and S.N. Naik, (2007), Biodiesel development from high acid value polanga seed oil and performance evaluation in CI engine, *Fuel*, **86**, 448–454.

S. Sakai. Y. Tsuchida, H. Nakamoto, *et al.*, (2007) Effect of lignocellulose – derived inhibitors on growth of and ethanol production by growth arrested *Corneybacterium glutamicum* R., *Appl. Environ. Microbiol.*, **73**, 2349–2353.

K. Sakamoto and W.N. Konings, (2003), Beer Spoilage bacteria and hop resistance, *Int. J. Food Microbiol.*, **89**, 105–124.

P. Sassner, C.G. Martensson, M. Galbe and G. Zacchi, (2008), Stem pretreatment of H(2)SO(4)-impregnated salix for the production of bioethanol, *Biores. Technol.*, **99**, 137–145.

S. Sawayama, T. Minowa and S.Y. Yokohama, (1999), Possibility of renewable energy production and CO_2 mitigation by thermochemical liquefaction of microalgae, *Biomass and Bioenergy*, **17**, 33–39.

J. Scheffran, (2010), Global demand for biofuels: Technologies, Markets and Policies, in *Biomass to Biofuels*, A.A Vertès, N. Qureshi, H. Blaschek and H. Yukawa (eds), pp. 27–50, John Wiley & Sons, Ltd, Chichester.

C.A. Sharp, S.A. Howell and J. Jobs, (2000), The effect of biodiesel fuels on transient emissions from modern diesel engine, Part 1: Regulated emissions and performance, *Society of Automotive Engineers Paper No.2000-01-1967*, SAE, Warrendale, PA.

J. Sheehan, T. Dunahay, J. Benemann and P. Roessler, (1998) A look back at the U.S. Department of Energy's Aquatic Species Program – biodiesel from algae, National Renewable Energy Laboratory, Golden, *COReport NREL/TP-580–24190*.

F. Shirai, K. Kunii, C. Sata, *et al.*, (1998), Cultivation of microalgae in the solution from the desalting process of soy sauce waste treatment and utilization of algal biomass for ethanol fermentation. *World J. Microbiol. Biotechnol.*, **14**, 839–843.

H.S. Sii, H. Masjuki and A.M. Zaki, (1995), Dynamometer evaluation and engine wear characteristics of palm oil diesel emulsions, *J. Am. Oil. Chem. Soc.*, **72**, 905–909.

A. Singh, P.S. Nigam and J.D. Murphy, (2011), Renewable fuels from algae: An answer to debatable land based fuels, *Bioresource Technology*, **102**(1), 10–16.

G.A. Stanley, T.J. Hobley and N.B. Pamment, (1997), Effect of acetaldehyde on *Saccharomyces cerevisiae* and *Zymomonas mobilis* subjected to environmental shocks, *Biotechnol. Bioeng.*, **53**, 71–78.

N. Stern, (2007), *Stern Review on the Economics of Climate Change*. The Department of Environment, Food and Rural Affairs, London, UK.

K.J. Suda, (1984), *Vegetable oil or diesel fuel – A flexible option*, SAE **84004**.

Y. Suen, J.S. Hubbard, G. Holzer and T.G. Tornabene, (1987), Total lipid production of green alga *Nannochloropsis sp.* Q11 under different nitrogen regimes, *J. Phycol.*, **23**, 289–296.

Y. Sun and J. Cheng, (2002), Hydrolysis of lignocellulosic materials for ethanol production: A review, *Biores Technol.*, **83**, 1–11.

M.J. Taherzadeh, L. Gustafsson, C. Niklasson and G. Liden, (2000), Inhibition effect of furfural in aerobic and anaerobic batch cultivation of *Saccharomyces cerevisiae* growing on ethanol and/or acetic acid, *J. Biosci. Bioeng.*, **90**, 374–380.

M. Takaji, K. Watanabe, K. Yamabert, T. Yoshida, (2000), Limited feeding of potassium nitrate for intracellular lipid and triglyceride accumulation of *Nannochloropsis sp* UTEX LB 1999, *Appl. Microbiol. Biotechnol*, **54**, 112–117.

K. Tsukahara and S. Sawayama, (2005), Liquid fuel production using microalgae, *J. Jpn. Petrol. Inst.* **48**(5), 251–259.

J.H. Van Gerpen, C.L. Peterson and C.E. Goering, (2007), Biodiesel: An alternative fuel for compression ignition engines, ASABE distinguished lecture series, Tractor Design No. 31. *Presented at the 2007 Agriculture Equipment Technology Conference*, Louisville, KY 11–14.

J.H. Van Gerpen, (2005), Biodiesel processing and production, *Fuel Process. Technol.*, **86**, 1097–1107.

J.H. Van Gerpen, R. Pruszko, D. Clements, *et al.*, (2006), Building a successful biodiesel business. *Biodiesel Basics*, Dubuque, IA.

D.I.C. Wang, I. Biocic, H.Y. Fang and S.D. Wang, (1997), Direct microbial conversion of cellulosic biomass to ethanol. In *Proceedings of The Third Annual Biomass Energy Systems Conference*, Solar Energy Research Institute, Department of Energy, Washington, DC.

P.S. Wang, J. Tat and M.E. Van Gerpen, (2005), The production of fatty acid isopropyl esters and their use as a diesel engine fuel, *Journal of the American Oil Chemists' Society (JAOCS)*, **82**(11), 845–849.

T.A. Warnick, B.A. Methe and S.B. Leschine, (2002), *Clostridium phytofermentas* sp. nov., a cellulolytic mesophile from forest soil, *Int. J. Syst. Evol. Microbiol.*, **52**, 1155–1160.

R.M. West, Z.Y. Liu, M. Peter and J.A. Dumensic, (2008), Liquid alkanes with targeted molecular weights from biomass- derived carbohydrates, *ChemSusChem*, **1**, 417–424.

World Biofuels: FAPRI (2008) *Agricultural Outlook*, Available at: http://www.fapri.org/ outlook (accessed February 20, 2013).

Worldwatch Institute, (2007), *Biofuels for Transport. Global Potential and Implications For Sustainable Agriculture and Energy*. Earthscan, London.

W. Xie, X. Huang and H. Li, (2007), Soybean oil methyl esters preparation using NaX zeolites loaded with KOH as a heterogeneous catalyst, *Bioresource Technol.*, **98**(4), 936–939.

A. Yahya and S.J. Marley, (1994), Physical and chemical characterization of methyl soy oil and methyl tallow esters as CI engine fuels, *Biomass Bioenergy*, **6**, 321–328.

F. Zhang and P. Cooke, (2008), *Future Prospects and Perspectives on Biofuels*, Available at: www.dime-eu.org/Cooke-08-Fang-Biofuels-up.pdf (accessed February 20, 2013).

M. Zhang, C. Eddy, K. Deanda, *et al.*, (1995), Metabolic engineering of a pentose metabolism pathway in ethanologenic *Zymomonas mobilis*, *Science*, **267**, 240–243.

8

Inside the Bioplastics World: An Alternative to Petroleum-based Plastics

Vincenzo Piemonte

Department of Engineering, University of Rome, Campus-Biomedico, Rome, Italy

8.1 Bioplastic Concept

Biodegradable plastics were introduced in the 1980s as possible renewable feedstock in producing non-petroleum-based plastics, as well as to reduce environmental problems due to the growth in landfill volume. Nowadays, the worldwide production of bioplastics is about 1 000 000 tons per year and is very modest when compared with 200 million tons per year of conventional (petroleum-based) plastics; it is estimated that, in the near future, the growth will be exponential reaching about 1 500 000 tons per year in 2012 (Widdecke *et al.*, 2008). The major manufacturers are Nature Works (USA) and PURAC (The Netherlands) who produce a bioplastic based on polylactic acid (PLA) (Tokiwa and Calabia, 2006), and Novamont (Italy): the largest producer of a starch-based bioplastic (Mater-Bi) in Europe (Bastioli, 2005).

The interest in development of biodegradable plastics noticed in recent years is due to motives of both an environmental and strategic nature (Anderson *et al.*, 1998; Zhang *et al.*, 2000; Gross and Kalra, 2002; Demirbas, 2007). In order to reduce the environmental impact of plastics (especially in terms of CO_2 released in the environment) some of the products obtained from agriculture (starch, cellulose, wood, sugar) are used as raw materials. By this way, the net balance of carbon dioxide is greatly reduced, since the CO_2

Sustainable Development in Chemical Engineering – Innovative Technologies, First Edition.
Edited by Vincenzo Piemonte, Marcello De Falco and Angelo Basile.
© 2013 John Wiley & Sons, Ltd. Published 2013 by John Wiley & Sons, Ltd.

released during production, utilization, and disposal of plastics is balanced by the CO_2 consumed during the growth cycle of the plant. Furthermore, petroleum, with constantly rising prices, is replaced by renewable raw materials obtained from agriculture.

The use of bioplastics was also stimulated by a second environmental motive, related to problems connected with the disposal of waste: in the 1990s, the main system for the disposal of municipal solid waste (with a fraction of plastics equal 20–30%) was disposal in landfill. Traditional plastics undergo degradation phenomena with very slow kinetics; hence, the volume required by these materials in landfills is virtually stable over time. On the contrary, bioplastics show faster degradation rates in landfill (Ishigaki *et al.*, 2004) and, therefore, volumes can be contained.

This highlights the complex meaning of the word *bioplastic*. In fact the word bioplastic can refer to the source (renewable) used as raw material for the bioplastic production or to the biodegradability and/or compostability that foreshadows a specific final destination of the bioplastic (Kyrikou and Demetres, 2007). Indeed, a plastic based on renewable sources (bio-based) may not be biodegradable (for instance, polyethylene: PE, is obtained from bioethanol, see Ragauskas *et al.*, 2006) and on the contrary may be biodegradable but made from non-renewable sources (e.g., polycaprolactone: see Demirbas, 2007). Of course, there are also bioplastics that are both biodegradable and bio-based, such as polylactic acid (PLA) (Tokiwa and Calabia, 2006) and poly-3-hydroxybutyrate (Harding *et al.*, 2007), or others obtained by copolymerization between biodegradable (or bio-based) and unbiodegradable polymers such as Mater-Bi (Bastioli, 2005).

Due to the aforementioned high versatility of bioplastics, the main developmental fields of these innovative polymers varies from containers for bio-waste collection and shopping bags to mulching films for agriculture; from products for catering to be collected along with food residues, to products for the packaging of prepacked food, as well as containers for beverages and specialized products to be used in various fields such as biomedicine. This confirms that nowadays bioplastics represent a great reality but, at the same time, they also represent a great challenge for real sustainable development.

Indeed, in order to assess the reliability of the bioplastics as "world green painters" careful evaluation of their impact on the environment in comparison with petroleum-based plastics is a key point. To this end, the most important tool to evaluate the environmental profile of a bioplastic and/or a petroleum-based plastic (conventional plastic) is the lifecycle assessment (LCA). The objectives of the LCA are to evaluate the effects of the interactions between a product and the environment, and therefore the environmental impact directly or indirectly caused by the use of a given product. It is very important to highlight that the LCA is not a tool for product marketing; that is, the results derived from an LCA cannot be taken as guidelines for the promotion of a product over another product; on the contrary the "mission" of the LCA studies is to drive scientists, farmers and others towards a constant and careful re-design of biopolymers to make possible the "growth of the newborn bioplastics".

The next few pages are dedicated to describing the production processes of PLA and starch-based bioplastics (such as Mater-Bi), which represent two different concepts: the first of a bio-based plastic and the second of biodegradability. Then, the third section of this chapter examines reliability from an environmental point of view of these two bioplastics, using the LCA technique, which will provide to the reader interesting insights into the bioplastic world toward a continuous improvement of a product. Bioplastics, which could be in the near future, very important protagonists in the plastic world.

8.2 Bioplastic Production Processes

8.2.1 PLA Production Process

The process reported in the following refers to the production of PLA in the NatureWorks plant situated in Nebraska (Erwin *et al.*, 2010).

The PLA production plant is divided into five major units:

1. Corn processing wet mill;
2. Corn processing and the conversion of starch into dextrose;
3. Conversion of dextrose into lactic acid;
4. Conversion of lactic acid into lactide;
5. Polymerization of lactide into PLA polymer pellets.

After harvest, the corn grain is transported to a corn wet mill (CWM), where the starch is separated from the other components of the corn kernel (proteins, fats, fibers, ash, and water) and hydrolyzed to dextrose using enzymes. The dextrose solution is transported by pipeline to Cargill's lactic acid fermentation process, which is situated adjacent to the CWM.

Lactic acid is produced by fermentation of dextrose received from the CWM. The process, (see Figure 8.1), combines dextrose and other media, adds a microbial inoculum, and produces crude lactic acid. The pH is controlled to near neutral by the addition of calcium hydroxide. The lactic acid broth is then acidified by adding sulfuric acid, resulting in the formation and precipitation of gypsum. The gypsum is removed by filtration, and the lactic acid concentrated by evaporation.

There are two major routes in producing polylactic acid from the lactic acid monomer: direct condensation polymerization of lactic acid, and ring-opening polymerization through the lactide intermediate. The first route involves the removal of water by condensation and the use of a solvent under high vacuum and temperature. With this route only low to intermediate molecular-weight polymers can be produced, mainly because of the presence of water and impurities. Other disadvantages of this route are the relatively large reactor required, and the need for evaporation, recovery of the solvent and increased color and racemization. For these reasons NatureWorks uses the second route: ring-opening polymerization through the lactide intermediate which enables the production of high molecular weight PLA without the use of an organic solvent (Kawashima *et al.*, 2002).

Therefore, after final purification, the lactic acid enters the lactide/PLA manufacturing process. PLA is prepared through the polymerization of lactide to make polylactide polymer in a continuous process. In the first step, water is removed in a continuous condensation reaction of aqueous lactic acid to produce a low molecular weight prepolymer. Next, the prepolymer is catalytically converted into the cyclic dimer, lactide, and vaporized.

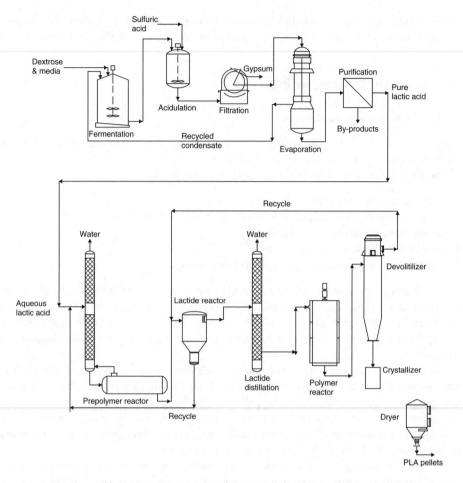

Figure 8.1   *NatureWorks PLA production process*

Production of the cyclic lactide dimer results in three potential forms: the D,D-lactide (called D-lactide), L,L-lactide (called L-Lactide) and L,D or D,L lactide called meso lactide. Meso lactide has different properties from D and L lactide. The D and L lactide are optically active, but the meso is not. Before polymerization the lactide stream is split into a low D lactide stream and a high D/meso lactide stream. Ring-opening polymerization of the optically active types of lactide can yield a "family" of polymers characterized by the molecular weight distribution and by the amount and the sequence of D-lactide in the polymer backbone. Polymers with high L-lactide levels can be used to produce crystalline polymers while the higher-D-lactide materials are more amorphous.

The lactide mixture is then purified by distillation. Finally, high molecular weight PLA is produced using a ring-opening lactide polymerization. The process does not use any solvents. After the polymerization is complete, any remaining lactide monomer is removed and recycled within the process. This polymer pellet production is the final stage of the PLA production plant.

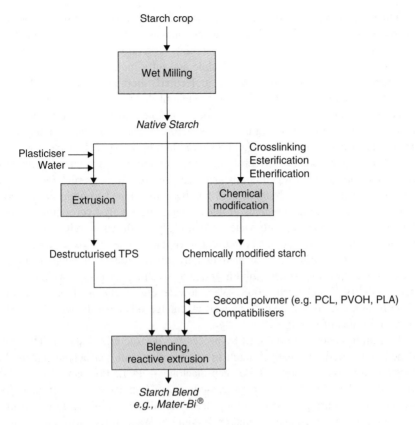

Figure 8.2   *Starch-based bioplastic production scheme*

8.2.2   Starch-based Bioplastic Production Process

The process reported in the following is common for the production of different types of starch-based bioplastics, among which Mater-Bi is the most important in Europe (Bastioli, 2005).

The process (see Figure 8.2) starts with the production of *native starch* obtained by extraction of starch from the maize kernel by wet milling process. The kernel is first softened by steeping it in a diluted acidic solution, the coarse fraction is ground to split the kernel and to remove the oil-containing germ. Finer milling separates the fiber from the endosperm which is then centrifuged to separate the less dense protein from the more dense starch. The starch slurry is then washed in a centrifuge, dewatered and dried prior to extrusion or granulation (Daniel *et al.*, 2000). A part of produced native starch is used to obtain *destructurized starch* and *chemically modified starch* that will be extruded together: the left over part of the native starch will be used to obtain the final starch-based bioplastic.

Destructurized starch (also referred as *thermoplastic starch*: TPS) is formed by processing native starch in an extruder (single or twin screw extruder) where under high temperature, high pressure, limited water, and sufficient time the native crystallinity and granular structures of amylase and amylopectin are almost completely destroyed. The increase in temperature during extrusion increases the mobility of starch granules and leads to the melting of the crystalline structures. The granules swell and take up the plasticizer, shear opens the granule and intra-molecular rearrangement takes place. Destructurized starch products are molecularly homogeneous (with both amylose and amylopectin dispersed uniformly throughout the material), have relatively high molecular-weight amylopectin, are not brittle or friable and have superior mechanical properties with respect to native starch. Fillers, additives and so on can be integrated into the extrusion process to provide the final resin product in one step. This includes also the addition of plasticizers such as glycerol, polyethers and urea, which have the function of reducing intermolecular hydrogen bonds and stabilize product properties (Weber, 2000).

As for chemically modified starch, this has been developed to address the problem that starch plastics with high contents of native starch are highly hydrophilic and readily disintegrate when contacted with water. Chemically modified starch is produced by treating native starch with chemicals in order to replace some hydroxyl groups by ester or ether groups. Crosslinking, in which two hydroxyl groups on neighboring starch molecules are chemically linked, inhibits granule swelling on gelatinization and gives increased stability (Foodstarch, 2008). Very low levels of chemical modification can significantly improve hydrophilicity, as well as change other rheological, physical, and chemical properties of starch.

Finally, starch blends are produced by processing destructurized starch (TPS), chemically modified starch, and native starch in combination with petrochemical, bio-based or inorganic compounds (such as Polycaprolactone – PCL, Polylactic Acid – PLA, Polyvinyl-alcohol – PVOH) into a (microscopically) homogenous material. Blending usually takes place during extrusion in a process called *reactive blending* which implies that the starch is bonded covalently and/or by Van der Waals forces with other polymers (Kalambur and Rizvi, 2006).

The starch content in a blend varies from 30–80% by mass depending on the end application. In the last years, the majority of co-polymers have been biodegradable polymers derived from fossil fuel feedstocks, such as PCL and PVOH. Thus, most starch blends are partially bio-based and fully biodegradable, such as Mater-Bi.

8.3   Bioplastic Environmental Impact: Strengths and Weaknesses

8.3.1   Life Cycle Assessment Methodology

From a general point of view, the LCA allows us to evaluate the interactions that a product or service has with the environment, considering its whole life cycle and includes the preproduction points (extraction and production of raw materials), production, distribution, use (including reuse and maintenance), recycling, and final disposal.

According to ISO 14040 and 14044, the LCA is achieved through four distinct phases. In the first phase, the *goal and scope* of study is formulated and specified in relation to the intended application. The object of study is described in terms of a so-called functional unit. Apart from describing the functional unit, the goal and scope should address the overall approach used to establish the system boundaries. The system boundary determines which unit processes are included in the LCA and must reflect the goal of the study.

The second phase, *inventory* involves data collection and modeling of the product system as well as description and verification of data. This phase encompasses all data related to environmental (e.g., CO_2) and technical (e.g., intermediate chemicals) quantities for all relevant unit processes within the study boundaries that compose the product system. The data must be related to the functional unit defined in the goal and scope phase. The results of the inventory are a lifecycle inventory (LCI), which provides information about all inputs and outputs in the form of elementary fluxes between the environment and all the unit processes involved in the study.

The third phase, the *Lifecycle Impact Assessment*, aims to evaluate the contribution to impact categories such as global warming and acidification. The first step of this phase is termed *characterization*. Here, impact potentials are calculated based on the LCI results. The next steps are *normalization* and *weighting*, but these are both voluntary according the ISO standard. Normalization provides a basis for comparing different types of environmental impact categories (all impacts get the same unit). Weighting implies assigning a weighting factor to each impact category depending on relative importance.

The last phase, *interpretation*, is an analysis of the major contributions, sensitivity analysis, and uncertainty analysis. This stage leads to the conclusion of whether the ambitions from the goal and scope can be met.

The LCA can be carried out by assessing the environmental footprint of a product from raw materials to its production (cradle to gate) or by analyzing the whole product life cycle, including product disposal (cradle to grave).

8.3.2 The Ecoindicator 99 Methodology: An End-Point Approach

Traditionally, in LCA, the emissions and resource extractions are expressed as 10 or more different impact categories (categories of midpoint level), such as acidification, ozone layer depletion, ecotoxicity, and resource extraction. For a panel of experts or nonexperts, it is very difficult to give meaningful weighting factors to such a large number of rather abstract impact categories. It was concluded that the panel should not be asked to weight the impact categories, but instead the different types of damage that are caused by these impact categories.

The Ecoindicator 99 method accounts for 11 impact categories of midpoint level: carcinogens, respiratory organics, respiratory inorganics, climate change, radiation, ozone layer, ecotoxicity, acidification/eutrophication, land use, minerals, and fossil fuels. The impact categories from carcinogens to ozone layer are then normalized and grouped in the macrocategory (end-point level or damage level), *Human Health* that takes in to account the overall impact (damage) of the emissions associated to the product analyzed on the human health (the damage to human health is expressed as the number of years of life lost and the number of years lived disabled). The categories ecotoxicity, acidification/eutrophication, and land use are included in the macrocategory *Ecosystem Quality* that accounts for the overall damage on the environment (the damage to Ecosystem Quality is expressed as the loss of species over a certain area, during a certain time), while the minerals and fossil fuels are grouped in the macrocategory *Resources* that accounts for the depletion of nonrenewable resources (the damage to Resources is expressed as the surplus energy needed for future extractions of minerals and fossil fuels). The impact category indicator results that are calculated in the characterization step are directly added to form damage categories. The damage categories (and not the impact categories) are then normalized on a European level (damage caused by 1 European per year), mostly based on 1993 as the base year, with some updates for the most important emissions (Goedkoop *et al.*, 2008).

Indeed, for a quantitative and objective comparison between the two products under investigation, on the basis of the three macrocategories of damage, a single overall impact index is needed. To this end, we must assign a weight to the individual macrocategory and define a global index of impact, I, given by:

$$I = \sum_i p_i c_i$$

where p_i is the assigned weight to the macrocategory of damage i, and c_i is the value of the macrocategory of damage.

The result in terms of advantage (i.e., a better overall impact score) of Product A or Product B is a function of the importance (i.e., the weight) that will be assigned to individual macrocategories. Representing the three categories in a *mixing diagram* (see Figure 8.3), each point within the triangle represents a weighting combination (Hofstetter *et al.*, 2008). In each point of the mixing triangle, the relative weights always add up

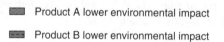

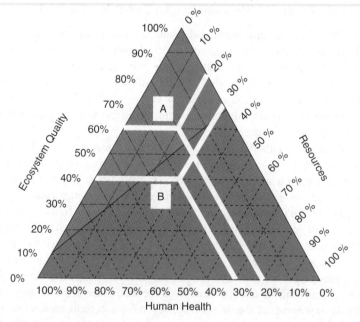

Figure 8.3 *Mixing triangle diagram. Reprinted by permission of the publisher (Taylor & Francis Ltd, https://www.tandf.co.uk/journals).*

to 100%. In a mixing triangle, each corner represents a weight of 100% for one damage category; in Figure 8.3, the top corner is the weighting combination where *Ecosystem Quality* is weighted 100%, and a 0% weighting is given to both *Human Health* and *Resources*. Any point on the base of the triangle of Figure 8.3 gives 0% weight to Ecosystem Quality, and the weights are split between 100% Human Health/0% Resources in the bottom-left corner to 0% Human Health/100% Resource in the bottom-right corner.

We only consider positive weightings; that is, a positive impact score always means damage: only points within the mixing triangle are taken as reasonable weighting sets. Therefore, it is possible to evaluate the value of I for the two products tested for each weighting set; in this way, we can locate the area of advantage for the Product A $(I_{PA} < I_{PB})$ or for the Product B depending on the set of weight, as shown in Figure 8.3.

Figure 8.3 shows two sets of weights corresponding to point A (60% Ecosystem Quality, 20% Human Health, and 20% Resources) and B (40% Ecosystem Quality, 30% Human Health, and 30% Resources). In condition A, Product A shows a better score than Product B, while in B, it is *preferable* (from an environmental point of view) to use the Product B; therefore, the focus on one or other material for the production of bottles is a function of the weights that will be assigned to individual macrocategories of impact. The border area between the two regions is the so-called line of indifference, that is, the set of values of the weights assigned to bring the two products tested to the same overall impact index, and thus a condition of substantial balance between the two products analyzed.

8.3.3   Case Study 1: PLA versus PET Bottles

This case study is devoted to examining environmental benefits of packages made from PLA in comparison with packages made from polyethylene terephthalate (PET), using the example of drinking water bottles.

The functional unit of this LCA is defined as 1000 U of 500 ml bottles to be used for drinking water. In particular, 12.58 kg of polymer granules will to produce 12.2 kg of bottles (1000 U) both for PLA and PET bottles. Furthermore, the same lifetime was considered for both types of bottles (Detzel and Krueger, 2006).

The system boundaries comprise:

- polymer production, starting from crude oil extraction for the PET and crop farming for PLA up to the final polymer;
- transportation of polymers to the bottle production plant. The transportation of PLA and PET pellets for subsequent processing in the finished product was carried out by rail. The transportation distance was assumed equal to 100 km. The environmental burden is calculated by taking into account the tons carried per kilometer (tkm);
- injection molding and blow stretching of the final bottles (bottle production processes).

Nature is not part of the system boundary, and therefore all the emissions (fertilizers, pesticides, etc.) relative to the area allocated for agricultural production are strictly taken into account.

Because PLA is produced from corn as a raw material through fermentation processing, and co-products, such as corn meal, are generated, the allocation step is needed. In this study, allocation procedures have been carried out via mass; the environmental credits

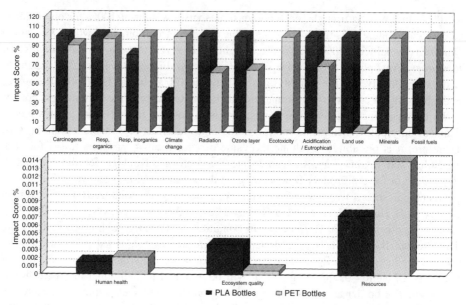

Figure 8.4   *Comparison between the LCAs cradle to gate on the production of PLA and PET bottles*

and burdens of co-products and recycled materials (see next section) are loaded for 0% (Ecoinvent criteria) and 100% to the systems under consideration, respectively.

For the implementation of the system models, SimaPro7 LCA software has been used and the Ecoindicator 99 methodology chosen, while the LCI has been obtained, both for PET and PLA, from data included in the Ecoinvent v.2.0 database (Frischknocht and Jungbluth, 2007).

The comparison between the LCAs cradle to gate of PET and PLA bottles is reported in Figure 8.4 both in terms of impact categories (mid-point level) and macrocategories of damage (end-point level). The figure shows that the production of PLA bottles has a strong impact in terms of damage to the ecosystem quality (it is worth noticing that this macrocategory includes the impact categories acidification, eutrophication, ecotoxicity, and land occupation). On the contrary, the effects on human health are roughly the same for PLA and PET. Finally, a greater damage in terms of consumption of non-renewable resources is associated with PET bottles production.

In order to give a quantitative comparison of these two products the obtained results are reported in a triangle mixing diagram (Figure 8.5). Using the set of weightings suggested by the LCA Suisse Group (40% Human Health, 40% Ecosystem Quality, and 20% Resources) (Hofstetter *et al.*, 2008), the PLA and PET bottles are roughly equivalent in terms of environmental impact. This result must be interpreted considering that if on one hand, the PLA can save in terms of fossil resources; on the other hand, it causes major damage in terms of human health and ecosystem quality. Indeed, for the production of raw material (corn), it is necessary to use intensive agriculture with the use of different chemicals (pesticides, herbicides, and fertilizers) that are harmful to humans and environment. The superiority of PLA to PET does not seem so obvious, but

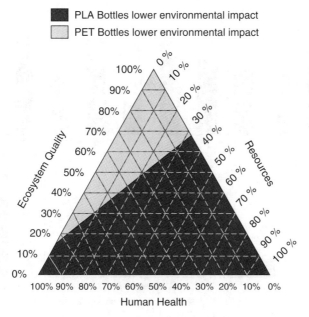

Figure 8.5   *Comparison between LCAs cradle to gate of PET and PLA bottles using the mixing triangle technique*

on the contrary, giving a high weighting to both Human Health and Ecosystem Quality the PLA failed' in comparison. The performed analysis corroborates the hypothesis that the true advantage of the PLA derives from the use of renewable resources, but the benefit is paid in terms of damage to human health and ecosystem quality (pesticides, consumption of land, and consumption of water).

8.3.4   Case Study 2: Mater-Bi versus PE Shoppers

In the following are reported the results obtained from the LCAs of shoppers made from Mater-Bi and PE using the Ecoindicator-99 methodology. The LCAs have been performed by using the SimaPro7 software, while the LCIs data have been taken both from the Ecoinvent v.2 and the Buwal 250 libraries. The data on Mater-Bi production (provided directly from Novamont, with data relating to production in Terni, Italy) refers to a co-polymer with a starch content of 35% and the remaining 65% is biodegradable polyester derived from nonrenewable sources (polyester type not specified). For the PE, the production process of granules was performed on average data, for various production sites throughout Europe, contained in the Eco-profiles prepared by the European Plastics Industry. The main assumptions made to perform the LCAs are:

- Nature is not part of the production system; this implies that all the emissions (fertilizers, pesticides, etc.) relative to the area allocated for agricultural production are strictly taken into account.
- As per the biodegradable polyester contained in the co-polymer Mater-Bi, the PCL has been considered (biodegradable polyester made from fossil fuels) (Demirbas, 2007).

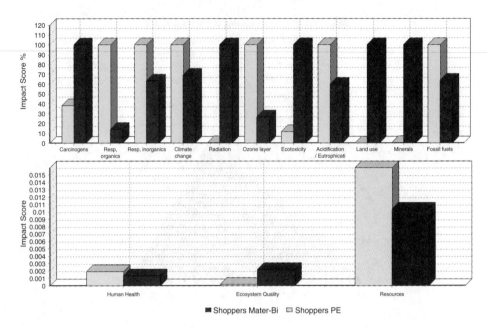

Figure 8.6   *Comparison LCA "cradle to gate" on the production of Mater-Bi and PE shoppers*

- The production of Mater-Bi and PE shoppers is achieved through three phases: production of granules, transportation of granules in the processing establishments, and process of production of the shoppers by blow foil extrusion.
- The LCAs have been realized on the basis of 1000 shoppers made from Mater-Bi (total weight of 14 kg) and PE (total weight of 12 kg), respectively. The different weights between the two shoppers is due to different mechanical properties of the Mater-Bi and PE, that is to obtain shoppers with the same mechanical characteristics, different thickness of polymer films are needed (Davis, 2003).

The results of the LCAs "cradle to gate" are reported in Figures 8.6 and 8.7. Figure 8.6 shows how, also in this case, the production of Mater-Bi shoppers has a strong impact in terms of damage to ecosystem quality, on the contrary, it is evident the greater damage, in terms of consumption of non-renewable resources determined by the shoppers made from PE with respect to that made from Mater-Bi. Finally, the effects on human health are roughly the same for both Mater-Bi and for PE.

Figure 8.4 underlines, once again, the great strength of bioplastics, which is the savings in non-renewable resources consumption, but, as heavy counterparts, there is a greater impact on the ecosystem quality. Of course, which of these two aspects should be considered dominant depends on the sensitivity of the reader, which is also based on the current conditions when the evaluation takes place. It is evident that if at any moment we were informed that all fossil resources are depleted, there is no much to choose. However, it would be very important to get to that moment with a mature technology and, if possible, with little or no weakness points.

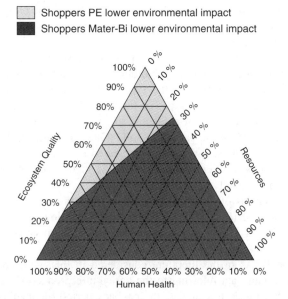

Figure 8.7   *Comparison between LCAs cradle to gate of Shoppers in PE and Mater-Bi using the mixing triangle technique*

8.3.5   Land Use Change (LUC) Emissions and Bioplastics

Recently, a discussion has emerged around greenhouse gas emissions due to direct and indirect land use change (LUC) of expanding agricultural areas dedicated to bio-products production. Converting rainforests, peatlands, savannas, or grasslands to produce food crop-based bio-products creates a "carbon debt" by releasing 9–170 times more CO_2 than the annual greenhouse gas (GHG) reductions that these bio-products (such as the bio-plastics) would provide by displacing petroleum-based products (conventional plastics). In fact, to produce bioplastics, farmers can directly plow up more forest or grassland, which releases to the atmosphere much of the carbon previously stored in plants and soils through decomposition or fire. The loss of maturing forests and grasslands also foregoes ongoing carbon sequestration as plants grow each year, and this foregone sequestration is the equivalent of additional emissions. Therefore, by excluding emissions from Land-Use Change, most previous accounts were one-sided because they counted the carbon benefits of using land for bioplastics but not the carbon costs, the carbon storage and sequestration sacrificed by diverting land from its existing uses (Righelato and Spracklen, 2007; Plevin and Mueller, 2008; Searchinger *et al.*, 2008).

Emissions from LUC can be divided in indirect land-use change emissions (iLUC) and direct land-use change emissions (dLUC). The former emissions occur when land currently used for feed or food crops is changed into bio-product feedstock production and the demand for the previous land use (i.e., feed, food) remains. Therefore, the displaced agricultural production will move to other places where unfavorable land-use change may occur (Fritsche, 2008). On the other hand, dLUC emissions occur when a new agricultural land is taken into production and feedstock for bio-products purposes displaces a prior land use, thereby generating possible changes to the carbon stock of that land (Cherubini, 2010).

In order to account for LUC emissions in the LCA studies, the method proposed by Searchinger *et al.* (2008) (applied to biofuels) can be followed. In particular, since greenhouse gas emissions depend on the type of land converted, new cropland was assigned in each region to different types of forest, savannah, or grassland on the basis of the proportion of each ecosystem converted to cultivation in the 1990s and assumed that conversion emits 25% of the carbon in soils (Guo and Gifford, 2002; Murty *et al.*, 2002) and all carbon in plants, which must be cleared for cultivation. Searchinger *et al.* (2008) in their work report an emission value of 351 metric tons of CO_{2eq} per hectares of land converted to cropland.

To carry out an analysis of the LUC emissions effect on the LCA results in terms of global warming potential (time horizon 100 yr), in the following, we added the LUC emissions to some LCAs of literature, considering the LUC emissions value reported by Searchinger *et al.*

In Table 8.1 the LCAs in the literature considered in this work are reported, along with the most important data referring to the agriculture production. The data reported in this table have been used to calculate the payback period (defined as the period over which the annual GHG savings due to the substitution of petroleum-based products with bio-based products equalize the emissions from land-use change) for all of the LCAs considered (Figure 8.8). For example, if we consider that for the production of 1 kg of PLA granules, 1.56 kg of corn are needed and that the production of 1 kg of corn requires $1.7\,m^2$ of arable land (Table 8.1, case H), and assuming an emission value of 351 metric tons of CO_{2eq} per ha of land converted to cropland (Searchinger *et al.*, 2008), the PLA would equalize and therefore pay-back carbon emissions from Land-Use Change in about 47.5 yr, meaning GHGs increase until the end of that period. Therefore, it is evident from Figure 8.8 that considering the LUC emissions in the LCA studies, the advantages achievable by displacing petroleum-based plastics with bio-based plastics are strongly reduced.

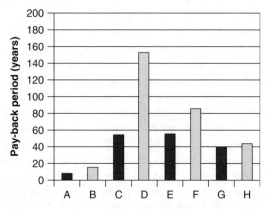

Figure 8.8 *Calculated pay-back period for the different LCAs of bioplastics considered. Reproduced with permission © 2010 American Institute of Chemical Engineers (AIChE).*

Table 8.1 *Key data on agricultural production in the LCA studies considered and GHG savings (Erwin et al., 2007). Reproduced with permission of John Wiley & Sons, Ltd, © 2008*

Study	Type of Polymers	GHG Savings kgCO$_{2eq}$/kg Polymer	Crop	Country	Crop Yield (Mg/[ha yr])	Crop Input (kg crop/kg Polymer)
Dinkel *et al.* (1996)	TPS vs LDPE Pellets (A)	3.90	Potato, corn	CH	Potato: 18.0 Corn: 12.5	2.23 + 0.385
Dinkel *et al.* (1996)	TPS vs LDPE Films (B)	1.99	Potato, corn	CH	Potato: 18.0 Corn: 12.5	2.23 + 0.385
Wurdinger *et al.* (2002)	TPS vs EPS Loose fills (C)	0.28	Corn	DE	Corn: 6.45	0.786
Estermann *et al.* (2000)	TPS vs EPS Loose fills (D)	0.76	Corn	FR	Corn: 8.20	0.971
Vink *et al.* (2003)	PLA vs LDPE Pellets (E)	1.20	Corn	US	Corn: 9.06	1.74
Wotzel *et al.* (1999)	Natural fiber/EPS vs ABS Automotive parts (F)	1.0	Hemp/ EPS	DE	n/a	0.49
Pervaiz and Sain (2003)	Natural fiber/PP vs fiberglass/PP Composite (G)	2.80	Hemp/ PP	CA	Hemp: 2	0.65
Piemonte and Gironi (2012)	PLA vs PET Granules (H)	1.95	Corn	US	Corn: 5.6	1.56

8.4 Conclusions

Bioplastics production for the replacement of a part of petroleum-based plastics seems to be a real and effective strategy toward sustainable development. In fact, the displacing of conventional plastics with bioplastics can lead to considerable energy and GHGs emissions savings. Overall, these advantages are paid for in terms of damage on the ecosystem quality because, for the production of raw material (corn), it is necessary to use intensive agriculture with use of different chemicals (pesticides, herbicides, fertilizers) that are harmful to the environment. Furthermore, accounting for the Land-Use Change emissions can limit the attractiveness of bioplastics for the displacement of petroleum-based plastics, at least from an environmental point of view. Therefore, in order to further develop the bioplastic potentialities it is very important to carefully manage both the feedstock chain for bioplastic production and waste disposal. Indeed, bioplastics from perennials grown on degraded cropland and from waste biomass would minimize habitat destruction, competition with food production, and carbon debts, all of which are associated with direct and indirect land clearing for bioplastic production. In this context, commercial availability of technologies for producing bioplastics based on lignocellulosic feedstocks (like biofuels of second generation) may induce changes in land use toward more lignocellulosic crops cultivation.

As for bioplastic disposal, mechanical recycling could be the best solution in order to maximize energy saving and reduce renewable resource consumption, therefore reducing the use of pesticides, herbicides, fertilizers and so on for their production.

To conclude, bioplastics are an important reality with huge potential, but the challenge to the total replacement of conventional plastics has just begun.

Acknowledgements

This chapter has been developed on the basis of a bigger work financial supported by Italian Consortium of Packaging (CONAI). The author many thanks Professor Fausto Gironi for their useful suggestions.

References

J.M. Anderson, A. Hiltner, M.J. Wiggins, *et al.* (1998). Recent advances in biomedical polyurethane biostability and biodegradation. *Polymer Intl.*, **46**, 163–171.

C. Bastioli (2005). Starch-based technology. In C. Bastioli (ed.), *Handbook of Biodegradable Polymers*, pp. 257–286, Italy: Rapra Technology Limited.

F. Cherubini (2010). GHG balances of bioenergy systems – Overview of key steps in the production chain and methodological concerns, *Renewable Energy*, **35**, 1565–1573.

J.R. Daniel, R.L. Whistler and H. Röper. (2000) Starch. In: *Ullmann's Encyclopaedia of Industrial Chemistry 2007*. Wiley-VCH Verlag GmbH & Co. KGaA

G. Davis (2003). Characterization and Characteristics of Degradable Polymer Sacks. *Materials Characterization*, **51**, 147–157.

A. Demirbas (2007). Biodegradable plastics from renewable resources. *Energy Sources Part A*, **29**, 419–424.

A. Detzel and M. Krueger (2006). Life cycle assessment of polylactide (PLA). *A comparison of food packaging made from NatureWorks PLA and alternative materials, IFEU report*, Heidelberg, Germany, July.

F. Dinkel, C. Pohl, M. Ros and B. Waldeck. (1996). Life-cycle assessment of starch-containing polymers, *Schriftenreiche Umwelt Nr. 271, Bern: BUWAL (Swiss Agency for the Environment, Forests, and Landscape)*.

T.H. Erwin, D.A. Glassner, J.J. Kolstad, *et al.* (2007). The eco-profiles for current and near-future NatureWorks polylactide (PLA) production, *Industrial Biotechnology*, **3**, 58–81.

T.H. Erwin, S. Davies and J.J. Kolstad, (2010). The eco-profile for current Ingeo® polylactide production, *Industrial Biotechnology*, **6**(4), 212–224.

R. Estermann, B. Schwarzwalder and B. Gysin (2000). *Life Cycle Assessment of Mater-Bi and EPS Loose Fills*, Olten, Switzerland: Composto.

Foodstarch (2008) Dictionary of food starch terms/crosslinked starch, *Website of the National Starch and Chemical Company*: Available at: http://eu.foodinnovation.com /dictionary/c.asp (accessed February 12, 2013).

R. Frischknocht and N. Jungbluth. (2007). Overview and Methodology, *Ecoinvent Report No. 1*, Swiss Centre of Life Cycle Inventories, Dubendorf.

U.R Fritsche. (2008). The "iLUC factor" as a means to hedge risk of GHG emissions from indirect land-use change associated with bioenergy feedstock provision. In *Background paper for the EEA Expert Meeting in Copenhagen*, June 10, 2008 (Draft version), Oko Institute.

M. Goedkoop, M. Oele, A. Schryver and M. Vieira. (2008). *SimaPro Database Manual: Methods Library*, pp. 22–25, The Netherlands: PRé Consultants.

R.A. Gross and B. Kalra (2002). *Biodegradable polymers for the environment*. Green Chem, **297**, 803–807.

L.B. Guo and R.M. Gifford (2002). Soil carbon stocks and land use change: A meta-analysis, *Global Change Biology*, **8**, 345.

K.G. Harding, J.S. Dennis, fix> Blottnitz, H. von and S.T.L. Harrison (2007). Environmental analysis of plastic production processes: Comparing petroleum-based polypropylene and polyethylene with biologically-based poly-b-hydroxybutyric acid using life cycle assessment, *Journal of Biotechnology*, **130**, 57–66.

P. Hofstetter, A. Braunschweig, M. Mettier, *et al.* (2008). The mixing triangle: Correlation and graphical decision support for LCA-based comparison, *Journal of Industrial Ecology*, **3**, 97–115.

T. Ishigaki, W. Sugano, A. Nakanishi, *et al.* (2004). The degradability of biodegradable plastics in aerobic and anaerobic waste landfill model reactors. *Chemosphere* **54**, 225–233.

ISO 14040: (2006), *Environmental management – Life cycle assessment – Principles and framework*. Requirements and guidelines.

S. Kalambur and S.S.H. Rizvi (2006) An overview of starch-based plastic blends from reactive extrusion. *Journal of Plastic Film and Sheeting* **22**, 39.

N. Kawashima, S. Ogawa, S. Obuchi, *et al.* (2002). Polylactic acid "LACEA". In: Y. Doi and A. Steinbuchel, (eds). *Biopolymers* in 10 volumes, Vol. **4**, Polyesters III Applications and Commercial Products. Weinheim: Wiley-VCH Verlag, 251–274.

I. Kyrikou and B. Demetres (2007). Biodegradation of agricultural plastic films: A critical review, *Journal of Polymer Environment*, **15**, 125–150.

D. Murty, M.U.F. Kirschbaum, R.E. McMurtie and H. McGilvray (2002). Does conversion of forest to agricultural land change soil carbon and nitrogen? *A review of the literature, Global Change Biology*, **8**, 105.

M. Pervaiz and M.M. Sain (2003). Carbon storage potential in natural fiber composites, *Resources, Conservation, and Recycling*, **39**, 325–340.

V. Piemonte and F. Gironi (2012) Bioplastics and GHGs saving: The land use change (LUC) emissions issue, *Energy Sources Part A: Energy, Recovery and Environmental Effect* **34** (21), 1995–2003(DOI: 10.1080/15567036.2010.497797).

R.J. Plevin and S. Mueller (2008). The effect of CO_2 regulations on the cost of corn ethanol production, *Environmental Research Letters*, **3**, 1–9.

A.J. Ragauskas, C.K. Williams, B.H. Davison, *et al.* (2006). The path forward for biofuels and biomaterials, *Science*, **331**, 484–489.

R. Righelato and D.V. Spracklen (2007). Carbon mitigation by biofuels or by saving and restoring forests? *Science*, **317**, 902.

T. Searchinger, R. Heimlich, R.A. Houghton, *et al.* (2008). Use of U.S. Croplands for biofuels increases greenhouse gases through emissions from land-use change, *Science*, **319**, 1237–1241.

Y. Tokiwa and B.P. Calabia (2006). Biodegradability and biodegradation of poly(lactide), *Energy Sources Part A*, **72**, 244–251.

E.T.H. Vink, K.R. Rabago, D.A. Glassner and P.R. Gruber (2003). Applications of life cycle assessment to NatureWorks(TM) polylactide (PLA) production, *Polymer Degradation and Stability*, **80**, 403–419.

R. Weber (2000) *Bio-Based Packaging Materials for the Food Industry: Status and Perspectives*. KVL Department of Dairy and Food Science, Frederiksberg, Denmark.

H. Widdecke, H. Otten, A. Marek and S. Apelt (2008). *Bioplastics 07/08. Processing Parameters and Technical Characteristics*, A Global Overview. CTC GmbH Fachhochschule Braunschweig/Wolfenbuttel.

K. Wotzel, R. Wirth, and R. Flake (1999). Life cycle studies on hemp fibre reinforced components and ABS for automotive parts, *Angewandte Makromolekulare Chemie*, **272**, 121–127.

E. Wurdinger, U. Roth, A. Wegener, *et al.* (2002). Biobased Polymers: Comparative Lifecycle Assessment of Loose-Fill-Packaging From Starch and Poly(Styrene). *Endbericht (DBU-Az. 04763)*. Bavarian Institute of Applied Environmental Research and Technology.

J.Y. Zhang, E.J. Beckman, N.P. Piesco, and S. Agarwal (2000). A new peptide-based urethane polymer: Synthesis, biodegradation, and potential to support cell growth in vitro. *Biomaterials*, **21**, 1247–1258.

9

Biosurfactants

Maria Giovanna Martinotti, Gianna Allegrone, Massimo Cavallo, and Letizia Fracchia
Dipartimento di Scienze del Farmaco, Università del Piemonte Orientale "Amedeo Avogadro", Novara, Italy

9.1 Introduction

Surfactants are SURFace ACTive AgeNTS with wide ranging properties including the lowering of surface and interfacial tensions of liquids. Some surfactants known as biosurfactants (BSs) are surface-active substances naturally synthesized by a variety of microorganisms on substrates including sugars, oils, alkanes, and wastes (Lin, 1994). They have the properties of reducing surface tension, stabilizing emulsions, promoting foaming, and are generally not toxic and biodegradable. BSs are amphiphilic compounds produced on living surfaces, mostly on microbial cell surfaces, or excreted extracellularly, and contain hydrophobic and hydrophilic moieties that confer the ability to accumulate between fluid phases, thus reducing surface and interfacial tension at the surface and interface respectively. The various kinds of BSs include: lipopeptides synthesized by many species of *Bacillus* and other species such as *Brevibacterium aureum*, *Nocardiopsis alba*, glycolipids by species of *Pseudomonas*, *Mycobacterium*, *Rhodococcus*, *Arthrobacter*, *Nocardia*, *Gordonia*, yeast *Candida*, *Yarrowia* and *Pseudozima*, and fungi such as *Ustilago scitaminea*, phospholipids by *Thiobacillus thiooxidans*, lipid-polysaccharide complexes by *Acinetobacter* species and lastly the microbial surfaces themselves (Desai and Banat, 1997; Bodour and Miller-Maier, 2002; Mulligan, 2005; Rivardo *et al.*, 2009; Franzetti *et al.*, 2009; Banat *et al.*, 2010; Morita *et al.*, 2011). Marine BS lipopeptides and glycolipids are too produced by some marine actinobacteria and have received more attention, particularly for bioremediation of the sea polluted by

Sustainable Development in Chemical Engineering – Innovative Technologies, First Edition.
Edited by Vincenzo Piemonte, Marcello De Falco and Angelo Basile.
© 2013 John Wiley & Sons, Ltd. Published 2013 by John Wiley & Sons, Ltd.

crude oil (Maneerat, 2005; Gandhimathi *et al.*, 2009; Karthik *et al.*, 2010; Kiran *et al.*, 2010a,b). A thorough review on these marine products has been done by Das *et al.* (2010) with particular attention to their environmental and industrial potentials and biological action.

In the last 20 years a large amount of investigations have been dedicated to BSs not only as potential replacement for synthetic surfactants in many environmental and industrial applications such as bioremediation, enhanced soil recovery, paint, textile, detergent applications, but also in cosmetic, food, agrochemical and more recently pharmaceutical and medical applications (Banat *et al.*, 2000, 2010; Fracchia *et al.*, 2011).

Despite their potential, the use of BSs is currently extremely limited, the major reason being the cost of production of chemical surfactants. This is the reason why there are no commercial large-scale processes for BS production (Mukherjee *et al.*, 2006). Solid state cultivation on renewable substrates has been recently suggested as alternative method to submerged cultivation, to increase the yield of the product and, in addition, to eliminate serious problems created by foam production (Krieger *et al.*, 2010).

In recent years, the development of new functional structures and/or systems using self-assembly of amphiphilic molecules has evolved into a dynamic and rapidly growing area of nanotechnology due to their ability to self-assemble into hierarchically ordered structures using hydrogen bonding, hydrophobic and van der Waals interactions (Kitamoto *et al.*, 2009). The BS mediated process and microbial synthesis of nanoparticles are emerging as clean, nontoxic, and environmentally acceptable "green chemistry" procedures (Xie *et al.*, 2006; Kasture *et al.*, 2008; Reddy *et al.*, 2009; Kiran *et al.*, 2010c, 2011). BS-mediated synthesis is superior to the methods using bacteria or fungi for nanoparticle synthesis, as these amphiphilic compounds reduce the formation of nanoparticles (Kiran *et al.*, 2011). BSs have the potential to be used a dispersant too (Raichur, 2007; Biswas and Raichur, 2008).

This chapter will focus on the most recent results obtained in the field of production, optimization, recovery and applications of BSs.

9.2   State of the Art

In the last two years two books have been dedicated to the subject of *Biosurfactants* (Sen, 2010; Soberón-Chávez, 2011), this leads to the conclusion that BSs are quickly becoming more and more interesting for the international scientific community working in founding sustainable alternatives to chemical surfactants. BSs are amphiphilic compounds produced on living surfaces, mostly, of microbial cells. They contain hydrophobic and hydrophilic domains which allow them to partition at the interfaces between liquid phases, be they solid, liquid or gasses thus reducing surface and interfacial tension (Mehta *et al.*, 2010). They have the ability to form molecular aggregates including micelles. The micellar aggregation of biosurfactants is originated at the critical micelle concentration (CMC) typically from $1-200$ mg L^{-1} and interestingly, about 10- to 40-fold lower than that of chemical surfactants.

BSs are classified by their chemical composition and microbial origin. They were divided by Rosenberg and Ron (1999) in two groups: low molecular weight polymers which efficiently lower surface and interfacial tension; and high molecular weight which are efficient emulsion-stabilizing agents. The low molecular mass BSs include glycolipids such as rhamnolipids, and sophorolipids or lipopeptides such as the well-known surfactin and polymyxin. The high molecular mass group includes lipoproteins, lipopolysaccharides, and amphipathic forms. Figure 9.1 shows the structures of some BSs and bio-emulsifyers (BEs).

9.2.1　Glycolipids

A great majority of biosurfactants are glycolipids, that is, carbohydrates in combination with long chain aliphatic acids or hydroxyaliphatic acids. Rhamnolipids are composed of one or two molecules of rhamnose linked to one or two molecules of β-hydroxydecanoic acid. Jarvis and Johnson (1949) were the first to describe the exact chemical nature of these biomolecules produced by *Pseudomonas aeruginosa*. Investigations have revealed a large diversity of congeners and homologs produced by this strain following different culture conditions and by other bacterial species (Ochsner *et al.*, 1996; Dubeau *et al.*, 2009; Abdel-Mawgoud *et al.*, 2010, 2011; Costa *et al.*, 2011). Trehalolipids are glycolipids in which trehalose is a non-reducing disaccharide with two glucose units linked in an α,α-1,1-glycosidic linkage. The most reported trehalose lipid is trehalose 6,6'-dimycolate, which is a α-branched chain mycolic acid esterified to the C6 position of each glucose. Different trehalose containing lipids are produced by *Arthrobacter parafineus*, *Nocardia* sp., *Rhodococcus erythropolis*, *Rhodococcus opacus*, *Micrococcus luteus*, *Mycobacterium paraffinicum* and *Gordonia amarae* (Kuyukina and Ivshina, 2010; Franzetti *et al.*, 2010b; Shao, 2011). Sophorolipids are synthesized by a selected number of yeast species. They consist of a hydrophobic fatty acid tail of 16 or 18 carbon atoms and a hydrophilic carbohydrate head, sophorose. The hydrophilic part sophorose is a di-glucose with an unusual β-1,2 bond and may contain acetyl groups at the 6'- and/or 6' positions (Van Bogaert *et al.*, 2007). *Rhodotorula bogoriensi*, *Candida bombicola*, *Wicherhamiella domercqiae*, *Pichia anomala* are yeast species producing sophorolipids (Thaniyavarn *et al.*, 2008; Van Bogaert *et al.*, 2011). Mannosylerythritol lipids (MELs) are glycolipids that contain a 4-*O*-β-D-manno pyranosyl-*meso*-erythritol as hydrophylic group and a fatty acid and/or an acetyl group as the hydrophobic moiety (Arutchelvi and Doble, 2011). Based on the degree of acetylation MELs (from *Pseudozyma aphidis*) have been classified as MEL-A when diacetylated, MEL-B and MEL-C when monoacetylated and MEL-D when nonacetylated (Rau *et al.*, 2005). MELs are produced by yeast strains of the genus *Pseudozima* sp., *Ustilago* sp., or *Kurtzmanomyces* sp. when grown on soybean oil or *n*-alkane (Kitamoto *et al.*, 1990; Kakugawa *et al.*, 2002; Morita *et al.*, 2009).

9.2.2　Lipopeptides

Lipopeptides are mainly produced by different *Bacillus* species, up to now they have been classified into three different families: surfactin, iturins and fengycin (or plipastatins). Surfactin family is composed of a cyclic heptapeptide with a chiral sequence LLDLLDL interlinked with a β-idroxy fatty acid which can contain 12–16 C atoms and show *n*, *iso*, and *anteiso* configurations. All iturins are heptapeptides and have the same LDDLLDL chiral sequence. The length of the fatty acid chain differs from one member to another. Fengycins (A and B) also called plipastatin are lipodecapeptides with an internal lactone ring in the peptide moiety between the carboxyl terminal amino

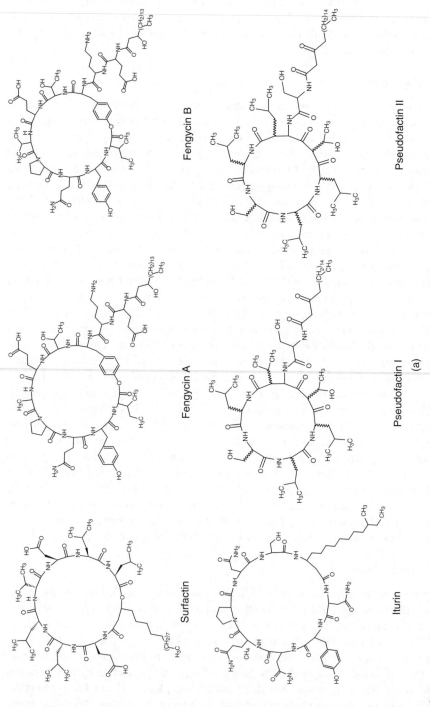

Figure 9.1 *Structures of some biosurfactants and bioemulsifiers. Reproduced from Bioresource Technology, 101/15, Janek et al., Isolation and charaterization of two new lipopeptide biosurfactants produced by Pseudomonas fluorescensBD5 isolated from water from the Artic Archipelago of Svalbard, 6118–6123, October 2011, with permission from Elsevier.*

(a)

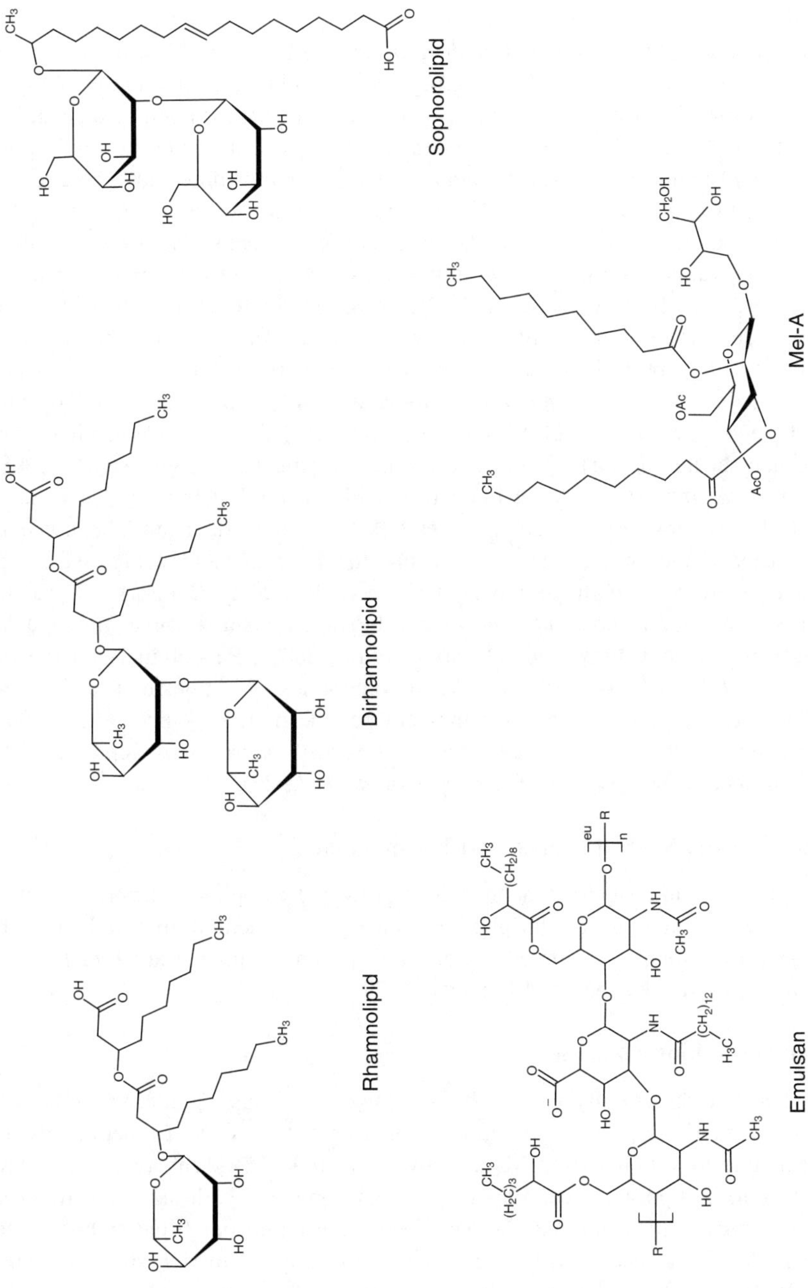

Sophorolipid

Dirhamnolipid

Rhamnolipid

Mel-A

Emulsan

(b)

Figure 9.1 *(continued)*

acid (Ile) and the hydroxyl group in the side chain of the tyrosine residue in position 3. The main representative fatty acid chains are C15, C16 and C17. Generally, *Bacillus* species co-produce various families of lipopeptides with different homologues and isoforms. Pecci *et al.* (2010) reported of three surfactin homologues and eight fengycin homologues produced by the strain *B. licheniformis* V9T14. Kurstakins, discovered by Hathout *et al.* (2000), make up the fourth group that should be included in the family of lipopeptides. Kurstakins are at least four and they have the same amino acid sequence, Thr-Gly-Ala-Ser-His-Gln-Gln, but different fatty acids are linked via an amide bond to the N-terminal amino acid residue. Bamilocyn A, a novel lipopeptide produced by the strain *Bacillus amiloliquefaciens* LP03 was characterized by Lee *et al.* (2007). The lipopeptide has a molecular mass of 1,022.6 Da and differed from surfactins in the substitution of leucine, valine and aspartic acid in positions 3, 4, and 5 by methionine, leucine, and proline, respectively. Jacques (2011) has reviewed in depth the high diversity of all these lipopeptides and the complex modular enzymes and regulation of their biosynthesis.

A number of lipopeptide BSs produced by Gram-negatives bacteria have been also described: Viscosin 1 from *Pseudomonas libanensis* M9-3 contains a total of nine amino acids, two of which (leu-glu) link to the fatty acid tail and the remaining seven [val-leu-ser-leu-ser-ile-allo(thr)] forming a cyclic oligopeptide (Saini *et al.*, 2008), Pseudophomins A and B, are cyclic lipodepsipeptides from *Pseudomonas fluorescens* strain BRG100 (Pedras *et al.*, 2003), Pseudofactin I and II, from *Pseudomonas fluorescens* BD5, have in their cyclic structure the palmitic acid connected to the terminal amino group of eight amino acids in peptide moiety (Janek *et al.*, 2010), Massetolides from a *Pseudomonas fluorescens* biovar II contains nine amino acids and a 10-carbon hydroxy fatty acid (De Souza *et al.*, 2003), Putisolvin I and II from *Pseudomonas putida* PCL1445 are cyclic lipodepsipeptides with a hexanoic lipid chain connected to the N-terminus of a 12-amino-acid peptide moiety (Kuiper *et al.*, 2004) and, finally, the non ionic BSs Serrawettins from *Serratia marcescens*, represented by the molecular species W1, W2 and W3 (Matsuyama *et al.*, 2011).

9.2.3   Fatty Acids, Neutral Lipids, and Phospholipids

Several fungi, yeasts and bacteria that are able to grow on hydrophobic substrates such as alkanes, secrete large amounts of phospholipids, fatty acids or neutral lipids to facility the uptake of the carbon source. Examples of producers of such compounds are *Acinetobacter* sp. and *Aspergillus* sp. (Käppeli and Finnerty, 1979).

9.2.4   Polymeric Biosurfactants

Polymeric BSs are structurally formed by several components. Emulsan, synthesized by *Acinetobacter calcoaceticus*, is the most studied one. It consists of a heteropolysaccharide backbone to which fatty acids are covalently linked (Rosenberg *et al.*, 1988). Another example is liposan, a carbohydrate-protein complex synthesized by the yeast *Yarrowia lipolytica* (Cirigliano and Carman, 1984). An emulsifier (lipid-carbohydrate-protein) complex associated with the cell wall of the tropical marine strain *Yarrowia lipolytica* NCIM 3589, produced in the presence of alkanes or crude oil, was also described by Zinjarde and Pant (2002).

9.2.5 Particulate Biosurfactants

Extracellular membrane vesicles that partitions hydrocarbons to form a microemulsion plays an important role in alkane uptake in *Pseudomonas nautica* (Husain *et al.*, 1997). Cell surface of *Sphingomonas* sp. was covered with extracellular vesicles when grown on poly-aromatic hydrocarbons (Gilewicz *et al.*, 1997).

A list of major types of biosurfactants and of their microbial sources is shown in Table 9.1.

9.3 Production Technologies

According to Syldatk and Hausmann (2010), the reasons for limited use of microbial surfactants in industry are the expensive cost of substrates, the limited product concentrations, low yields and formation of product mixtures rather than pure compounds. Therefore, a number of reviews were published on the development of strategies for high-throughput and low cost production on many renewable substrates (Fiechter, 1992; Makkar and Cameotra, 1999, 2002; Helmy *et al.*, 2011; Makkar *et al.*, 2011).

9.3.1 Use of Renewable Substrates

Vegetable oil from canola, corn, sunflower, safflower, olive, rapeseed, grape seed, palm, coconut, fish, and soy bean oil, dairy, and sugar industrial wastes, lignocellulosic waste, starch rich substrates or unconventional substrates are reported to be good resources for BSs production (Rahman *et al.*, 2002; Benincasa *et al.*, 2004; Nitschke and Pastore, 2004; Rodrigues *et al.*, 2006a; Sarubbo *et al.*, 2007; Thavasi *et al.*, 2007; Barros *et al.*, 2008; Joshi *et al.*, 2008; Portilla *et al.*, 2008). Sunflower seed oil and oleic acid were used for rhamnolipids production by the thermophilic strain *Thermus thermophilus* HB8 (Pantazaki *et al.*, 2010). In this work, sunflower seed oil was directly hydrolyzed by secretion of lipases and became a favorable carbon source for rhamnolipid production. Yields of rhamnolipids in oleic acid and sunflower seed oil were as high as 250 and 300 mg L^{-1}, respectively.

Solid state cultivation involves the growth of microorganisms on moist organic solid particles within beds in which there is a continuous gas phase between the particles (Mitchell *et al.*, 2006). Solid state cultivation on rice husks alone or added with defatted rice bran both enriched with either soy oil or diesel oil in a fixed bed column bioreactor was used to produce BSs from *Aspergillus fumigatus* and *Phialemonium* sp. (Martins *et al.*, 2006). Sugarcane bagasse impregnated with a solution containing glycerol was suggested as a valuable means to produce rhamnolipids from *Pseudomonas aeruginosa* UFPEDA 614 by Neto *et al.*, (2008, 2009). BS yields reached 40 g of rhamnolipid per kilogram of dry initial substrate after 12 days and the surface properties were maintained after autoclaving of the fermented solids. Molasses as whole medium were also used by Saimmai *et al.* (2011) to produce lipopeptide BSs by *Bacillus licheniformis* TR 7 and *Bacillus subtilis* SA9 isolated from mangrove sediment. Under optimized conditions, the yields produced by TR7 and SA9 were found to be 3.30 and 3.78 g L^{-1}, respectively.

Table 9.1 Microbial sources and major types of microbial surfactants

Class	Biosurfactants/bioemulsifiers	Microorganism(s)	Reference(s)
Low molecular weight	Glycolipids	Pseudomonas aeruginosa, Pseudomonas sp., Serratia rubidaea	Matsuyama et al. (1990); Banat et al. (2000); Benincasa et al. (2004); Mulligan (2005); Mehta et al. (2010)
	Rhamnolipids (Hydroxyalkanoyloxy)alkanoic acids (HAAs) Rhamnolipid precursor	Pseudomonas aeruginosa	Déziel et al. (2003)
	Glucose lipids	Alcanivorax borkumensis, Alcaligenes sp.	Poremba et al. (1991); Abraham et al. (1998)
	Glycolipids	Alcanivorax borkumensis, Arthrobacter sp, Corynebacterium sp., Rhodococcus erythropolis, Serratia marcescens, Tsukamurella sp.,	Singer et al. (1990a); Singer et al. (1990b); Pruthi and Cameotra (1997); Vollbrecht et al. (1998); Mulligan (2005)
	Trehalolipids	Rhodococcus erythropolis, Mycobacterium sp., Arthrobacter sp., Arthrobacter parafifineus, Corynebacterium sp., Nocardia sp., Rhodococcus ruber, Rhodococcus wratislaviensis	Uchida et al. (1989); Schulz et al. (1991); Philp et al. (2002); Tuleva et al. (2008); Franzetti et al. (2010b)
	Sophorolipids	Candida bombicola, Candida apicola, Candida lipolytica, Candida bogoriensis, Torulopsis petrophilum	Hommel et al. (1994); Brakemeier et al. (1995); Mulligan (2005)
	Mannosylerythritol lipids	Candida antarctica	Kitamoto et al. (1993)
	Cellobiolipids	Ustilago maydis	Fiechter (1992)
	Polyol lipids	Rhodotorula glutinis, Rhodotorula graminis	Yoon and Rhee (1983)
	Diglycosyl diglycerides	Rhizobium trifolii, Lactobacillus fermentii	Orgambide et al. (1992); Mulligan (2005)
	Flavolipids	Flavobacterium sp.	Bodour et al. (2004)

Lipopeptides and lipoproteins

Compound	Source organism	Reference
Surfactin/iturin/fengicin	*Bacillus subtilis, Bacillus licheniformis, Bacillus pumilus*	Thimon *et al.* (1995); Carrillo *et al.* (2003); Rivardo *et al.* (2010)
Bacillomycin L, plipastatin, surfactin	*Bacillus subtilis*	Roongsawang *et al.* (2002)
Bamylocin A	*Bacillus amyloliquefaciens*	Lee *et al.* (2007)
Lichenysin	*Bacillus licheniformis*	Lin *et al.* (1994); Yakimov *et al.* (1995)
Subtilisin	*Bacillus subtilis*	Bott *et al.* (1988)
Gramicidin S	*Bacillus brevis*	Azuma and Demain (1996)
Polymyxins	*Bacillus polymyxa*	Falagas and Kasiakou (2005)
Arthrofactin	*Arthrobacter sp.*	Morikawa *et al.* (1993)
Pelgipeptins C and D	*Paenibacillus elgii*	Ding *et al.* (2011)
Kurstakins	*Bacillus thuringiensis*	Hathout *et al.* (2000)
Serrawettin	*Serratia marcescens*	Li *et al.* (2005)
Viscosin	*Pseudomonas fluorescens*	Neu *et al.* (1990)
Syringomycins and syringopeptins	*Pseudomonas syringae pv. syringae*	Hutchison and Gross (1997)
Putisolvin I and II	*Pseudomonas putida*	Kuiper *et al.* (2004)
Pseudophomins A and B	*Pseudomonas fluorescens*	Pedras *et al.* (2003)
Lokisin	*Pseudomonas koreensis*	Hultberg *et al.* (2010a)
Ornithine lipids	*Myroides sp.*	Maneerat *et al.* (2005a)

Fatty acids, neutral lipids, phospholipids

Compound	Source organism	Reference
Fatty acids (corynomycolic acids, spiculisporic acids, etc.)	*Corynebacterium lepus, Nocardia erythropolis, Arthrobacter parafineus, Corynebacterium lepus, Penicillium spiculisporum, Talaromyces trachyspermus, Capnocytophaga sp. Corynebacterium insidibasseosum*	Makkar and Cameotra (2002); Nitschke and Costa (2007)
Neutral lipids	*Nocardia erythropolis*	Macdonald *et al.* (1981)
Phospholipids	*Thiobacillus thiooxidans, Acinetobacter sp.,*	Lemke *et al.* (1995); Mulligan (2005)
Bile acids	*Myroides sp.*	Maneerat *et al.* (2005b)

Aminoacids and peptides

Compound	Source organism	Reference
Ornithine, lysine peptides	*Gluconobacter cerinus, Thiobacillus thiooxidans, Streptomyces sioyaensis*	Mulligan (2005)
Streptofactin	*Streptomyces tendae*	Richter *et al.* (1998)

(continued overleaf)

Table 9.1 *(Continued)*

Class	Biosurfactants/bioemulsifiers	Microorganism (s)	Reference (s)
High molecular weight	Emulsan	*Acinetobacter calcoaceticus*	Rosenberg (1993)
	Alasan	*Acinetobacter radioresistens*	Navon-Venezia et al. (1995)
	Biodispersan	*Acinetobacter calcoaceticus*	Rosenberg et al. (1988)
	Mannan-lipid-protein	*Candida tropicalis*	Metha et al. (2010)
	Liposan	*Candida lipolytica*	Cirigliano and Carman (1984)
	Carbohydrate-protein-lipid	*Psudomonas fluorescens, D. polymorphis, Yarrowia lipolytica, Pseudomonas nautica*	Husain et al. (1997); Zinjarde and Pant (2002); Mehta et al. (2010)
	Lipopolysaccharides	*Acinetobacter calcoaceticus, Pseudomonas sp., Candida lipolytica*	Mulligan (2005)
	Protein PA	*Pseudomonas aeruginosa*	Metha et al. (2010)
	Food emulsifier	*Candida utilis*	Shepherd et al. (1994)
	Insecticide emulsifier	*Pseudomonas tralucida*	Anu Appaiah and Karanth (1991)
	Sulfated polysaccharide	*Halomonas eurihalina*	Martinez Checa et al. (2002)
	Acetyl heteropolysaccharide	*Sphingomonas paucimobilis*	Ashtaputre and Shah (1995)
	N-acetyl and O-pyruvil heteropolysaccharide	*Pseudomonas fluorescens*	Bonilla et al. (2005)
Particulate bio-surfactants	Vesicles and fimbriae	*Acinetobacter calcoaceticus*	Mehta et al. (2010)
	Whole microbial cells	Cyanobacteria, variety of other bacteria	Denger and Schink (1995); Gilewicz et al. (1997); Zinjarde and Pant (2002)
	Particulate surfactant (PM)	*Pseudomonas marginalis*	Burd and Ward (1996)

The yeast *Candida bombicola* was shown to produce $63.6 \, g \, L^{-1}$ of sophorolipids when grown on a cheap fermentative medium containing sugarcane molasses, yeast extract, urea and soybean oil (Daverey and Pakshirajan, 2009). The same authors reported the production of sophorolipids by the yeast *Candida bombicola* grown on medium containing de-proteinized whey, glucose, yeast extract and oleic acid with a yield of up to $34 \, g \, L^{-1}$ (Daverey and Pakshirajan, 2010).

An unconventional substrate such as potato peel was used by Das *et al.* (2007) to produce BSs from the strains *Bacillus subtilis* DM03 and DM04 in submerged and solid state fermentation. Both strains produced appreciable and equal amounts of crude lipopeptide BSs demonstrating also that the two fermentation systems were equally efficient.

Thavasi *et al.* (2011), by culturing *Lactobacillus delbrueckii* on peanut oil cake as carbon source, observed a yield of $5.35 \, mg \, ml^{-1}$ of BS at 144 h when the cell reached their early stationary phase. De-oiled cake of mahua (*Madhuca indica*) was used by Hazra *et al.* (2011) to grow *Pseudomonas aeruginosa* AB4 and produce glycolipid BSs. A maximum rhamnolipid production of $40 \, g \, L^{-1}$ was achieved at C : N ratio of 10 to 0.6 at pH 8.5 and with a temperature of 40°C.

Sugarcane juice was used by Morita *et al.* (2011) for the production of mannosylerythritol lipid B (MEL B) by *Ustilago scitaminea* NBRC 32730. Yield of MEL B on sugarcane alone was $12.7 \, g \, L^{-1}$ and after the supplementation of $1 \, g \, L^{-1}$ of urea, the yield was $25.1 \, g \, L^{-1}$ in seven days.

Patil *et al.* (2011) produced MELs by replacing conventional carbon sources with glycerol for the growth of *Pseudozyma antarctica* (ATCC 32657). The biomass and concentration of MELs ($0.77 \, g \, L^{-1}$) was maximal on glycerol at 10% (w v^{-1}) and when glycerol was used in combination with soybean oil, the yield increased up to $3.62 \, g \, L^{-1}$.

9.3.2 Medium Optimization

Medium optimization is one of the primary approaches used for obtaining yields of BSs in fermentative production. The most effective method applied for optimization of factors is the statistical approach (Sen, 1997). Experimental designs such as Plackett–Burman (Jacques *et al.*, 1999; Mukherjee *et al.*, 2008); the response surface methodology (RSM) (Mutalik *et al.*, 2008); factorial designs (Rodrigues *et al.*, 2006b); and artificial neural network modeling coupled to genetic algorithms (Pal *et al.*, 2009; Sivapathasekaran *et al.*, 2010), are statistical methods used for improved optimization of BS production.

The Plackett–Burman-based statistical screening have been used by many authors to optimize the yield of different BSs. Fermentation optimization experiments to improve iturin A and surfactin S1 production by *Bacillus subtilis* S499 were performed by Jacques *et al.* (1999). The amount of lipopeptides in the supernatant of a culture carried out in optimized medium was about five times higher than that obtained in non-optimized rich medium and chemical analysis revealed a third family of lipopeptides: the fengycins. Time dependent analysis of the production of these three families of lipopeptides showed that surfactin S1 was produced during the exponential phase while iturin A and fengycin during the stationary phase. The same statistical approach was used to optimize the growth of a marine *Bacillus* sp. and produce increased yield of BS by Mukherjee *et al.* (2008). The modified media combination showed a significant increase in the BS yield by 84.7% over the control medium and the quality and property of the BS remained unchanged.

An attempt to optimize production of sophorolipids by using *Starmerella bombicola* through a two stage optimization process was used by Parekh and Pandit (2011). In stage one, the Plackett– Burman screening experiments were applied to evaluate the significant variables that influenced the production of sophorolipids. In stage two, the Box–Behnken design of RSM was used to evaluate the effects of factors and predicting optimum conditions. It was found that pH, yeast extract, and oleic acid concentrations were the most influential variables that affected the production of sophorolipids. Under the optimized conditions, the sophorolipids production reached $18.2\,g\ L^{-1}$. The same experimental design was used to study the effect of nutrient ratios on BS production by *Serratia marcescens* (Roldán-Carrillo *et al.*, 2011). Results showed that ratios of C : N = 5, C : Fe = 26 000 and C : Mg = 30 reduced surface tension to $30\,mN\,m^{-1}$ and scale up of BS production in a 30 l bioreactor at the conditions of the best treatment gave a yield of $6.1\,g\ L^{-1}$ of pure BS.

A central composite rotatable design of RSM for media optimization to enhance BS production by *Rhodococcus* spp. MTCC 2574 was applied by Mutalik *et al.* (2008). Using the RSM, modeling of the individual and interactive effects of media components on BS yields was achieved. The medium optimization by RSM effectively enhanced the BS yield by 3.4-fold.

In order to improve BS production by *Yarrowia lypolitica* IMUFRJ, a factorial design was carried out (Cardoso Fontes *et al.*, 2010). A 2^4 full factorial design was used to investigate the effects of nitrogen sources on maximum variation of surface tension (ΔST) and emulsification index (EI). Results showed that glycerol and glucose were the best substrates to increase BS production. The experimental design optimization enhanced EI by 110.7% and ΔST by 108.1% in relation to the standard process.

Parekh *et al.* (2011) used statistical experimental methodology to optimize the fermentative production of sophorolipids by *Starmerella bombicola* NRRL Y-17069 at the shake flask scale. The Placket–Burman screening experiment was applied to evaluate the significant variables that influence the production of sophorolipids. The pH and concentrations of yeast extract and of oleic acid were found to be the most influential variables. The optimum levels of these three variables were achieved by using a Box–Behnken design of the response surface methodology. Under the optimized conditions, the sophorolipid production reached $18.2\,g\ L^{-1}$.

Optimization of the medium for BSs production by the probiotic bacteria *Lactococcus lactis* 53 and *Streptococcus thermophilus* A was carried out by Rodrigues *et al.* (2006b). The optimization of cellular growth of the probiotic bacteri was achieved using a 2^{6-2} fractional factorial central composite design and surface modeling method, after establishing that their BS were growth associated. By determining the relation between cellular growth and surface-activity of the BS in time (as a measure of its production) before and after the optimization procedure, an increase of about two times in the mass of produced cell-bound BS (milligram) per gram cell dry weight was observed.

Sivapathasekaran *et al.* (2010) optimized BS production from the marine strain *Bacillus circulans* MTCC 8281 by using artificial neural network (ANN) modeling. The critical medium components were identified by a single-factor-at-a-time strategy. The model was linked with a genetic algorithm (GA) to find the maximum production level and the optimum concentrations of the critical medium components that affect significantly the production process. The higher BS yield of $4.40\,g\ L^{-1}$, obtained in shake flasks, was significantly close to the value predicted by the ANN-GA model.

BS production by the marine actinobacterium *Brevibacterium aureum* MSA13 was optimized by Kiran *et al.* (2010b) using industrial and agro-industrial solid waste residues as substrates in solid state culture (SSc). Based on the optimization experiments, the BS production by MSA13 was increased to threefold over the original isolate under SSc conditions with pre-treated molasses as substrate and olive oil, acrylamide, FeCl3 and inoculums size as critical control factors.

9.3.3   Immobilization

Immobilization of living cells in porous support seems to offer enormous advantages in production of BS and cell protection. It is an efficient way to reduce or avoid the factors that causes low product yields. Entrapment in polyvinyl alcohol beads was used as a rapid, nontoxic, inexpensive, versatile method for BS production by Jeong *et al.* (2004). In this work, the marine bacterial strain *Pseudomonas aeruginosa* BYK-2 (KCTC 18012P) was immobilized by entrapment in 10% polyvinyl alcohol beads and optimized for the continuous production of rhamnolipids. Continuous culture, performed in a 1.8 l airlift bioreactor in fish oil (pollock liver) at a dilution rate of $0.018\,h^{-1}$, at 25°C, with initial pH 7, yielded 0.1 g rhamnolipids h^{-1}.

Enhanced production of surfactin from *Bacillus subtilis* ATCC 21332 was obtained by Yeh *et al.* (2005) by adding to fermentation broth small quantity of solid porous carriers such as activated carbon. Addition of $25\,g\,L^{-1}$ of activated carbon into fermentation broth triggered a remarkable enhancement in surfactin production from 100 mg to $3.6\,g\,L^{-1}$. Ninety percent pure surfactin product was successfully recovered from the experimental broth.

The study of Gancel *et al.* (2009) concerned surfactin and/or fengycin batch production by immobilized cells of *Bacillus subtilis* ATCC 21332. Light carriers designed for a three phase inverse fluidized bed biofilm reactor (TPFIBR) were used and a new support based on iron grafting onto polypropylene foams was proposed. A suspension solid-state grafting process to graft ferric acetylacetonate onto polypropylene (PP) foams was applied. Influence of particles on lipopeptide production was analyzed in two kinds of experiments: preliminary colonization step of particles, followed by a production step in modified culture medium (colonization step) or direct addition of pellets in culture medium (production step). All PP plus iron pellets promoted biomass enhancement. The immobilized cultures produced 2–4 times more BSs than planktonic cells. In production experiments addition of carriers seemed to modify the ratio between surfactin and fengycin with an enhancement of the fengycin production. The same strain *Bacillus subtilis* ATCC 21332 was used by Chtioui *et al.* (2010) in experiments of immobilization with polypropylene foamed with powder activated carbon (PPch) for lipopeptide production. Results showed that the immobilized strain on PPch support produced, in time course, 2–4 times more surfactin than the free cells. During the aerobic production, the immobilized cells of *B. subtilis* secreted both surfactin and fengycin molecules, and the synthesis of fengycin was greatly enhanced. The kinetics of biofilm formation and the rate of lipopeptide production were affected by the presence of lipopeptides and their concentration.

Investigation on magnetic immobilizates of *Pseudomonas aeruginosa* for application in continuous rhamnolipid production by foam fractionation and retention of entrained immobilizates by high-gradient magnetic separation from foam was carried on by Heyd *et al.* (2009). Good magnetic separation was achieved at 5% (w w^{-1}) magnetite loading.

Best homogeneous embedding, good diffusion properties, and stability enhancement was achieved with alginate beads with embedded Bayoxide® magnetite of MagPrep® silica particles.

9.4   Recovery of Biosurfactants

BS recovery depends on their ionic charge, water solubility and location (extracellular, intracellular or cell-bound). Downstream processing in many biotechnological processes are responsible for up to 60% of the total production cost. A list of BS downstream/recovery processes with relative dis-/advantages has been reported by Helmy *et al.* (2011).

The most widely used technique in BS downstream process is solvent extraction with a mixture of at least two solvents in various ratios which facilitates adjustment of the polarity of extraction agent to the target extractable material. The demand for large and expensive amounts of solvents is a disadvantage and some of the solvents are toxic too. Ethyl-acetate is frequently used, since it is inexpensive, for extraction of lipopeptide BSs from culture filtrates of *Bacillus subtilis* (Rivardo *et al.*, 2009; Canova *et al.*, 2010) *Pseudomonas putida* (Kuiper *et al.*, 2004), *Candida lipolytica* (Sarubbo *et al.*, 2007), *Candida glabrata* (de Gusmão *et al.*, 2010). Extraction with the even less toxic solvent methyl tertiary-butyl ether (MTBE) was used by Kuyukina *et al.* (2001) for recovery of BS from *Rhodococcus ruber* IEGM 231 bacterial cultures.

Membrane ultra-filtration is a fast one-step recovery process, which gives high level of purity and was used for purification of a lipopeptide BS from *Bacillus subtilis* by Lin *et al.* (1997). Sivapathasekaran *et al.* (2011) directly concentrated and purified extracellular BSs from the fermentation broth of the marine bacterium *Bacillus circulans* DMS-2 (MTCC 8281) using a single step of purification by ultrafiltration (UF) employing YM 30 kDa (UF-I) and Omega 10 kDa (UF-II) polyethersulfone membranes in a stirred cell UF system. The optimum operating pressure required for both membranes, UF-I and UF-II, were found to be 30 and 35 psi, respectively. The BS was recovered in higher amounts using UF-II (89%) and showed a CMC of 15 mg L^{-1}. The ultrafiltered products characterized using Fourier transform infrared spectroscopy and matrix-assisted laser desorption ionization time of flight mass spectral analysis, demonstrated the presence of the surfactin and fengycin lipopeptides.

Foam fractionation requires specially designed bioreactors which facilitate foam recovery during fermentation and it has been used by Chen *et al.* (2006a, 2006b) for continuous and batch surfactin production. Foam fractionation in batch mode was also used by Sarachat *et al.* (2010) to recover excreted rhamnolipid biosurfactant by free-cell culture medium of *P. aeruginosa* SP4. Operating conditions were optimized with an air flow rate of 30 ml min^{-1}, an initial foam height of 60 cm, a pore size of the sintered glass disk in the range of 160–250 μm, an initial liquid volume of 25 ml, and an operation time of 4 h and provided a BS recovery of 97%.

Biofilm production of lipopeptides coupled with an adapted separation can be an important process strategy. Chtioui *et al.* (2010) produced the lipopeptides surfactin and fengycin by immobilizing aerobic cells of *Bacillus subtilis* ATCC 1332 on polypropylene particles foamed with powder activated carbon. Foaming was avoided by air injected over the surface of cultural medium. Isolation from fermentation broth was obtained by a perspective continuous separation based on a liquid membrane technique (known also as pertraction). Pertraction was carried out in a rotating disc contactor. Even if surfactin could be successfully extracted from fermentation broth by using pertraction with *n*-heptane as liquid membrane and a phosphate buffer solution at pH 7.3 as a receiving phase at agitation of $10\,min^{-1}$, its accumulation was relatively slow.

Ultrasound separation technology (UES) was used to separate microbial cells of *Candida bombicola* ATCC 22214 from sophorolipids by Palme *et al.* (2010). The study reports the use of ultrasonically enhanced settling of *C. bombicola* in order to establish an integrated production process where cells are separated during production for simplified product recovery. Results showed that separation efficiency of about 99% *C. bombicola* could be achieved, leading to $8\,g\,L^{-1}$ of nearly cell-free sophorolipid product.

When distillery wastewater (DW) was used by Dubey *et al.*, (2005) as the nutrient medium for BS production by *Pseudomonas aeruginosa* strain BS2, the usual methods for BS recovery could not be used because recovery by any of these methods imparted color to the BS. Therefore, a new downstream technique for recovery from fermented DW comprised of adsorption-desorption processes using wood-based activated carbon (WAC) was developed. WAC was the most efficient adsorbent compared to silica gel, activated alumina and zeolite. The WAC concentration of 1% w v^{-1} was optimum to achieve 99.5% adsorption efficiency and that of 4% w v^{-1} facilitated complete removal of the BS from collapsed foam. Acetone was a specific viable eluant because it selectively facilitated maximum recovery, that is, 89%. Complete regeneration of WAC by acetone was feasible and it could be reused for BS recovery for up to three cycles.

9.5 Application Fields

BSs have an unlimited number of applications that involves every industry and every aspect of life, such as enhanced oil and metal recovery, crude oil drilling, lubricants, surfactant-aided bioremediation of water-insoluble pollutants (Kosaric, 2001; Blais *et al.*, 2010; Franzetti *et al.*, 2009, 2010a; Pacwa-Plociniczak *et al.*, 2011; Daverey and Pakshirajan, 2011), hygiene and cosmetics (Lourit and Kanlayavattanakul, 2009; Williams, 2009; Kanlayavattanakul and Lourith, 2010), textiles (Kesting *et al.*, 1996), paint (Flemming and Wingender, 2001), detergents and cleaning (Futura *et al.*, 2002; Rai and Mukherjee, 2010), food processing (Nitschke and Costa, 2007; Zezzi do Valle Gomes and Nitschke, 2012) and pharmaceutical and medical applications (Rodrigues *et al.*, 2006c; Seydlová and Svobodová, 2008; Rivardo *et al.*, 2009; Rodrigues and Teixeira, 2010; Gharaei-Fathabad, 2011; Seydlová *et al.*, 2011; Fracchia *et al.*, 2011). Shete *et al.* (2006) and Banat *et al.* (2010) reviewed interesting marketable products and patents issued in many fields of environmental application and, more recently, in the health care and cosmetic industries.

9.5.1 Environmental Applications

The main industrial application for BSs is in the field of oil recovery and processing. BSs that have been extensively studied include rhamnolipids produced by *Pseudomonas aeruginosa*, sophorolipids produced by *Candida bombicola* (formerly *Torulopsis bombicola*) from vegetable oils and sugars, and surfactin produced by *Bacillus subtilis* (Mulligan, 2005). Many properties of BSs such as emulsification/de-emulsification, dispersion, foaming, wetting and coating are useful for physicochemical and biological remediation technologies applied to decontaminate both organic and metal polluted sites (Mulligan, 2009). A comprehensive overview of advances in the applications of BSs and BS-producing microorganisms in hydrocarbon and metal remediation technologies have been recently done by Pacwa-Plociniczak *et al.* (2011).

The activity of an efficient rhamnolipid BS producer and n-hexadecane utilizing *Pseudomonas sp* isolated from oil contaminated soil was studied by Cameotra and Singh (2009). BS action brought about the dispersion of hexadecane to droplets smaller than 0.22 μm increasing the availability of the hydrocarbon to the degrading organism. BS formed an emulsion with hexadecane thereby facilitating increased contact between hydrocarbon and the degrading bacteria. Interestingly, by electron microscopy studies, it was observed that *internalization* of *BS layered hydrocarbon droplets* was taking place suggesting a process similar in appearance to active pinocytosis.

Franzetti *et al.*, (2010b) have studied and proposed the role of trehalose lipid BSs with respect to their interaction with microorganisms and hydrocarbons and suggested that during growth stages, BSs can change hydrocarbon accession modes. In a previous work (Franzetti *et al.*, 2009), it was observed that the *Gordonia* sp. strain BS29, while growing on hydrocarbons, produced cell-bound glycolipid and extracellular emulsifier and during the growth on hexadecane the surface hydrophobicity changed.

Sorption is a common treatment for removing pollutants from natural environments (Pavan *et al.* 2000). Chemical surfactants or BSs intercalated into layered double hydroxides (LDHs) are considered as potential organic pollutant sorbents. Chuang *et al.* (2010) evaluated the effects of using either calcined or uncalcined LDH and various solid/solution ratios, the concentrations of the chemical surfactant sodium dodecyl sulfate (SDS) or the BS rhamnolipid (RL) from *Pseudomonas aeruginosa*, and the reaction temperatures and times, on the synthesis of SDS-LDH and RL-LDH and compared the naphthalene removal efficiency of RL-LDH and SDS-LDH. Analysis by small angle X-ray diffraction and X-ray diffraction patterns revealed that using 2 : 1 LDH and 1000 mg L^{-1} RL, reacted at solid/solution ratio of 0.5 g L^{-1}, with a reaction time of three days and a temperature of 65°C, a RL-LDH was synthesized having high carbon contents and larger interlayer space. RH-LDH has higher potential than SDS-LDH as a sorbent for removal organic contamination. RH-LDH has the highest sorption ability for naphthalene. The authors suggested that RL-LDH could be potentially employed as a sorbent for removal organic contaminants from aqueous solutions, and particularly in optimizing aquatic environment remediation technologies.

Groboillot *et al.* (2011) observed that decontamination of dredged harbor sediments by bioremediation or electromigration processes by adding the cyclolipopeptide amphisin could enhance the bioavailability or mobility of polycyclic aromatic hydrocarbons (PAH) in an aqueous phase. They showed that the BS produced by the strain *Pseudomonas fluorescens* DSS73 seemed very promising with regard to mobilization in an aqueous phase of strongly adsorbed PAHs on the finest sediment particles. It was almost as efficient as conventional synthetic surfactants, such as SDS, for solubilizing the heaviest and most strongly adsorbed PAHs. Mobilization by amphisin needed much lower quantities than those required with the synthetic surfactant. The study also proved that the strain was able to produce the cyclolipopeptide in hostile growth conditions corresponding to real environment, that is, in estuarine water feeding a dredged material disposal site. *P. fluorescens* DSS73 was also able to produce amphisin, in the presence of a relatively high PAH contamination, and even with low oxygen growth conditions suggesting that in situ bioaugmentation of the biosurfactant producer could be conceivable.

A patent related to the strain *Pseudomonas aeruginosa* NY3 and compounds produced by the strain having BS activity has been applied by Xihou *et al.* (2011). Also provided are methods of treating environmental materials contaminated with hydrocarbons, heavy metals, or pesticides with such compounds and methods of inhibiting microbial growth.

Investigation on the effect of short-chain organic acids (SCOAs) on BS-enhanced mobilization of phenanthrene in soil-water system was performed by An *et al.* (2011). The desorption characteristics of phenanthrene by soils were assessed in the presence of rhamnolipid and four SCOAs, including acetic acid, oxalic acid, tartaric acid and citric acid. The removal of phenanthrene using rhamnolipid and SCOAs gradually increased as the SCOA concentration increased up to 300 mmol l^{-1}. The extent of phenanthrene desorption was more significant at pH 6 and 9. Desorption amounts of phenanthrene differentiated for varied combinations, in which the effect of ternary and binary SCOAs were more remarkable. The phenanthrene desorption was higher in the ditch soil compared with the under-plant soil. The authors suggest that soil property might be an important factor influencing the mobility behaviors of phenanthrene with combined use of biosurfactant and organic acids. In addition, adjusting the pH and ionic strength could control the effect of biosurfactant and organic acids, allowing higher remediation efficiency at specific values.

Comparison of surface abilities and oil recovery of BSs produced by the bacterial strains *Pseudomonas aeruginosa* WJ-1, *Bacillus subtilis* H10 and *Rhodococcus erythropolis* Z25, isolated from the formation water of a Chinese petroleum reservoir was assessed by Xia *et al.* (2011). Maximum BS production reached to about 2.66 g L^{-1} and the surface tension of liquid decreased from 71.2 to 22.56 mN m^{-1} using *P. aeruginosa*. The three strains showed a good ability to emulsify the crude oil, and the *P. aeruginosa* BS attained the greatest emulsification index (80%) for crude oil. Results showed an excellent resistance of all BSs to retain their surface-active properties at extreme conditions, a good stability above pH of 5 and capability to support the salinity up to 20 g L^{-1}. All BSs were stable up to 120°C. Oil recovery experiments in physical simulation showed 7.2–14.3% recovery of residual oil after water flooding when the BSs of three strains were added.

The efficiency of combinations of iodide (I^-) ligand and BSs produced by different bacterial species in the simultaneous removal of cadmium (Cd^{2+}) and phenanthrene in a Haplustox soil sample was investigated by Lima *et al.* (2011). Four BSs produced by the strains *Bacillus* sp. LBBMA 111A (mixed lipopeptide), *Bacillus subtilis* LBBMA 155 (lipopeptide), *Flavobacterium* sp. LBBMA 168 (mixed of flavolipids) and *Arthrobacter oxydans* LBBMA 201 (lipopeptide) and the synthetic surfactant Triton X-100 were tested with different concentrations of ligand. Soil samples contaminated with Cd^{2+} and phenanthrene underwent consecutive washings with a surfactant/ligand solution. Cd^{2+} removal increased significantly with the number of washes and the ligand concentration for most of the surfactants evaluated, particularly in solutions containing the lipopeptide

BS produced by *Bacillus subtilis* LBBMA155 and the mixture of flavolipids by *Flavobacterium* sp. LBBMA168 and Triton X-100. Maximum Cd^{2+} removal efficiency of 99.2% was observed for the lipopeptide by *Arthrobacter oxydans* LBBMA 201 and 99.2% for the mixed lipopeptide by *Bacillus* sp. LBBMA111A in the presence of 0.336 mol iodide l^{-1}. The BS solutions removed 80–88.0% of phenanthrene in soil, and the removal was not influenced by the presence of the ligand.

Batch experiments were conducted to investigate Arsenic (As) mobilization from mine tailings in the presence of a BS (JBR425, mixed rhamnolipids) and to evaluate the feasibility of using it in remediating As contaminated mine tailings/soils (Wang and Mulligan, 2009). As mobilization by the BS was greatest after 24 h when the mass ratio was 10 mg BS per gram mine tailings at pH 11, arsenic mobilization was found to be positively correlated with the mobilization of Fe and other metals (i.e., Cu, Pb and Zn), which might further enhance As mobilization by helping incorporate it into soluble complexes or micelles.

Foam fractionation is a cost-effective and simple separation process ideal for the removal of heavy metals from contaminated sites, especially when the heavy metal concentration is low (Melek *et al.* 2006). The separation of mercury ions from artificially contaminated water by the foam fractionation process using surfactin and surfactants (SDS and Tween-80) was investigated by Chen *et al.* (2011). The recovery efficiency of mercury ions was highly dependent on the surfactin concentration. The maximum recovery of mercury ions (10.4%) was obtained with a surfactin concentration 10 times its CMC. The mercury enrichment value corresponding to the highest metal recovery by surfactin was 1.53. Diluted solution (2 mg L^{-1} Hg^{2+}) resulted in better separation (36.4%). An increase in the digestion time of the metal solution with surfactin yielded better separation as compared with a freshly-prepared solution, and an increase in the airflow rate increased bubble production, resulting in higher metal recovery but low enrichment. Basic solutions yielded higher mercury separation as compared with acidic solutions due to the precipitation of surfactin under acidic conditions. Overnight incubation of the surfactin–mercury solution, increased the percentage recovery four-fold.

Nguyen and Sabatini (2011) have recently reviewed the characterization of rhamnolipid and sophorolipid BSs based on their hydrophilicity/hydrophobicity and their ability to form microemulsions with a range of oils without additives. In addition, they have discussed the use of rhamnolipid and sophorolipid in detergency and vegetable oil extraction for biodiesel application. BS mixtures were used in vegetable oil extraction for biofuel application (Nguyen *et al.*, 2010a). In this work, reverse-micellar microemulsions of diesel were used as extraction solvent for vegetable oil extraction. Diesel-based reverse-micellar microemulsions of lecithin/sophorolipid/rhamnolipid were shown to extract vegetable oil more effectively than diesel itself and even conventional hexane. The same BS mixture was studied in microemulsion formations of a range of oil types and oil hydrophobicity (Nguyen *et al.*, 2010b). The results showed the robust performance of the mixture in microemulsion formation and interfacial tension (IFT) reduction for all the oils tested. The microemulsions for limonene and isopropyl myristate were studied in fish phase behavior diagrams as potential applications for hard surface cleansers, cosmetics and pharmaceuticals. It was found that isopropyl myristate microemulsions using

BS mixtures of lecithin/sophorolipid/rhamnolipid were virtually temperature-insensitive and not significantly affected by the change in electrolyte concentration. These properties make biosurfactant formulations desirable in cosmetics and drug delivery applications.

Potential environmental applications of BEs produced by the strain *Variovorax paradoxus* 7bCT5 have been recently evaluated by Franzetti *et al.* (2012). The strain *V. paradoxus* 7bCT5 produces a mixture of high molecular weight polysaccharides. They found that the extracellular BEs were able to produce a thick stable oil/water emulsion with emulsification activity unaffected by both boiling and low temperatures. Biodegradability, toxicity and soil sorption of the emulsifiers were assessed. Respirometric tests showed moderate biodegradability occurred by soil bacterial inoculum. Furthermore, the produced compounds did not show any toxic properties through different ecotoxicological tests. The compounds showed high sorption affinity to soil particles. The soil sorption affinity likely affected the *V. paradoxus* BEs ability to remove hydrocarbon from contaminated soils; in fact, they significantly increased the removal of crude-oil from sandy soil compared to water.

9.5.2   Biomedical Applications

Several interesting features of BSs have led to a wide range of potential applications in the pharmaceutical and biomedical applications. Their antibacterial, antifungal, and antiviral activities make them relevant molecules for applications in combating many diseases and as therapeutic agents. Furthermore, BSs are generally considered safer than synthetic pharmaceuticals, due to their biological origin. Biomedical applications of BSs have been described in many reviews (Cameotra and Makkar, 2004; Rodrigues *et al.*, 2006c; Seydlová and Svobodová, 2008; Banat *et al.*, 2010; Rodrigues and Teixeira, 2010; Gharaei-Fathabad, 2011; Kalyani *et al.*, 2011; Seydlová *et al.*, 2011; Fracchia *et al.*, 2011).

9.5.2.1   Antimicrobial Activity

The antimicrobial activity of a glycolipid produced by a strain of *Rhodococcus erythropolis* isolated from contaminated sites in Riyadh area was analyzed by Abdel-Megeed *et al.* (2011). The BS exhibited high inhibitory activity against *Escherichia coli*, *Pseudomonas aeruginosa*, *Aspergillus niger* and *Aspergillus flavus*. The glycolipid effects, examined by scanning electronic microscope, showed that bacteria were totally deformed and exhibited severe destruction. Abnormal cell division was observed at high frequencies among cells that tried to divide in the presence of the glycolipids. Many cells were enlarged, elongated, empty hosts, or fragmented, consistent with the extremely low viability. The authors suggested that these glycolipids could be of considerable interest to the development of new anti-microbial materials for medical applications as water purification plants, dental surgery equipment and pharmaceutical purposes.

The antimicrobial effects of a mixture of a BS obtained by cultivating a strain of *Bacillus subtilis* on cassava wastewater and an alkaline lipase (AL) from *Fusarium oxysporum* (AL/BS mix) on several types of microorganisms, as well as their abilities to remove *Listeria innocua* ATCC 33093 biofilm from stainless steel coupons were analyzed by Pereira de Quadros *et al.* (2011). The AL/BS mix had a surface tension of around $30\,\mathrm{mN\,m^{-1}}$, indicating that the presence of alkaline lipase did not interfere in the surface activity properties of the tensoactive component. The antimicrobial activity of the AL/BS mix was determined by minimum inhibitory concentration (MIC) micro-assays. The mixture only affected the growth of *B. subtilis* CCT 2576, *B. cereus* and *L. innocua* ATCC 33090. The most sensitive strain was *B. cereus* ATCC 10876 with a MIC of 0.013 mg ml^{-1}. The mixture was also tested against *L. innocua* attached to stainless steel coupons and results showed that the presence of the AL/BS mix improved the removal of adhered cells relative to treatment done without the mix.

Ding *et al.* (2011) isolated and identified two lipopeptide antibiotics, pelgipeptins C and D from the strain *Paenibacillus elgii* B69. Results of the MIC of the two lipopep-tides against a number of Gram-positive and Gram-negative bacteria and against the pathogenic fungus *Candida albicans* showed that they were active against all the tested microorganisms. Time-kill assays demonstrated that pelgipeptin D exhibited rapid and effective bactericidal action against a methicilin resistant strain of *S. aureus*. Regarding toxicity, the intraperitoneal LD50 value showed that pelgipeptin D toxicity was slightly higher than that of the structurally related antimicrobial agent polymyxin B.

The antimicrobial activity of the strain *Bacillus subtilis* B38 against human pathogenic *Candida albicans* of different origins.was studied by Tabbene *et al.* (2011). Specific PCR primers revealed the presence of the *bam*C gene, which is involved in the biosynthesis of bacillomycin D. Three anti-Candida compounds designated a1, a2, and a3 were identified as analogues of bacillomycin D-like lipopeptides of 14, 15, and 16 carbon fatty acid long chains, respectively. The compound a3 displayed the strongest fungicidal activity against pathogenic *C. albicans* strains and it was even more active than amphotericin B against the pathogenic strain *C. albicans* sp. 311 isolated from fingernail.

9.5.2.2   Anti-Adhesive Activity

Microbial biofilms formation on medical and technical equipment is an important and mostly hazardous occurrence, especially as bacteria within such biofilms usually become highly resistant to antibiotics and adverse environmental challenges. Current biofilm preventive strategies are essentially aimed at coating medical surfaces with antimi-crobial agents, a process not always successful (von Eiff *et al.*, 2005; Francolini and Donelli, 2010). Several reports have suggested that, in addition to their direct action against pathogens, biosurfactants are able to interfere with biofilm formation, modulat-ing microbial interaction with interfaces and, consequently, their adhesion (Rasmussen and Givskov, 2006; Rodrigues *et al.*, 2007; Kiran *et al.*, 2010a).

The analysis of the effect of a protein-like BS produced by the strain *Lactobacillus acidophilus* DSM 20079 on adherence and on the expression level of the genes *gtf*B and *gtf*C in *Streptococcus mutans* biofilm cells showed that the BS was able to interfere in the adhesion and biofilm formation of the *S. mutans* to glass slide. It also could make streptococcal chains shorter. The authors also showed that several properties of *S. mutans* cells (the surface properties, biofilm formation, adhesion ability and gene expression) were changed after *L. acidophilus* BS treatment (Tahmourespour *et al.*, 2011). The anti-biofilm potential of a glycolipid surfactant produced by a tropical marine strain isolated from the hard coral *Symphyllia* sp. and identified as *Serratia marcescens* was analyzed by Dusane *et al.* (2011). The bacterium showed antimicrobial activity towards the pathogens *Candida albicans* and *Pseudomonas aeruginosa* and the marine biofouling bacterium

Bacillus pumilus and produced a glycolipid BS composed of glucose and palmitic acid that reduced the surface tension to $27\,\text{mN m}^{-1}$. The compound prevented adhesion of the strains *C. albicans* BH, *P. aeruginosa* PAO1 and *B. pumilus* TiO and also disrupted preformed biofilms of these bacterial cultures. Confocal laser scanning microscopy and electron microscopy confirmed the effective removal of biofilms from glass surfaces.

Biofilm inhibition and antimicrobial activity of a lipopeptide BS produced by a soil strain of *Bacillus cereus* resistant to the heavy metals iron, lead and zinc was described by Sriram *et al.* (2011). Critical micelle concentration of the BS was $45\,\text{mg L}^{-1}$ and occurred at the surface tension value of $36\,\text{mN m}^{-1}$. The compound was very stable and retained its surfactant activity even at extremes of pH, temperature and NaCl concentration showing higher antibacterial activity against Gram-positive bacteria than Gram-negative. In addition, the BS strongly inhibited the biofilm formation by pathogenic strains of *Pseudomonas aeruginosa* and *Staphylococcus aureus*.

Antimicrobial and anti-adhesive activities of two BSs, produced by *Candida lipolytica* UCP0988 and *Candida sphaerica* UCP0995 and respectively named Rufisan and Lunasan, against Gram-positive, Gram-negative and *Candida albicans* pathogenic bacteria were reported by Rufino *et al.* (2011) and Luna *et al.* (2011).

Rivardo *et al.* (2009) demonstrated that two lipopeptide BSs produced by the strains *B. subtilis* V9T14 and *B. licheniformis* V19T21 were able to selectively inhibit biofilm formation produced by pathogenic strains on polystyrene. In particular, biofilm formation of *S. aureus* ATCC 29213 and *E. coli* CFT073 were decreased by 97 and 90%, respectively. The V9T14 BS was active on the Gram-negative strain yet ineffective against the Gram-positive while the opposite was observed for the V19T21 BS. These effects were observed either by coating the polystyrene surface with these compounds or by adding the BSs to the inoculum. The chemical characterization of V9T14 lipopeptide BS carried out by LC/ESI-MS/MS revealed that it was composed of 77% of surfactin and 23% of fengycin (Pecci *et al.*, 2010). The lipopeptide BS produced by the strain *Bacillus subtilis* V9T14 in association with AgNO$_3$ was tested against a preformed *E. coli* biofilm (Rivardo *et al.*, 2010). Results indicated that the activity of silver could be synergistically enhanced by the presence of V9T14 BS, allowing a reduction in the quantity of silver used to achieve greater antimicrobial impact. The concentrations of silver in the silver–BS solutions were 129–258 fold less than the concentrations needed when silver was used alone. In another study, by Rivardo *et al.* (2011), the V9T14 BS in association with antibiotics led to a synergistic increase in the efficacy of antibiotics in uropathogenic *E. coli* CFT073 biofilm inhibition and, in some combinations, to the total biofilm eradication of the pathogenic strain. An international patent on this application has been published (Ceri *et al.*, 2010).

Again, anti-adhesive activity against two *C. albicans* pathogenic biofilm-producing strains (CA-2894 and DSMZ 11225) was shown by Fracchia *et al.* (2010) with a BS produced by the strain *Lactobacillus* sp. CV8LAC, isolated from fresh cabbage. The CV8LAC BS significantly inhibited the adhesion of fungal pathogens to polystyrene both in pre-coating and co-incubation experiments. In pre-coating assays, biofilm formation of strain CA-2894 was reduced by 82% while that of strain DSMZ 11225 by 81%. In co-incubation, biofilm formation of the two strains was inhibited by 70% and 81% respectively. No inhibition of both *C. albicans* planktonic cells was observed, thus indicating that the BS displayed specific anti-biofilm formation but not antimicrobial activity. Prevention of *Candida albicans* biofilm formation on resins and silicon materials by biosurfactants produced by an isolate of *Bacillus* sp. have been recently shown by Rimondini *et al.* (2012).

Zeraik and Nitschke (2010) analyzed the effective importance of temperature and hydrophobicity on adhesion reduction to polystyrene of pathogenic bacteria produced by BSs. Polystyrene surfaces were conditioned with surfactin and rhamnolipid and then the attachment of *Staphylococcus aureus*, *Listeria monocytogenes*, and *Micrococcus luteus* to these surfaces was assessed at 35°, 25°, 4°C. Results showed that surfactin was able to inhibit adhesion in all the conditions analyzed with a 63.66% of adhesion reduction at 4°C. Rhamnolipid promoted only a slight decrease in the attachment of *S. aureus*.

9.5.3   Agricultural Applications

Another important field of application for BSs is the biological control of fungal phytopathogen plant diseases. Biological control, that is, the use of natural antagonists organisms or of their secreted products to combat pests or suppress plant diseases, offers interesting alternatives to the use of chemicals Both glycolipid and lipopeptide BSs have been shown to possess antagonistic activity against fungal phytopathogens (Ongena and Jacques, 2008; Droby *et al.*, 2009; D'aes *et al.*, 2010; Raaijmakers *et al.*, 2010; Pérez-Garcia *et al.*, 2011). Plant-associated species include many plant beneficial and plant pathogenic *Pseudomonas* spp. which produce primarily cyclic lipopeptide and rhamnolipid type biosurfactants (Fogliano *et al.*, 2002; De Souza *et al.*, 2003; De Jonghe *et al.*, 2005; Debode *et al.*, 2007).

A BS indicated as lokisin produced by the strain *Pseudomonas koreensis* 2.74 was able to induce a significant reduction of the infection produced by the oomycetes *Pythium ultimum* in hydroponic tomato cultivation (Hultberg *et al.*, 2010a). In addition, suppression of potato late blight disease induced by the zoospore-producing pathogen *Phytophtora infestans* was observed when the same BS was used and a direct effect of the BS on the zoospores of *P. infestans* was observed (Hultberg *et al.*, 2010b)

Kruijt *et al.* (2009) characterized the BSs produced by the strain *Pseudomonas putida* 267 as putisolvin-like cyclic lipopeptides and demonstrated that these compounds were essential in swarming motility and biofilm formation. *P. putida* 267 suppressed *Phytophthora capsici* damping-off of cucumber. The putisolvin-like BSs exhibited zoosporicidal and antifungal activities, yet they do not contribute to biocontrol of *P. capsici* and colonization of cucumber roots by *P. putida* 267.

Three lipopeptides BS (compound 1, 2, 3) were isolated from the organic extract of *Bacillus amyloliquefaciens* strain (BO7) and were characterized as belonging to the surfactin family by Romano *et al.* (2011). The lipopeptides, particularly compound 3, displayed both *in vivo* and *in vitro* strong and dose-dependent antifungal activity against the plant pathogenic fungus *Fusarium oxysporum*.

Antifungal compounds produced by the strains *Bacillus amyloliquefaciens* $ARP_2 3$ and $MEP_2 18$ have been characterized by Alvarez *et al.* (2012). Mass spectra from RPHPLC eluted fractions showed the presence of surfactin C15, fengycins A (C16–C17) and B (C16) isoforms in supernatants from strain ARP23 cultures, whereas the major lipopeptide produced by strain MEP218 was iturin A C15. When the purified BSs were tested against *Sclerotinia sclerotiorum*, alterations in mycelial morphology and sclerotial germination were observed. In addition, foliar application of *Bacillus amyloliquefaciens* strains on soybean plants prior to *S. sclerotiorum* infection resulted in significant protection against sclerotinia stem rot compared with non-inoculated plants or plants inoculated with a non-lipopeptide producing *B. subtilis* strain.

More interestingly, by combined electrospray and imaging mass spectrometry-based approaches, Nihorimbere *et al.* (2012) have recently determined the detailed pattern of surfactins, iturins and fengycins produced *in planta* by the strain *Bacillus amyloliquefaciens* S499. Very different production rates were observed for the three cyclo-lipopeptides families. Whereas surfactin accumulated in significant amounts, much lower quantities of iturins and fengycins were detected in the environment of colonized roots in comparison with laboratory medium. In addition, the surfactin pattern produced by strain S499 evolving on roots was enriched in homologues with long fatty acid chains (C15) compared with the chains typically secreted under *in vitro* conditions. These results demonstrated that lipopeptide production by root-associated S499 cells depended on the specific nutritional context of the rhizosphere (exudates enriched in organic acids, oxygen limitation) but also on the formation of biofilm-related structures around root hairs.

9.5.4 Biotechnological and Nanotechnological Applications

In Table 9.2 the latest discoveries in the biotechnological and nanotechnological fields applicable to BSs are listed. In particular, the latest successful results of mannosylerythritol lipids (MELs) applications in the enhancement of gene trasfection efficiency of cationic liposomes as well as of glycolipids and other BSs in drug delivery and gene therapy.

The BS mediated process and microbial synthesis of nanoparticles are emerging as clean, non-toxic and environmentally acceptable "green chemistry" procedures (Sastry *et al.*, 2004; Kasture *et al.*, 2008; Kitamoto *et al.*, 2009; Kiran *et al.*, 2011). BS mediated nanomaterial synthesis and/or stabilization is a recent development in the field of nanotechnology.

Surfactin was used to stabilize silver nanoparticles by Reddy *et al.* (2009) while Kiran *et al.* (2010c) used the glycolipid brevifactin produced by the marine strain *Brevibacterium casei* MSA19. These last Authors synthesized silver nanoparticles using a borohydrate reduction at different pH and temperatures. The nanoscale silver could be synthesized in reverse micelles using the glycolipid as stabilizer and nanoparticles were uniform and stable for two months. SEM micrographs of the synthesized silver nanoparticles are shown in Figure 9.2.

Sophorolipids too have been used as nanoparticle capping agents for the production of water-dispersible cobalt and silver nanoparticles (Kasture *et al.*, 2007; Singh *et al.*, 2009). Sophorolipids were also used to covalently attach silver nanoparticles to polymer scaffolds (D'Britto *et al.*, 2011). Antibacterial activity of these silver studded scaffolds against different human bacterial pathogenic strains was already observed within 6 h of exposure. CHO-K1 cell survival studies performed on these scaffolds showed that while cells thrived on them with enhanced attachment and proliferation, bacteria struggled to grow on them.

Calcium phosphates are bioactive materials that are a major constituent of human teeth and bone tissue. Maity *et al.* (2011) used a reverse microemulsion technique with surfactin to build up nanocrystalline brushite particles of calcium phosphate. The particle

Table 9.2 *Examples of recent biosurfactant applications in biotechnological and nanotechnological fields (Fracchia et al., 2011)*

Biosurfactant type	Activity/application	Study
Mannosylerythritol lipids-A	Ligand system for immunoglobulin G and M and lectins	Konishi et al. (2007); Ito et al. (2007); Imura et al. (2007, 2008)
	DNA capsulation and membrane fusion with anionic liposomes	Ueno et al. (2007a)
	In vitro promotion of gene transfection mediated by cationic liposomes	Igarashi et al. (2006); Ueno et al. (2007b); Inoh et al. (2010)
	In vivo promotion of liposome-mediated gene transfection	Inoh et al. (2009)
	Herpes simplex virus thymidine kinase gene therapy	Maitani et al. (2006)
	Water-in-oil microemulsions	Worakitkanchanakul et al. (2008)
	Increase membrane fluidity of monolayers composed of L-α-dipalmitoylphosphatidylcholine (DPPC)	Kitamoto et al. (2009)
Mannosylerythritol lipids-B	Self-assembling and vesicle-forming activity	Worakitkanchanakul et al. (2008)
Rhamnolipids and sophorolipids	Deuterated rhamnolipids and sophorolipids	Smyth et al. (2010)
Rhamnolipids	Cadmium sulfide nanoparticles	Singh et al. (2011)
	Biocompatible microemulsions of lecithin/rhamnolipid/sophorolipid biosurfactants	Nguyen et al. (2010b)
	Silver nanoparticles with antibioticmicrobial activity	Kumar et al. (2010)
	Nickel oxide nanoparticles by microemulsion technique	Palanisamy and Raichur (2009)
	Silver nanoparticles	Xie et al. (2006)
	ZnS nanoparticles	Narayanan and Sakthivel (2010)
	Microemulsions	Xie et al. (2007)
	Alcohol-free microemulsions	Nguyen and Sabatini (2009)
Sophorolipids	Cobalt nanoparticles	Kasture et al. (2007)
	Silver nanoparticles	Kasture et al. (2008)
	Sophorolipid-coated silver and gold nanoparticles with antibacterial activity	Singh et al. (2009, 2010)
	Biocompatible microemulsions of lecithin/rhamnolipid/sophorolipid biosurfactants	Nguyen et al. (2010b)
Glycolipid biosurfactant	Silver nanoparticles	Kiran et al. (2010c)
Fengycin and surfactin	Enhancers for the skin accumulation of aciclovir	Nicoli et al. (2010)
Surfactin	Surfactin-mediated synthesis of gold nanoparticles	Reddy et al. (2009)
	Cadmium sulfide nanoparticles	Singh et al. (2011)

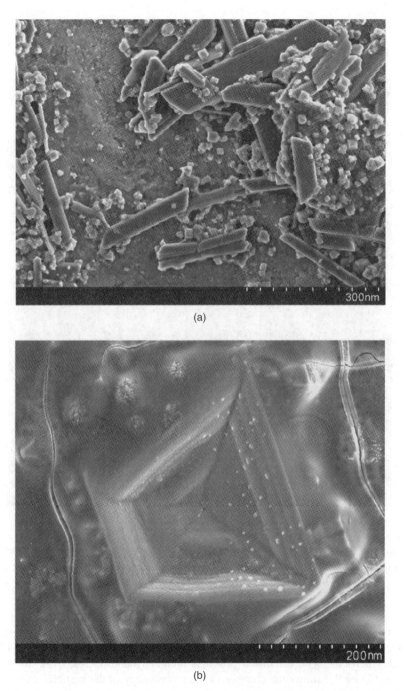

(a)

(b)

Figure 9.2 *SEM micrographs of silver nanoparticles stabilized by brevifactin (A) showing the nanoparticles (B) silver nanoparticles capped by brevifactin remain dispersed in solution for a longer time. Dimensions can be viewed with the footer-scale. Reproduced from Journal of Biotechnology, 148/4, Kiran, Sabu and Selvin, Synthesis of silver nanoparticles by glycolipid biosurfactant produced from marine Brevibacterium caseiMSA19, 221–225, August 2010, with permission from Elsevier.*

Figure 9.3 *SEM images of nanoparticles obtained by the microemulsion process with different water to surfactin ratios (W : S): (a) 250, (d) 500, (g) 1000 and (j) 40 000 without calcination; (b) 250, (e) 500, (h) 1000 and (k) 40 000 with calcination at 800°C. TEM images of nanoparticles obtained by the microemulsion process with different water to surfactin ratios (W : S): (c) 250, (f) 500, (i) 1000 and (l) 40 000 without calcination (Maity et al., 2011)*

sizes were found to be in the 16–200 nm range. Morphological variety was observed in the as-synthesized microemulsions: which consisted of nanospheres and needle-like non calcinated particles. The calcinated products included nanospheres, oval, and nanorod particles. Figure 9.3 shows the changes of the morphology of the particles with different water to surfactin ratios.

9.6   Future Prospects

The demand for new specialty surfactants in the environmental, cosmetic, agriculture, food, and pharmaceutical industries is on the rise and BSs perfectly meet this demand owing to their effectiveness and environmentally friendly characteristics (Mukherjee *et al.*, 2006; Banat *et al.*, 2010; Fracchia *et al.*, 2011).

Prospects of sophorolipid-based products include several types of facial cosmetics, lotions, beauty washes and hair products (Shete *et al.*, 2006). Concerning surfactin validation for applications in biomedical and health-related areas, further research needs to be carried out into its interaction with animal cellular membranes and its global effect on the macroorganism and natural microbiota.

The proven antimicrobial, anti-adhesives, immune modulating properties of some biosurfactants (lipopeptides, glycolipids) and the recent successful applications of mannosylerythritol lipids in gene therapy, immunotherapy, and medical insertion safety suggest it is worth persisting in this field.

Bacterial surfactants can alter the environment; in particular, plants can be affected by them in very different ways, the knowledge of which is important for insight into pathogenesis as well as into plant growth promoting and disease suppressing effects (D'aes *et al.*, 2010).

Among the programs aiming at constructing a sustainable society, the introduction of "green technology" into all areas of industry is one of the most important challenges (Kiran *et al.*, 2011).

Some problems have to be solved to make production of BS more profitable and economically feasible by (1) defining protocols to cultivate BS producing bacterial strains or hyper-producing mutants on renewable cheap substrates, (2) optimizing growth/production conditions and (3) implementing large scale production and recovery processes in order to compete economically with the chemical surfactants.

Finally, more in-depth studies of the natural role of BS in microbial competitive interactions, cell-to-cell communication, pathogenesis, motility and biofilm formation and maintenance could improve and suggest many other interesting applications.

References

A.M. Abdel-Mawgoud, F. Lépine and E. Déziel, Rhamnolipids: diversity of structures, microbial origins and roles, *Appl. Microbiol. Biotechnol.*, **86**, 1323–1336 (2010).

A.M. Abdel-Mawgoud, R. Hausmann, F. Lépine and E. Déziel, Rhamnolipids: detection, analysis, biosynthesis, genetic regulation, and bioengineering of production, *in Biosurfactants: From Genes to Applications*, G. Soberón-Chávez (ed.), Microbiology Monographs 20, Springer-Verlag, Berlin/Heidelberg, (2011).

A. Abdel-Megeed, A.N. Al-Rahma1, A.A. Mostafa and K. Husnu Can Baser, Biochemical characterization of anti-microbial activity of glycolipids produced by *Rhodococcus erythropoli*, *Pak. J. Bot.*, **43**, 1323–1334 (2011).

W-R. Abraham, H. Meyer and M. Yakimov, Novel glycine containing glucolipids from the alkane using bacterium *Alcanivorax borkumensis*, *Biochim. Biophys. Acta*, **1393**, 57–62 (1998).

F. Alvarez, M. Castro, A. Príncipe, *et al.*, The plant-associated *Bacillus amyloliquefaciens* strains $MEP_2 18$ and $ARP_2 3$ capable of producing the cyclic lipopeptides iturin or surfactin and fengycin are effective in biocontrol of sclerotinia stem rot disease, *J. Appl. Microbiol.*, **112**, 159–174 (2012).

C.J. An, G.H. Huang, J. Wei and H. Yu, Effect of short-chain organic acids on the enhanced desorption of phenanthrene by rhamnolipid biosurfactant in soil-water environment, *Water Res.*, **45**, 5501–5510 (2011).

K.A. Anu Appaiah and N.G.K. Karanth, Insecticide specific emulsifier production by hexachlorocyclohexane utilizing *Pseudomonas tralucida* Ptm strain, *Biotechnol. Lett.*, **13**, 371–374 (1991).

J. Arutchelvi and M. Doble, Mannosylerythritol lipids: microbial production and their applications, *in Biosurfactants: from Genes to Applications*, G. Soberón-Chávez (ed.), *Microbiology Monographs 20*, Springer-Verlag, Berlin/ Heidelberg (2011).

A.A. Ashtaputre and A.K. Shah, Studies on a viscous, gel-forming exopolysaccharide from *Sphingomonas paucimobilis* GS1, *Appl. Environ. Microbiol.*, **61**, 1159–1162 (1995).

T. Azuma and A.L. Demain, Interactions between gramicidin S and its producer, *Bacillus brevis*, *J. Ind. Microbiol. Biotechnol.*, **17**, 56–61 (1996).

I.M. Banat, R.S. Makkar and S.S. Cameotra, Potential commercial applications of microbial surfactants, *Appl. Microbiol. Biotechnol.*, **53**, 495–508 (2000).

I.M. Banat, A. Franzetti, I. Gandolfi, *et al.*, Microbial biosurfactants production, applications and future potential, *Appl. Microbiol. Biotechnol.*, **87**, 427–444 (2010).

F.F.C. Barros, A.N. Ponezi and G.M. Pastore, Production of biosurfactant by *Bacillus subtilis* LB5a on pilot scale using cassava wastewater as substrate, *J. Ind. Microbiol. Biotechnol.*, **35**, 1071–1078 (2008).

M. Benincasa, A. Abalos, I. Oliveira and A. Manresa, Chemical structure, surface properties and biological activities of the biosurfactant produced by *Pseudomonas aeruginosa* LBI from soapstock, *Antonie van Leeuwenhoek*, **85**, 1–8 (2004).

M. Biswas and A.M. Raichur, Electrokinetic and rheological properties of nano-zirconia in the presence of rhamnolipid biosurfactant, *J. Am. Ceram. Soc.*, **91**, 3197–3201 (2008).

J.F. Blais, N. Meunier and G. Mercier, New technologies for toxic metals removal from contaminated sites, *Recent Pat. Eng.*, **4**, 1–6 (2010).

A.A. Bodour and R.M. Maier, Biosurfactants: types, screening methods, and applications, *in Encyclopedia of Environmental Microbiology*, G. Bitton (ed.), John Wiley & Sons, Inc., Hoboken, NJ, (2002).

A.A. Bodour, C. Guerrero-Barajas, B.V. Jiorle, *et al.*, Structure and characterization of flavolipids, a novel class of biosurfactants produced by *Flavobacterium* sp. strain MTN11, *Appl. Environ. Microbiol.*, **70**, 114–120 (2004).

M. Bonilla, C. Olivaro, M. Corona, *et al.*, Production and characterization of a new bioemulsifier from *Pseudomonas putida* ML2, *J. Appl. Microbiol.*, **98**, 456–463 (2005).

R. Bott, M. Ultsch, A. Kossiakoff, *et al.*, The three-dimensional structure of *Bacillus amyloliquefaciens* subtilisin at 1.8 Å and an analysis of the structural consequences of peroxide inactivation, *J. Biol. Chem.*, **263**, 7895–7906 (1988).

A. Brakemeier, S. Lang, D. Wullbrandt, *et al.*, Novel sophorose lipid from microbial conversion of 2-alkanols, *Biotechnol. Lett.*, **17**, 1183–1188 (1995).

G. Burd and O.P. Ward, Physiochemical properties of PM factor a surface active agent produced by *Pseudomonas marginalis*, *Can. J. Microbiol.*, **42**, 243–251 (1996).

S.S. Cameotra and R.S. Makkar, Recent applications of biosurfactants as biological and immunological molecules, *Curr. Opin. Microbiol.*, **7**, 262–266 (2004).

S.S. Cameotra and P. Singh, Synthesis of rhamnolipid biosurfactant and mode of hexadecane uptake by *Pseudomonas* species, *Microb. Cell Fact.*, **8**, doi:10.1186/1475-2859-8-16 (2009).

S.P. Canova, T. Petta, L.F. Reyes *et al.*, Characterization of lipopeptides from *Paenibacillus* sp. (IIRAC30) suppressing *Rhizoctonia solani*, *World. J. Microbiol. Biotechnol.*, **26**, 2241–2247 (2010).

G. Cardoso Fontes, P.F. Amaral, M. Nele and M.A. Coelho, Factorial design to optimize biosurfactant production by *Yarrowia lipolytica*, *J. Biomed. Biotechnol.*, doi:10.1155/2010/821306 (2010).

C. Carrillo, J.A. Teruel, F.J. Aranda and A. Ortiz, Molecular mechanism of membrane permeabilization by the peptide antibiotic surfactin, *Biochim. Biophys. Acta.*, **1611**, 91–97 (2003).

H. Ceri, R. Turner, M.G. Martinotti, *et al.*, Biosurfactant composition produced by a new *Bacillus licheniformis* strain, uses and products thereof. International patent filed PCT/IB2009/055334, published WO/2010/067345 (2010).

C.Y. Chen, S.C. Baker and R.C. Darton, Continuous production of biosurfactant with foam fractionation, *J. Chem. Technol. Biotechnol.*, **81**, 1915–1922 (2006a).

C.Y. Chen, S.C. Baker and R.C. Darton, Batch production of biosurfactant with foam fractionation, *J. Chem. Technol. Biotechnol.*, **81**, 1923–1931 (2006b).

H.R. Chen, C.C. Chen, A.S. Reddy, *et al.*, Removal of mercury by foam fractionation using surfactin, a biosurfactant, *Int. J. Mol. Sci.*, **12**, 8245–8258 (2011).

O. Chtioui, K. Dimitrov, F. Gancel and I. Nikov, Biosurfactants production by immobilized cells of *Bacillus subtilis* ATCC 21332 and their recovery by pertraction, *Proc. Biochem.*, **45**, 1795–1799 (2010).

Y.H Chuang., C.H. Liu, Y.M. Tzou, J *et al.*, Comparison and characterization of chemical surfactants and bio-surfactants intercalated with layered double hydroxides (LDHs) for removing naphthalene from contaminated aqueous solutions, *Colloids Surf. A Physicochem. Eng. Aspects*, **366**, 170–177 (2010).

M.C Cirigliano and G.M. Carman, Isolation of a bioemulsifier from *Candida lipolytica*, *Appl. Environ. Microbiol.*, **48**, 747–750 (1984).

S.G.V.A.O. Costa, E. Déziel and F. Lépine, Characterization of rhamnolipid production by *Burkholderia glumae*, *Lett. Appl. Microbiol.*, **53**, 620–627 (2011).

J. D'aes, K. De Maeyer, E. Pauwelyn and M. Hôfte, Biosurfactants in plant-*Pseudomonas* interactions and their importance to biocontrol, *Environ. Microbiol*. Rep., **2**, 359–372 (2010).

V. D'Britto, H. Kapse, H. Babrekar, *et al.*, Silver nanoparticles studded porous polyethylene scaffolds: bacteria struggle to grow on them while mammalian cells thrive, *Nanoscale*, **3**, 2957–2963 (2011).

K. Das and A.K. Mukherjee, Comparison of lipopeptide biosurfactants production by *Bacillus subtilis* strains in submerged and solid state fermentation systems using a cheap carbon source: Some industrial application of biosurfactants, *Proc. Biochem.*, **42**, 1191–1199 (2007).

P. Das, S. Mukherjee, C. Sivapathasekaran and R. Sen, Microbial surfactants of marine origin: potential and prospects, *in Biosurfactants: Advances in Experimental Medicine and Biology*, R. Sen (ed.), Landes Bioscience, Austin, TX (2010).

A. Daverey and K. Pakshirajan, Production, characterization and properties of sophorolipids from the yeast *Candida bombicola* using low-cost fermentative medium, *Appl. Biochem. Biotechnol.*, **158**, 663–674 (2009).

A. Daverey and K. Pakshirajan, Sophorolipids from *Candida bombicola* using mixed hydrophilic substrates: production, purification and characterization, *Colloids Surf. B Biointerfaces*, **79**, 246–253 (2010).

A. Daverey and K. Pakshirajan, Recent advances in bioremediation of contaminated soil and water using microbial surfactants, *in Microbes and Microbial Technology: Agricultural and Environmental Applications*, I. Ahmad, F. Ahmed, and J. Pichtel (eds), Springer, New York (2011).

C.A.B. De Gusmão, R.D. Rufino and L.A. Sarubbo, Laboratory production and characterization of a new biosurfactant from *Candida glabrata* UCP1002 cultivated in vegetable fat waste applied to the removal of hydrophobic contaminant, *World J. Microbiol. Biotechnol.*, **26**, 1683–1692 (2010).

K. De Jonghe, I. De Dobbelaere, R. Sarrazyn and M. Hôfte, Control of *Phytophthora cryptogea* in the hydroponing forcing of witloof chicory with the rhamnolipid-based biosurfactant formulation PRO1, *Plant. Pathol.*, **54**, 219–226 (2005).

J.T. De Souza, M. de Boer, P. de Waard, *et al.*, Biochemical, genetic, and zoosporicidal properties of cyclic lipopeptide surfactants produced by *Pseudomonas fluorescens*, *Appl. Environ. Microbiol.*, **69**, 7161–7172 (2003).

J. Debode, K. De Maeyer, M. Perneel, *et al.*, Biosurfactants are involved in the biological control of *Verticillium* microsclerotia by *Pseudomonas* spp., *J. Appl. Microbiol.*, **103**, 1184–1196 (2007).

K. Denger and B. Schink, New halo- and thermotolerant fermenting bacteria producing surface-active compounds, *Appl. Microbiol. Biotechnol.*, **44**, 161–166 (1995).

J.D. Desai and I.M. Banat, Microbial production of surfactants and their commercial potential, *Microbiol. Mol. Biol. Rev.*, **61**, 47–64 (1997).

E. Déziel, F. Lépine, S. Milot and R. Villemur, *rhlA* is required for the production of a novel biosurfactant promoting swarming motility in *Pseudomonas aeruginosa*: 3-(3-hydroxyalkanoyloxy)alkanoic acids (HAAs), the precursors of rhamnolipids, *Microbiology*, **149**, 2005–2013 (2003).

R. Ding, X.C. Wu, C.D. Qian, *et al.*, Isolation and identification of lipopeptide antibiotics from *Paenibacillus elgii* B69 with inhibitory activity against methicillin-resistant *Staphylococcus aureus*, *J. Microbiol.*, **49**, 942–949 (2011).

S. Droby, M. Wisniewski, D. Macarisin and C. Wilson, Twenty years of postharvest biocontrol research: It is time for a new paradigm, *Postharvest Biol. Technol.*, **52**, 137–145 (2009).

D. Dubeau, E. Déziel, D. Woods and F. Lépine, *Burkholderia thailandensis* harbors two identical *rhl* gene clusters responsible for the biosynthesis of rhamnolipids, *BMC Microbiol.*, **9**, 263–275 (2009).

K.V. Dubey, A.A. Juwarkar and S.K. Singh, Adsorption-desorption process using wood-based activated carbon for recovery of biosurfactant from fermented distillery wastewater, *Biotechnol. Progress*, **21**, 860–867 (2005).

D.H. Dusane, V.S. Pawar, Y.V. Nancharaiah, *et al.*, Anti-biofilm potential of a glycolipid surfactant produced by a tropical marine strain of *Serratia marcescens*, *Biofouling*, **27**, 645–654 (2011).

M.E. Falagas and S.K. Kasiakou, Colistin: the revival of polymyxins for the management of multidrug-resistant Gram-negative bacterial infections, *Clin. Infect. Dis.*, **40**, 1333–1341 (2005).

A. Fiechter, Biosurfactants: moving towards industrial application, *Trends Biotechnol.*, **10**, 208–217 (1992).

H.C. Flemming, J. Wingender, Relevance of microbial extracellular polymeric substances (EPSs): Part II: Technical aspects, *Water Sci. Technol.*, **43**, 9–16 (2001).

V. Fogliano, A. Ballio, M. Gallo, *et al.*, *Pseudomonas* lipodepsipeptides and fungal cell wall-degrading enzymes act synergistically in biological control, *Mol. Plant Microbe Interact.*, **15**, 323–333 (2002).

L. Fracchia, M. Cavallo, G. Allegrone and M.G. Martinotti, A *Lactobacillus*-derived biosurfactant inhibits biofilm formation of human pathogenic *Candida albicans* biofilm producers, *in Current Research, Technology and Education Topics in Applied Microbiology and Microbial Biotechnology*, A. Mendez Vilas (ed.), *Formatex*, Spain (2010).

L. Fracchia, M. Cavallo, M.G. Martinotti and I.M. Banat, Biosurfactants and bioemulsifiers: biomedical and related applications-present status and future potentials, *in Biomedical Science, Engineering and Technology*, D.N. Ghista (ed.), InTech, Rijeka (2011).

I. Francolini and G. Donelli, Prevention and control of biofilm-based medical-device-related infections, *FEMS Immunol. Med. Microbiol.*, **59**, 227–238 (2010).

A. Franzetti, P. Caredda, C. Ruggeri, *et al.*, Potentials applications of surface active compounds by Gordonia sp. strain BS29 in soil-remediation technologies, *Chemosphere*, **75**, 801–807 (2009).

A. Franzetti, E. Tamburini and I.M. Banat, Application of biological surface active compounds in remediation technologies, in *Biosurfactants*, R. Sen (ed.), Advances in Experimental Medicine and Biology Series 672, Springer, Berlin (2010a).

A. Franzetti, I. Gandolfi, G. Bestetti, *et al.*, Production and application of trehalose lipid biosurfactants, *Eur. J. Lipid. Sci. Tech.*, **112**, 617–627 (2010b).

A. Franzetti, I. Gandolfi, C. Raimondi, *et al.*, Environmental fate, toxicity, characteristics and potential applications of novel bioemulsifiers produced by *Variovorax paradoxus 7bCT5*, *Biores. Technol.* doi: 10.1016/j.biortech.2012.01.005 (2012).

T. Futura, K. Igarashi and Y. Hirata, Low-foaming detergent compositions. World patent 03/002700 (2002).

F. Gancel, L. Montastruc, T. Liu, *et al.*, Lipopeptide overproduction by cell immobilization on iron-enriched light polymer particles, *Proc. Biochem.*, **44**, 975–978 (2009).

R. Gandhimathi, G.S. Kiran, T.A. Hema, *et al.*, Production and characterization of lipopeptide biosurfactant by a sponge-associated marine acinomycetes *Nocardiopsis alba* MSA 10, *Bioprocess Biosyst. Eng.*, **32**, 825–835 (2009).

E. Gharaei-Fathabad, Biosurfactants in pharmaceutical industry: a mini review, *Am. J. Drug Dis. Dev.*, **1**, 58–69 (2011).

M. Gilewicz, T. Ni'matuzahroh, H. Nadalig, *et al.*, Isolation and characterization of a marine bacterium capable of utilizing 2-methylphenanthrene, *Appl. Microbiol. Biotechnol.*, **48**, 528–533 (1997).

A. Groboillot, F. Portet-Koltalo, F. Le Derf, *et al.*, Novel application of cyclolipopeptide amphisin: feasibility study as additive to remediate polycyclic aromatic hydrocarbon (PAH) contaminated sediments, *Int. J. Mol. Sci.*, **12**, 1787–1806 (2011).

Y. Hathout, Y.P. Ho, V. Ryzhov, *et al.*, Kurstakins: a new class of lipopeptides isolated from *Bacillus thuringiensis, J. Nat Prod.*, **63**, 1492–1496 (2000).

C. Hazra, D. Kundu, P. Ghosh, *et al.*, Screening and identification of *Pseudomonas aeruginosa* AB4 for improved production, characterization and application of a glycolipid biosurfactant using low-cost agro-based raw materials, *J. Chem. Technol. Biotechnol.*, **86**, 185–198 (2011).

Q. Helmy, E. Kardena, N. Funamizu and Wisjnuprapto, Strategies toward commercial scale of biosurfactant production as potential substitute for its chemically counterparts, *Int. J. Biotechnol.*, **12**, 66–86 (2011).

M. Heyd, P. Weigold, M. Franzreb and S. Berensmeier, Influence of different magnetites on properties of magnetic *Pseudomonas aeruginosa* immobilizates used for biosurfactant production, *Biotechnol. Prog.*, **25**, 1620–1629 (2009).

R.K. Hommel, L. Weber, A. Weiss, *et al.*, Production of sophorose lipid by *Candida (Torulopsis) apicola* grown on glucose, *J. Biotechnol.*, **33**, 147–155 (1994).

M. Hultberg, T. Alsberg, S. Khalil and B. Alsanius, Suppression of disease in tomato infected by *Pythium ultimum* with a biosurfactant produced by *Pseudomonas koreensis, BioControl*, **55**, 435–444 (2010a).

M. Hultberg, T. Bengtsson and E. Liljeroth, Late blight on potato is suppressed by the biosurfactant producing strain *Pseudomonas koreensis* 2.74 and its biosurfactant, *BioControl*, **55**, 543–550 (2010b).

D.R. Husain, M. Goutx, M. Acquaviva, *et al.*, The effect of temperature on eicosane substrate uptake modes by a marine bacterium *Pseudomonas nautica* strain 617: relationship with the biochemical content of cells and supernatants, *World J. Microbiol. Biotechnol.*, **13**, 587–590 (1997).

M.L. Hutchison and D.C. Gross, Lipopeptide phytotoxins produced by *Pseudomonas syringae* pv. *syringae*: comparison of the biosurfactant and ion channel-forming activities of syringopeptin and syringomycin, *Mol. Plant Microbe Interact.*, **10**, 347–354 (1997).

S. Igarashi, Hattori Y. and Y. Maitani, Biosurfactant MEL-A enhances cellular association and gene transfection by cationic liposome, *J. Control. Release*, **112**, 362–368 (2006).

T. Imura, S. Ito, R. Azumi, *et al.*, Monolayers assembled from a glycolipid biosurfactant from *Pseudozyma (Candida) antarctica* serve as a high-affinity ligand system for immunoglobulin G and M, *Biotechnol. Lett.*, **29**, 865–870 (2007).

T. Imura, Y. Masuda, S. Ito, *et al.*, Packing density of glycolipid biosurfactant monolayers give a significant effect on their binding affinity toward immunoglobulin, G. *J. Oleo Sci.*, **57**, 415–422 (2008).

Y. Inoh, T. Furuno, N. Hirashima and M. Nakanishi, Nonviral vectors with a biosurfactant MEL-A promote gene transfection into solid tumors in the mouse abdominal cavity, *Biol. Pharm. Bull.*, **32**, 126–128 (2009).

Y. Inoh, T. Furuno, N. Hirashima, D. Kitamoto and M. Nakanishi, The ratio of unsaturated fatty acids in biosurfactants affects the efficiency of gene transfection, *Int. J. Pharm.*, **398**, 225–230 (2010).

S. Ito, T. Imura, T. Fukuoka, *et al.*, Kinetic studies on the interactions between glycolipid biosurfactant assembled monolayers and various classes of immunoglobulins using surface plasmon resonance, *Colloids Surf. B. Biointerfaces.*, **58**, 165–171 (2007).

P. Jacques, C. Hbid, J. Destain, *et al.*, Optimization of biosurfactant lipopeptide production from *Bacillus subtilis* S499 by Plackett-Burman Design, *Appl. Biochem. Biotechnol.*, **77**, 223–233 (1999).

P. Jacques, Surfactin and other lipopeptides from *Bacillus* spp., in *Biosurfactants: from Genes to Applications*, G. Soberón-Chávez (ed.), Microbiology Monographs, Vol. **20**, Springer-Verlag, Berlin/Heidelberg (2011).

T. Janek, M. Lukaszewicz, T. Rezanka and A. Krasowska, Isolation and characterization of two new lipopeptide biosurfactants produced by *Pseudomonas fluorescens* BD5 isolated from water from the Arctic Archipelago of Svalbard, *Bioresour. Technol.*, **101**, 6118–6123 (2010).

F.G. Jarvis and M.J. Johnson, A glyco-lipide produced by *Pseudomonas aeruginosa*, *J. Am. Chem. Soc.*, **71**, 4124–4126 (1949).

H.S. Jeong, D.J. Lim, S.H. Hwang, *et al.*, Rhamnolipid production by Pseudomonas aeruginosa immobilized in polyvinyl alcohol beads, *Biotechnol. Lett.*, **26**, 35–39 (2004).

S. Joshi, C. Bharucha, S. Jha, *et al.*, Biosurfactant production using molasses and whey under thermophilic conditions, *Bioresour. Technol.*, **99**, 195–199 (2008).

K. Kakugawa, M. Tamai, K. Imamura, *et al.*, Isolation of yeast *Kurtzmanomyces* sp.1-11, novel producer of mannosylerythritol lipid, *Biosci. Biotechnol. Biochem.*, **66**, 188–191 (2002).

R. Kalyani, M. Bishwambhar and V. Seneetha, Recent potential usage of biosurfactant from microbial origin in pharmaceutical and biomedical arena: a perspective, *Int. Res. J. Pharm.*, **2**, 11–15 (2011).

M. Kanlayavattanakul and N. Lourith, Lipopeptides in cosmetics, *Int. J. Cosmet. Sci.*, **32**, 1–8 (2010).

O. Käppeli and W.R. Finnerty, Partition of alkane by an extracellular vesicle derived from hexadecane-grown *Acinetobacter*, *J. Bacteriol.*, **140**, 707–712 (1979).

L. Karthik, G. Kumar and K.V. Bhaskara Rao, Comparison of methods and screening of biosurfactant producing marine actinobacteria isolated from Nicobar marine sediment, *IIOAB Journal*, **1**, 34–38 (2010).

M.B. Kasture, S. Singh, P. Patel, *et al.*, Multiutility sophorolipids as nanoparticle capping agents: synthesis of stable and water dispersible Co nanoparticles, *Langmuir*, **23**, 11409–11412 (2007).

M.B. Kasture, P. Patel, A.A. Prabhune, *et al.*, Synthesis of silver nanoparticles by sophorolipids: effect of temperature and sophorolipid structure on the size of particles, *J. Chem. Sci.*, **120**, 515–520 (2008).

W. Kesting, M. Tummuscheit, H. Schacht and E. Schollmeyer, Ecological washing of textiles with microbial surfactants, *Progr. Colloid. Polym. Sci.*, **101**, 125–130 (1996).

G.S. Kiran, B. Sabarathnam and J. Selvin, Biofilm disruption potential of a glycolipid biosurfactant from marine *Brevibacterium casei, FEMS Immunol. Med. Microbiol.*, **59**, 432–438 (2010a).

G.S. Kiran, T.A. Thomas, J. Selvin, *et al.*, Optimization and characterization of a new lipopeptide biosurfactant produced by marine *Brevibacterium aureum* MSA13 in solid state culture, *Bioresour. Technol.*, **101**, 2389–2396 (2010b).

G.S. Kiran, A. Sabu and J. Selvin, Synthesis of silver nanoparticles by glicolipid biosurfactant produced from marine *Brevibacterium casei* MSA 19, *J. Biotechnol.*, **148**, 221–225 (2010c).

G.S. Kiran., J. Selvin, A. Manilal and S. Sujith, Biosurfactants as green stabilizers for the biological synthesis of nanoparticles, *Crit. Rev. Biotechnol.*, **31**, 354–364 (2011).

D. Kitamoto, K. Haneishi, T. Nakahara and T. Tabuchi, Production of mannosylerythritol lipids by *Candida antartica* from vegetable oils, *Agric. Biol. Chem.*, **54**, 31–36 (1990).

D. Kitamoto, H. Yanagishita, T. Shinbo, *et al.*, Surface active properties and antimicrobial activities of mannosylerythritol lipids as biosurfactant produced by *Candida Antarctica, J. Biotechnol.*, **29**, 91–96 (1993).

D. Kitamoto, T. Morita, T. Fukuoka, *et al.*, Self-assembling properties of glycolipid biosurfactants and their potential applications, *Curr. Opin. Colloid Interface Sci.*, **14**, 315–328 (2009).

M. Konishi, T. Imura, T. Fukuoka, *et al.*, A yeast glycolipid biosurfactant, mannosylerythritol lipid, shows high binding affinity towards lectins on a self-assembled monolayer system, *Biotechnol. Lett.*, **29**, 473–480 (2007).

N. Kosaric, Biosurfactants and their application for soil bioremediation, *Food Technol. Biotechnol.*, **39**, 295–304 (2001).

N. Krieger, D.C. Neto and D.A. Mitchell, Production of microbial biosurfactants by solid-state cultivation, *Adv. Exp. Med. Biol.*, **672**, 203–210 (2010).

M. Kruijt, H. Tran and J.M. Raaijmakers, Functional, genetic and chemical characterization of biosurfactants produced by plant growth-promoting *Pseudomonas putida* 267, *J. Appl. Microbiol.*, **107**, 546–556 (2009).

I. Kuiper, E.L. Lagendijk, R. Pickford, *et al.*, Characterisation of two *Pseudomonas putida* lipopeptides biosurfactants, putisolvin I and II, which inhibit biofilm formation and break down existing biofilms, *Mol. Microbiol.*, **54**, 97–113 (2004).

C.G. Kumar, S.K. Mamidyala, B. Das, *et al.*, Synthesis of biosurfactant-based silver nanoparticles with purified rhamnolipids isolated from *Pseudomonas aeruginosa* BS-161R, *J. Microbiol. Biotechnol.*, **20**, 1061–1068 (2010).

M.S. Kuyukina, I.B. Ivshina, J.C. Philp, *et al.*, Recovery of *Rhodococcus* biosurfactants using methyl tertiary-butyl ether extraction, *J. Microbiol. Methods*, **46**, 149–156 (2001).

M.S. Kuyukina and I.B. Ivshina, *Rhodococcus* biosurfactants: biosynthesis, properties and potential applications biology of *Rhodococcus, in Biology of Rhodococcus*, H.M. Alvarez (ed.), Microbiology Monographs Series 16, Springer, Berlin/Heidelberg (2010).

S.-C. Lee, S.-H. Kim, I.-H. Park, S.-Y. Chung and Y.-L. Choi, Isolation and structural analysis of bamylocin A, novel lipopeptide from *Bacillus amyloliquefaciens*

LP03 having antagonistic and crude oil-emulsifying activity, *Arch. Microbiol.*, **188**, 307–312 (2007).

M.J. Lemke, P.F. Churchill and R.G. Wetzel, Effect of substrate and cell surface hydrophobicity on phosphate utilization in bacteria, *Appl. Environ. Microbiol.*, **61**, 913–919 (1995).

H. Li, T. Tanikawa, Y. Sato, *et al.*, *Serratia marcescens* gene required for surfactant *Serrawettin* W1 production encodes putative aminolipid synthetase belonging to non-ribosomal peptide synthetase family, *Microbiol. Immunol.*, **49**, 303–310 (2005).

T.M.S. Lima, L.C. Procópio, F.D. Brandão, *et al.*, Simultaneous phenanthrene and cadmium removal from contaminated soil by a ligand/biosurfactant solution, *Biodegradation*, **22**, 1007–1015 (2011).

S.C. Lin, M.A. Minton, M.M. Sharma and G. Georgiou, Structural and immunological chracterization of a biosurfactant produced by a *Bacillus licheniformis* JF-2, *Appl. Environ. Microbiol.*, **60**, 31–38 (1994).

S.C. Lin and H.J. Jiang, Recovery and purification of the lipopeptide biosurfactant of *Bacillus subtilis* by ultrafiltration, *Biotechnol. Tech.*, **11**, 413–416 (1997).

S.C. Lin, Biosurfactants: recent advances, *J. Chem. Tech. Biotechnol.*, **66**, 109–120 (1996).

N. Lourith and M. Kanlayavattanakul, Natural surfactants used in cosmetics: glycolipids, *Int. J. Cosmetic Sci.*, **31**, 255–261 (2009).

J.M. Luna, R.D. Rufino, L.A. Sarubbo, *et al.*, Evaluation antimicrobial and antiadhesive properties of the biosurfactant Lunasan produced by *Candida sphaerica* UCP 0995, *Curr. Microbiol.*, **62**, 1527–1534 (2011).

R.C. Macdonald, D.G. Cooper and J.E. Zajic, Surface-Active lipids from *Nocardia erythropolis* grown on hydrocarbons, *Appl. Environ. Microbiol.*, **41**, 117–123 (1981).

Y. Maitani, S. Yano, Y. Hattori, *et al.*, Liposome vector containing biosurfactant-complexed DNA as herpes simplex virus thymidine kinase gene delivery system, *J. Liposome Res.*, **16**, 359–372 (2006).

J.P. Maity, T.J. Lin, H.P.H. Cheng, *et al.*, Synthesis of brushite particles in reverse microemulsions of biosurfactant surfactin, *Int. J. Mol. Sci.*, **12**, 3821–3820 (2011).

R.S. Makkar and S.S. Cameotra, Biosurfactant production by microorganisms on unconventional carbon sources, *J. Surf. Det.*, **2**, 237–241 (1999).

R.S. Makkar and S.S. Cameotra, An update on the use of unconventional substrates for biosurfactant production and their new applications, *Appl. Microbiol. Biotechnol.*, **58**, 428–434 (2002).

R.S. Makkar, S.S. Cameotra and I.M. Banat, Advances in utilization of renewable substrates for biosurfactant production, *AMB Express*, **1**, 5–23 (2011).

S. Maneerat, Biosurfactants from marine microorganisms, *Songklanakarin J. Sci. Technol.*, **27**, 1263–1272 (2005).

S. Maneerat, T. Bamba, K. Harada, *et al.*, A novel crude oil emulsifier excreted in the culture supernatant of a marine bacterium, *Myroides* sp. strain SM1, *Appl. Microbiol. Biotechnol.*, **70**, 254–259 (2005a).

S. Maneerat, T. Nitoda, H. Kanzaki and F. Kawai, Bile acids are new products of a marine bacterium, Myroides sp. strain SM1, *Appl. Microbiol. Biotechnol.*, **67**, 679–683 (2005b).

F. Martinez Checa, F.L. Toledo, R. Vilchez, *et al.*, Yield production, chemical composition, and functional properties of emulsifier H28 synthesized by *Halomonas eurihalina* strain H-28 in media containing various hydrocarbons, *Appl. Microbiol. Biotechnol.*, **58**, 358–363 (2002).

V.G. Martins, S.J. Kalil, T.E. Bertolin and J.A. Costa, Solid state biosurfactant production in a fixed-bed column bioreactor, *Z. Naturforsch. C.*, **61**, 721–726 (2006).

T. Matsuyama, K. Kaneda, I. Ishizuka, *et al.*, Surface active novel glycolipid and linked 3-hydroxy fatty acids produced by *Serratia rubidaea*, *J. Bacteriol.*, **172**, 3015–3022 (1990).

T. Matsuyama, T. Tanikawa and Y. Nakagawa, Serrawettins and other surfactants produced by *Serratia, in Biosurfactants: From Genes to Applications*, G. Soberón-Chávez (ed.), Microbiology Monographs, Vol. **20**, Springer-Verlag, Berlin/Heidelberg (2011).

S.T. Mehta, S. Sharma, N. Mehta and S.S. Cameotra, Biomimetic amphiphiles: properties and potentials use, *in Biosurfactants: Advances in Experimental Medicine and Biology*, R. Sen (ed.), Landes Bioscience, Austin (2010).

D.Y. Melek, Recovery of Zn(II), Mn(II) and Cu(II) in aqueous solutions by foam fractionation with sodium dodecyl sulfate in combination with chelating agents, *Sep. Sci. Technol.*, **41**, 1741–1756 (2006).

A.D. Mitchell, N. Krieger and M. Berovic, *Solid-State Fermentation Bioreactors: Fundamentals of Design and Operation*, Springer, Hiedelberg (2006).

M. Morikawa, H. Daido, T. Takao, *et al.*, A new lipopeptide biosurfactant produced by *Arthrobacter* sp. strain MIS38, *J. Bacteriol.*, **175**, 6459–6466 (1993).

T. Morita, Y. Ishibashi, T. Fukuoka, *et al.*, Production of glycolipid biosurfactants, mannosylerythritol lipids by smut fungus *Ustilago scitaminea* NBRC 32730, *Biosci. Biotechnol. Biochem.*, **73**, 788–792 (2009).

T. Morita, Y. Ishibashi, N. Hirose, *et al.*, Production and characterization of a glycolipid biosurfactant, mannosylerythritol lipid B, from sugarcane juice by *Ustilago scitaminea* NBRC 32730, *Biosci. Biotechnol. Biochem.*, **75**, 1371–1376 (2011).

S. Mukherjee, P. Das and R. Sen, Towards commercial production of microbial surfactants, *Trends Biotechnol.*, **24**, 509–515 (2006).

S. Mukherjee, P. Das, C. Sivapathasekaran and R. Sen, Enhanced production of biosurfactant by a marine bacterium on statistical screening of nutritional parameters, *Biochem. Eng. J.*, **42**, 254–260 (2008).

C.N. Mulligan, Environmental applications for biosurfactants, *Environ. Pollut.*, **133**, 183–198 (2005).

C.N. Mulligan, Recent advances in the environmental applications of biosurfactants, *Curr. Opin. Colloid Interface Sci.*, **14**, 372–378 (2009).

S.R. Mutalik, B.K. Vaidya, R.M. Joshi, *et al.*, Use of response surface optimization for the production of biosurfactant from *Rhodococcus* spp. MTCC 2574, *Biores. Technol.*, **99**, 7875–7880 (2008).

K.B. Narayanan and N. Sakthivel, Biological synthesis of metal nanoparticles by microbes, *Adv. Colloid Interface*, **156**, 1–13 (2010).

S. Navon-Venezia, Z. Zosim, A. Gottlieb, *et al.*, Alasan, a new bioemulsifier from Acinetobacter radioresistens, *Appl. Environ. Microbiol.*, **61**, 3240–3244 (1995).

D.C. Neto, J.A. Meira, J.M. De Araújo, *et al.*, Optimization of the production of rhamnolipids by *Pseudomonas aeruginosa* UFPEDA 614 in solid-state culture, *Applied Microbiol. Biotechnol.*, **81**, 441–448 (2008).

D.C. Neto, J.A. Meira, E. Tiburtius, *et al.*, Production of rhamnolipids in solid-state cultivation: characterization, downstream processing and application in the cleaning of contaminated soils, *Biotechnol. J.*, **4**, 748–755 (2009).

T.R. Neu and K. Poralla, Emulsifying agents from bacteria isolated during screening for cells with hydrophobic surfaces, *Appl. Microbiol. Biotechnol.*, **32**, 521–525 (1990).

T.T. Nguyen and D.A. Sabatini, Formulating alcohol-free microemulsions using rhamnolipid biosurfactant and rhamnolipid mixtures, *J. Surfact. Deterg.*, **12**, 109–115 (2009).

T.T. Nguyen, L. Do and D.A. Sabatini, Biodiesel production *via* peanut oil extraction using diesel-based reverse-micellar microemulsions, *Fuel*, **89**, 2285–2291 (2010a).

T.T. Nguyen, A. Edelen, B. Neighbors and D.A. Sabatini, Biocompatible lecithin-based microemulsions with rhamnolipid and sophorolipid biosurfactants: Formulation and potential applications, *J. Colloid Interface Sci.*, **348**, 498–504 (2010b).

T.T. Nguyen and D.A. Sabatini, Characterization and emulsification properties of rhamnolipid and sophorolipid biosurfactants and their applications, *Int. J. Mol. Sci.*, **12**, 1232–1244 (2011).

S. Nicoli, M. Eeman, M. Deleu, *et al.*, Effect of lipopeptides and iontophoresis on aciclovir skin delivery, *J. Pharm. Pharmacol.*, **62**, 702–708 (2010).

V. Nihorimbere, H. Cawoy, A. Seyer, *et al.*, Impact of rhizosphere factors on cyclic lipopeptide signature from the plant beneficial strain *Bacillus amyloliquefaciens* S499, *FEMS Microbiol. Ecol.*, **79**, 176–191 (2012).

M. Nitschke and G.M. Pastore, Biosurfactant production by *Bacillus subtilis* using cassava-processing effluent, *Appl. Biochem. Biotechnol.*, **112**, 163–172 (2004).

M. Nitschke and S.G.V.A.O. Costa, Biosurfactants in food industry, *Trends Food Sci. Technol.*, **18**, 252–259 (2007).

U.A. Ochsner, T. Hembach and A. Fiecthter, Production of rhamnolipids by a *Pseudomonas alcaligenes* strain, *Adv. Biochem. Eng. Biotechnol.*, **53**, 89–118 (1996).

M. Ongena and P. Jacques, Bacillus lipopeptides: versatile weapons for plant disease control, *Trends Microbiol.*, **16**, 115–125 (2008).

G.G. Orgambide, R.I. Hollingsworth and F.B. Dazzo, Structural characterization of a novel diglycosyl diacylglyceride glycolipid from *Rhizobium trifolii* ANU843, *Carbohydr. Res.*, **233**, 151–159 (1992).

M. Pacwa-Plociniczak, G.A. Plaza, Z. Piotrowska-Seget and S.S. Cameotra, Environmental applications of biosurfactants: recent advances, *Int. J. Mol. Sci.*, **12**, 633–654 (2011).

M.P. Pal, B.K. Vaidya, K.M. Desai, *et al.*, Media optimization for biosurfactant production by *Rhodococcus erythropolis* MTCC 2794: artificial intelligence versus a statistical approach, *J. Ind. Microbiol. Biotechnol.*, **36**, 747–756 (2009).

P. Palanisamy and A.M. Raichur, Synthesis of spherical NiO nanoparticles through a novel biosurfactant mediated emulsion technique, *Mater. Sci. Eng. C*, **29**, 199–204 (2009).

O. Palme, G. Comanescu, I. Stoineva, *et al.*, Sophorolipids from *Candida bombicola*: Cell separation by ultrasonic particle manipulation, *Eu. J. Lipid Sci. Technol.*, **112**, 663–673 (2010).

A.A. Pantazaki, M.I. Dimopoulou, O.M. Simou and A.A. Pritsa, Sunflower seed oil and oleic acid utilization for the production of rhamnolipids by *Thermus thermophilus* HB8, *Appl. Microbiol. Biotechnol.*, **88**, 939–951 (2010).

V.J. Parekh and A.B. Pandit, Optimization of fermentative production of sophorolipid biosurfactant by *Starmerella bombicola* NRRL Y-17069 using response surface methodology, *Int. J. Pharm. Biol. Sci.*, **1**, 103–116 (2011).

S.V. Patil, S.D. Wadekar, S.B. Kale, *et al.*, Effect of glycerol and soybean oil as a carbon source on the production of mannosylerythritol lipids by *Pseudozyma antarctica* (ATCC 32657), *J. Lipid Sci. Technol.*, **43**, 16–20 (2011).

P.C. Pavan and E.B.V. João, Sorption of anionic surfactants on layered double hydroxides, *J. Colloid Interface Sci.*, **229**, 346–352 (2000).

Y. Pecci, F. Rivardo, M.G. Martinotti and G. Allegrone, LC/ESI-MS/MS characterisation of lipopeptide biosurfactants produced by the *Bacillus licheniformis* V9T14 strain, *J. Mass Spectrom.*, **45**, 772–778 (2010).

M.S.C. Pedras, N. Ismaila, J.W. Quail and S.M. Boyetchko, Structure, chemistry, and biological activity of pseudophomins A and B, new cyclic lipodepsipeptides isolated from the biocontrol bacterium *Pseudomonas fluorescens*, *Phytochemistry*, **62**, 1105–1114 (2003).

C. Pereira de Quadros, M.C. Teixeira Duarte and G.M. Pastore, Biological activities of a mixture of biosurfactant from *Bacillus subtilis* and alkaline lipase from *Fusarium oxysporum*, *Braz. J. Microbiol.*, **42**, 354–361 (2011).

A. Pérez-Garcia, D. Romero and A. de Vicente, Plant protection and growth stimulation by microorganisms: biotechnological applications of Bacilli in agriculture, *Curr. Opin. Biotechnol.*, **22**, 187–193 (2011).

J.C. Philp, M.S. Kuyukina, I.B. Ivshina, *et al.*, Alkanotrophic *Rhodococcus ruber* as a biosurfactant producer, *Appl. Microbiol. Biotechnol.*, **59**, 318–324 (2002).

K. Poremba, W. Gunkel, S. Lang and F. Wagner, Marine biosurfactants, III. Toxicity testing with marine microorganisms and comparison with synthetic surfactants, *Z. Naturforsch*, **46c**, 210–216 (1991).

O.M. Portilla, B. Rivas, A. Torrado, *et al.*, *J. Sci. Food Agric.* **88**, 2298–2308 (2008).

V. Pruthi and S.S. Cameotra, Production of a biosurfactant exhibiting excellent emulsification and surface active properties by *Serratia marcescens*, *World J. Microbiol. Biotechnol.*, **13**, 133–135 (1997).

J.M. Raaijmakers, I. de Bruijn, O. Nybroe and M. Ongena, Natural functions of lipopeptides from *Bacillus* and *Pseudomonas*: more than surfactants and antibiotics, *FEMS Microbiol. Rev.*, **34**, 1037–1062 (2010).

K.S.M. Rahman, T.J. Rahman, S. McClean, *et al.*, Rhamnolipid biosurfactant production by strains of *Pseudomonas aeruginosa* using low-cost raw materials, *Biotechnol. Prog.*, **18**, 1277–1281 (2002).

S.K. Rai and A.K. Mukherjee, Statistical optimization of producing, purification and industrial application of a laundry detergent and organic solvent-stable subtilisin-like serine protease (Alzwiprase) from *Bacillus subtilis* DM-04, *Biochem. Eng. J.*, **41**, 173–180 (2010).

A.M. Raichur, Dispersion of colloidal alumina using a rhamnolipid biosurfactant, *J. Dispers. Sci. Technol.*, **28**, 1272–1277 (2007).

T.B. Rasmussen and M. Givskov, Quorum-sensing inhibitors as anti-pathogenic drugs, *Int. J. Med. Microbiol.*, **296**, 149–161 (2006).

U. Rau, L.A. Nguyen, S. Schulz, *et al.*, Formation and analysis of mannosylerythritol lipid secreted by *Pseudozyma aphidis*, *Appl. Microbiol. Biotechnol.*, **66**, 551–559 (2005).

A.S. Reddy, C.Y. Chen, S.C. Baker, *et al.*, Synthesis of silver nanoparticles using surfactin: a biosurfactant as stabilizing agent, *Mat. Letters*, **63**, 1227–1230 (2009).

M. Richter, J.M. Willey, R. Süβmuth, *et al.*, Streptofactin, a novel biosurfactant with aerial mycelium inducing activity from *Streptomyces tendae* Tü 901/8c, *FEMS Microbiol. Lett.*, **163**, 165–171 (1998).

L. Rimondini, A. Cochis, M.G. Martinotti and L. Fracchia, Biosurfactants prevent in-vitro *C. albicans* biofilm formation on resins and silicon materials for prosthetic devices, *Oral Surg. Oral Med. Oral Pathol. Oral Radiol. Endod.*, **113**, 755–761 (2012).

F. Rivardo, R.J. Turner, G. Allegrone, *et al.*, Anti-adhesion activity of two biosurfactants produced by *Bacillus* spp. prevents biofilm formation of human bacterial pathogens, *Appl. Microbiol. Biotechnol.*, **8**, 541–553 (2009).

F. Rivardo, M.G. Martinotti, R.J. Turner and H. Ceri, The activity of silver against *Escherichia coli* biofilm is increased by a lipopeptide biosurfactant, *Can. J. Microbiol.*, **56**, 272–278 (2010).

F. Rivardo, M.G. Martinotti, R.J. Turner and H. Ceri, Synergistic effect of lipopeptide biosurfactant with antibiotics against *Escherichia coli* CFT073 biofilm, *Int. J. Antimicrob. Agents.*, **37**, 324–331 (2011).

L.R. Rodrigues, J.A. Teixeira and R. Oliveira, Low-cost fermentative medium for biosurfactant production by probiotic bacteria, *Biochem. Eng. J.*, **2**, 135–142 (2006a).

L.R. Rodrigues, J.A. Teixeira, R. Oliveira and H.C. van der Mei, Response surface optimization of the medium components for the production of biosurfactants by probiotic bacteria, *Proc. Biochem.*, **41**, 1–10 (2006b).

L.R. Rodrigues, I.M. Banat, J.A. Teixeira and R. Oliveria, Biosurfactants: Potential applications in medicine, *J. Antimicrob. Chemoter.*, **57**, 609–618 (2006c).

L.R. Rodrigues, I.M. Banat, J. Teixeira and R. Oliveira, Strategies for the prevention of microbial biofilm formation on silicone rubber voice prostheses, *J. Biomed. Mater. Res. Part B: Appl. Biomater.*, **81B**, 358–370 (2007).

L.R. Rodrigues and J. A. Teixeria, Biomedical and therapeutic applications of biosurfactants, *Adv. Exp. Med. Biol.*, **672**, 75–87 (2010).

T. Roldán-Carrillo, X. Martinez-García, I. Zapata-Peñasco *et al.*, Evaluation of the effect of nutrient ratios on biosurfactant production by *Serratia marcescens* using a Box-Behnken design, *Colloids Surf. B Biointerfaces*, **86**, 384–389 (2011).

A. Romano, D. Vitullo, A. Di Pietro, *et al.*, Antifungal lipopeptides from *Bacillus amyloliquefaciens* strain BO7, *J. Nat. Prod.*, **74**, 145–151 (2011).

N. Roongsawang, J. Thaniyavarn, S. Thaniyavarn, *et al.*, Isolation and characterization of a halotolerant *Bacillus subtilis* BBK-1 which produces three kinds of lipopeptides: bacillomycin L, plipastatin, and surfactin, *Extremophiles*, **6**, 499–506 (2002).

E. Rosenberg, C. Rubinovitz, R. Legmann and E.Z. Ron, Purification and chemical properties of *Acinetobacter calcoaceticus* A2 biodispersan, *Appl. Environ. Microbiol.*, **54**, 323–326 (1988).

E. Rosenberg, Exploiting microbial growth on hydrocarbons – new markets, *Trends Biotechnol.*, **11**, 419–424 (1993).

E. Rosenberg and E.Z. Ron, High- and low-molecular-mass microbial surfactants, *Appl. Microbiol. Biotechnol.*, **52**, 154–162 (1999).

R.D. Rufino, J.M. Luna, L.A. Sarubbo, *et al.*, Antimicrobial and anti-adhesive potential of a biosurfactant Rufisan produced by *Candida lipolytica* UCP 0988, *Colloids Surf. B Biointerfaces.*, **84**, 1–5 (2011).

A. Saimmai, V. Sobhon and S. Maneerat, Molasses as a whole medium for biosurfactants production by *Bacillus* strains and their application, *Appl. Biochem. Biotechnol.*, **165**, 315–335 (2011).

H.S. Saini, B.E. Barragán-Huerta, A. Lebrón-Paler, *et al.*, Efficient purification of the biosurfactant viscosin from *Pseudomonas libanensis* strain M9-3 and its physicochemical and biological properties, *J. Nat. Prod.*, **71**, 1011–1015 (2008).

T. Sarachat, O. Pornsunthorntawee, S. Chavadej and R. Rujiravant, Purification and concentration of a rhamnolipid biosurfactant produced by *Pseudomonas aeruginosa* SP4 using foam fractionation, *Bioresour. Technol.*, **101**, 324–330 (2010).

L.A. Sarubbo, C.B. Farias and G.M. Campos-Takaki, Co-utilization of canola oil and glucose on the production of a surfactant by *Candida lipolytica*, *Curr. Microbiol.*, **54**, 68–73 (2007).

M. Sastry, A. Ahmad, M.I. Khan and R. Kumar, Microbial nanoparticles production, *in Nanobiotechnology*, C.M. Niemeyer and C.A. Mirkin (eds), Wiley-VCH Verlag, Weinheim (2004).

D. Schulz, A. Passeri, M. Schmidt, *et al.*, Marine biosurfactants, Screening for biosurfactants among crude oil degrading marine microorganisms from the North Sea, *Z. Naturforsch.*, **46**, 197–203 (1991).

R. Sen, Response surface optimization of the critical media components for the production of surfactin, *J. Chem. Technol. Biotechnol.*, **68**, 263–270 (1997).

R. Sen, *Biosurfactants. Advances in Experimental Medicine and Biology*, Landes Bioscience, Austin, TX (2010).

G. Seydlová and J. Svobodová, Review of surfactin chemical properties and the potential biomedical applications, *Cent. Eur. J. Med.*, **3**,123–133 (2008).

G. Seydlová, R. Čabala and J. Svobodová. Surfactin - Novel Solutions for Global Issues, *in Biomedical Engineering, Trends, Research and Technologies*, M.A. Komorowska and S. Olsztynska-Janus (eds), InTech, Rijeka (2011).

Z. Shao, Trehalolipids, *in Biosurfactants: From Genes to Applications*, G. Soberón-Chávez (ed.) Microbiology Monographs Series 20, Springer-Verlag, Berlin/Heidelberg (2011).

R. Shepherd, J. Rockey, I.W. Sutherland and S. Roller, Effect of *C. utilis* physiology on the functionality of the bioemulsifier utilisan, in *Proceedings of the Conference on Fermentation Physiology*, Anon (ed.), Institution of Chemical Engineers Symposium Series, Institute of Chemical Engineers, Rugby (1994).

A.M. Shete, G. Wadhawa, I.M. Banat and B.A. Chopade, Mapping of patents on bioemulsifiers and biosurfactant: A review, *J. Sci. Ind. Res.*, **65**, 91–115 (2006).

M.E. Singer, W.R. Finnerty and A. Tunelid, Physical and chemical properties of a biosurfactant synthesized by *Rhodococcus* species H13A, *Can. J. Microbiol.*, **36**, 746–750 (1990).

M.E. Singer and W.R. Finnerty, Physiology of biosurfactant synthesis by *Rhodococcus* species H13-A, *Can. J. Microbiol.*, **36**, 741–745 (1990).

B.R. Singh, S. Dwivedi, A.A. Al-Khedhairy and J. Musarrat, Synthesis of stable cadmium sulfide nanoparticles using surfactin produced by *Bacillus amyloliquifaciens* strain KSU-109, *Colloids Surf. B Biointerfaces*, **85**, 207–213 (2011).

S. Singh, P. Patel, S. Jaiswal, *et al.*, A direct method for the preparation of glycolipid-metal nanoparticle conjugates: sophorolipids as reducing and capping agents for the synthesis of water re-dispersible silver nanoparticles and their inhibitory activity, *New J. Chem.*, **33**, 646–652 (2009).

S. Singh, V. D'Britto, A.A. Prabhune, *et al.*, A cytotoxic and genotoxic assessment of glycolipid-reduced and -capped gold and silver nanoparticles, *New J. Chem.*, **34**, 294–301 (2010).

C. Sivapathasekaran, S. Mukherjee, R. Sen, *et al.*, Single step concomitant concentration, purification and characterization of two families of lipopeptides of marine origin, *Bioprocess Biosyst. Eng.*, **34**, 339–346 (2010).

C. Sivapathasekaran, S. Mukherjee, A. Ray, *et al.*, Artificial neural network modeling and genetic algorithm based medium optimization for the improved production of marine biosurfactant, *Bioresour. Technol.*, **101**, 2884–2887 (2011).

T.J. Smyth, A. Perfumo, R. Marchant, *et al.*, Directed microbial biosynthesis of deuterated biosurfactants and potential future application to other bioactive molecules, *Appl. Microbiol. Biotechnol.*, **87**, 1347–1354 (2010).

G. Soberón-Chávez, *Biosurfactants: from Genes to Applications*, Springer-Verlag, Berlin/Heidelberg (2011).

M.I. Sriram, K. Kalishwaralal, V. Deepak, *et al.*, Biofilm inhibition and antimicrobial action of lipopeptide biosurfactant produced by heavy metal tolerant strain *Bacillus cereus* NK1, *Colloids Surf. B Biointerfaces*, **85**, 174–181 (2011).

C. Syldatk and R. Hausmann, Microbial biosurfactants, *Eur. J. Lipid Sci. Technol.*, **112**, 615–616 (2010).

O. Tabbene, L. Kalai, I. Ben Slimene, *et al.*, Anti-*Candida* effect of bacillomycin D-like lipopeptides from *Bacillus subtilis* B38, *FEMS Microbiol. Lett.*, **316**, 108–114 (2011).

A. Tahmourespour, R. Salehi and R.K. Kermanshahi, *Lactobacillus acidophilus*-derived biosurfactant effect on *gtf*B and *gtf*C expression level in *Streptococcus mutans* biofilm cells, *Braz. J. Microbiol.*, **42**, 330–339 (2011).

J. Thaniyavarn, T. Chianguthai, P. Sangvanich, *et al.*, Production of sophorolipid biosurfactant by *Pichia anomala*, *Biosci. Biotechnol. Biochem.*, **72**, 2061–2068 (2008).

R. Thavasi, S. Jayalakshmi, T. Balasubramanian and I.M. Banat, Biosurfactant production by *Corynebacterium kutscheri* from waste motor lubricant oil and peanut oil cake, *Lett. Appl. Microbiol.*, **45**, 686–691 (2007).

R. Thavasi, S. Jayalakshmi and I.M. Banat, Application of biosurfactant produced from peanut oil cake by *Lactobacillus delbrueckii* in biodegradation of crude oil, *Bioresour. Technol.*, **102**, 3366–3372 (2011).

L. Thimon, F. Peypoux, J. Wallach and G. Michel, Effect of the lipopeptide antibiotic, iturin A, on morphology and membrane ultrastructure of yeast cells, *FEMS Microbiol. Lett.*, **128**, 101–106 (1995).

B. Tuleva, N. Christova, R. Cohen, *et al.*, Production and structural elucidation of trehalose tetraesters (biosurfactants) from a novel alkanothrophic *Rhodococcus wratislaviensis* strain, *J. Appl. Microbiol.*, **104**, 1703–1710 (2008).

Y. Uchida, R. Tsuchiya, M. Chino, *et al.*, Extracellular accumulation of mono- and di-succinoyl trehalose lipids by a strain of *Rhodococcus erythropolis* grown on *n*-alkanes, *Agric. Biol. Chem.*, **53**, 757–763 (1989).

Y. Ueno, N. Hirashima, Y. Inoh, *et al.*, Characterization of biosurfactant-containing liposomes and their efficiency for gene transfection, *Biol. Pharm. Bull.*, **30**, 169–172 (2007a).

Y. Ueno, Y. Inoh, T. Furuno, *et al.*, NBD conjugated biosurfactant (MEL-A) shows a new pathway for transfection, *J. Control. Release.*, **123**, 247–253 (2007b).

I.N.A. Van Bogaert, K. Saerens, C. De Muynck, *et al.*, Microbial production and application of sophorolipids, *Appl. Microbiol. Biotechnol.*, **76**, 23–34 (2007).

I.N.A. Van Bogaert, S. Fleurackers, S. Van Kerrebroeck, *et al.*, Production of new-to-nature sophorolipids by cultivating the yeast *Candida bombicola* on unconventional hydrophobic substrates, *Biotechnol. Bioeng.*, **108**, 734–741 (2011).

E. Vollbrecht, R. Heckmann, V. Wray, *et al.*, Production and structure elucidation of di- and oligosaccharide lipids (biosurfactants) from *Tsukamurella* sp. nov., *Appl. Microbiol. Biotechnol.*, **50**, 530–537 (1998).

C. von Eiff, W. Kohnen, K. Becker and B. Jansen, Modern strategies in the prevention of implant-associated infections, *Int. J. Artif. Organs*, **28**, 1146–1156 (2005).

S. Wang and C.N. Mulligan, Arsenic mobilization from mine tailings in the presence of a biosurfactant, *Appl. Geochem.*, **24**, 928–935 (2009).

K. Williams, Biosurfactants for cosmetic application: Overcoming production challenges, *MMG 445 Basic Biotechnol.*, **5**, 78–83 (2009).

W. Worakitkanchanakul, T. Imura, T. Fukuoka, *et al.*, Aqueous-phase behavior and vesicle formation of natural glycolipid biosurfactant, mannosylerythritol lipid-B, *Colloids Surf. B Biointerfaces*, **65**, 106–112 (2008).

W.J. Xia, H.P. Dong, L. Yu and D.F. Yu, Comparative study of biosurfactant produced by microorganisms isolated from formation water of petroleum reservoir, *Colloids Surf. A Physicochem. Eng. Aspects*, **392**, 124–130 (2011).

Y.W. Xie, R.Q. Ye and H.L. Liu, Synthesis of silver nanoparticles in reverse micelles stabilized by natural biosurfactant. *Colloids Surf. A Physicochem. Eng. Aspects*, **279**, 175–178 (2006).

Y.W. Xie, R.Q. Ye and H.L. Liu Microstructure studies on biosurfactant-rhamnolipid/nbutanol/water/n-heptane microemulsion system, *Colloids Surf. A Physicochem. Eng. Aspects*, **292**, 189–195 (2007).

Y. Xihou, N. Maiqian and S. Qirong, Rhamnolipid biosurfactant from *Pseudomonas aeruginosa* strain NY3 and methods of use, U. S. Patent 20110306569, (2011).

M.M. Yakimov, K.N. Timmis, V. Wray and H.L. Fredrickson, Characterization of a new lipopeptide surfactant produced by thermotolerant and halotolerant subsurface *Bacillus licheniformis* BAS 50, *Appl. Environ. Microbiol.*, **61**, 1706–1713 (1995).

M-S. Yeh, Y-H. Wei and J-S. Chang, Enhanced production of surfactin from *Bacillus subtilis* by addition of solid carriers, *Biotechnol. Prog.*, **21**, 1329–1334 (2005).

S.H. Yoon and J.S. Rhee, Lipid from yeast fermentation: effects of cultural conditions on lipid production and its characteristics of *Rhodotorula glutinis*, *J. Am. Oil Chem. Soc.*, **60**, 1281–1286 (1983).

A.E. Zeraik and M. Nitschke, Biosurfactants as agents to reduce adhesion of pathogenic bacteria to polystyrene surfaces: effect of temperature and hydrophobicity, *Curr. Microbiol.*, **61**, 554–559 (2010).

M. Zezzi do Valle Gomes and M. Nitschke, Evaluation of rhamnolipid and surfactin to reduce the adhesion and remove biofilms of individual and mixed cultures of food pathogenic bacteria, *Food Control*, **25**, 441–447 (2012).

S.S. Zinjarde and A. Pant, Emulsifier from a tropical marine yeast, *Yarrowia lipolytica* NCIM 3589, *J. Basic Microbiol.*, **42**, 67–73 (2002).

10

Bioremediation of Water: A Sustainable Approach

Sudip Chakraborty[1,2], Jaya Sikder[3], Debolina Mukherjee[4], Mrinal Kanti Mandal[3], and D. Lawrence Arockiasamy[5]

[1]*Department of Chemical Engineering, Jadavpur University, West Bengal, India*
[2]*Department of Chemical Engineering and Materials, University of Calabria, Rende, Italy*
[3]*Department of Chemical Engineering, National Institute of Technology Durgapur, West Bengal, India*
[4]*Department of Geology, University of Calabria, Rende, Italy*
[5]*King Abdullah Institute for Nanotechnology, King Saud University, Riyadh, Saudi Arabia*

10.1 Introduction

Bioremediation is a combination of two words, *bio*, which means 'living', and *remediate* which means 'to solve a problem', or to bring things into their original state. It therefore refers to the use of microorganisms to degrade contaminants that pose environmental problems. Microorganisms have come into the limelight for safety and convenience of the environment and mankind. They degrade those contaminants obtained from polluted soil and water that pose threats to human race and its environment. This ability of microbes makes bioremediation a technology applicable to different areas. The process encompasses different microbes acting in parallel or sequentially to complete the degradation process (Thakur, 2006). *Biostimulation* involves stimulating a group of organisms in order to shift the microbial ecology towards the desired process. It can be achieved through changes in pH, moisture, aeration, or nutrient additions. *Bioaugmentation* is

Sustainable Development in Chemical Engineering – Innovative Technologies, First Edition.
Edited by Vincenzo Piemonte, Marcello De Falco and Angelo Basile.
© 2013 John Wiley & Sons, Ltd. Published 2013 by John Wiley & Sons, Ltd.

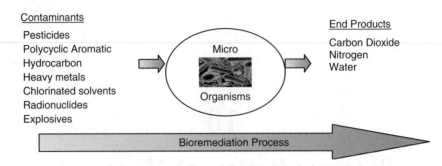

Figure 10.1   Bioremediation process

another approach where organisms selected for high degradation abilities are used to inoculate the contaminated site (Sylvia *et al.*, 2005). Though these two approaches are not mutually exclusive, they can be used simultaneously.

The goal of bioremediation is lowering toxicity levels and ensuring that anthropogenic waste gets accumulated more quickly into the natural biological cycle. Microbes can be degraded into different ways, generally classified under aerobic and anaerobic metabolism. Aerobic respiration is a process that involves microorganisms using oxygen to oxidize a carbon source that may exist in a contaminant. Many microorganisms use this mode of respiration to destroy organic contaminants (Sylvia *et al.*, 2005). Under these conditions, a drop in oxygen concentration occurs when microbes are active, which happens to be instrumental in the reproduction of some living organisms. Bacteria that are able to degrade alkanes and polyaromatic compounds in the presence of oxygen are *Pseudomonas*, *Alcaligenes*, *Sphingomonas*, *Rhodococcus* and *Mycobacterium*. They have been applied to bioremediation of pesticides, nitrates, phosphorous (Fig. 10.1). This removal is an important aerobic degradation carried out by bacteria. Anaerobic treatment can take place via methanogenesis, where organic matter is reduced to methane. It has disadvantages like the low growth of microorganisms (they have usually doubling times of several days), odor production and higher buffer requirements for pH control. Municipal wastewater sludge and solid waste stabilization are both carried out by this process. Another important aspect of anaerobic degradation is the low nutrient requirement.

Advantages of using bioremediation processes compared with other remediation technologies;

1. Instead of merely transferring contaminants from one environmental medium to another medium: bioremediation detoxifies hazardous substances.
2. Excavation-based processes are much more disruptive to the environment than biologically-based remediation; and
3. The cost of treating can be considerably lower than that for conventional treatment methods such as: vacuuming, absorbing, burning, dispersing, or moving the material.

10.2   State-of-the-Art: Recent Development

Bioremediation approaches are generally classified as in situ or ex situ processes. In situ bioremediation involves treating the polluted material at the site while ex situ involves the removal of the polluted material to be treated elsewhere. In situ bioremediation can be described as the process whereby organic pollutants are biologically degraded under natural conditions to either carbon dioxide and water or an attenuated transformation product. It is a low-cost, low maintenance, environmentally friendly and sustainable approach for the cleanup of polluted sites. With the need for excavation of the contaminated samples for treatment, the cost of ex situ bioremediation approaches can be high, relative to in situ methods. In addition, both the rate of biodegradation and the consistency of the process outcome differ between the in situ and ex situ methods (Thakur, 2006). Both remediation methods depend essentially on microbial metabolism, but in situ processes are preferred to those of ex situ for ecological restoration of contaminated soil and water environments (Sylvia *et al.*, 2005). Several microbiological methods are currently employed for the general soil quality assessment. The monitoring and efficacy testing for bioremediation require a careful selection from them, besides using specific information on abundance of microbial members or genes, and microbial processes and activities since factors such as water content, temperature and many others determine the course of attenuation (Table 10.1).

Persistent organic pollutants and emerging contaminants have been described in several reviews (Eljarrat, 2003; Purnomo, 2010) based on their relative toxic potential in environment. DDT [1,1,1-trichloro-2,2-bis (4-chlorophenyl) ethane] is a persistent pesticide that has been widely used for controlling mosquito-borne malaria and is still used for that purpose in some tropical countries (Turusov *et al.*, 2002; Purnomo *et al.*, 2008). The ability of brown-rot fungi to degrade DDT has been studied in Japan (Purnomo *et al.*, 2010). The study also included brown-rot fungi such as *Gloeophyllum trabeum*, *Fomitopsis pinicola* and *Daedalea dickinsii* showing these have a good ability to degrade

Table 10.1   *Factors affecting bioremediation of organic pollutants in soil & aquatic systems (Adapted from Michels et al., 2000)*

Factors	Effects
Auxiliary (co-)substrates	Enable co-metabolic transformation of contaminants
Co-contaminants	Influence bioavailability and enhance/inhibit biodegradation
Microbial communities	Pollutant-tolerant members within communities determine the rate of degradation
Nutrients	Growth and reproduction of microorganisms
Organic matter	Influence degradation and sorption/entrapment
pH	Microorganisms and enzymes exhibit pH-dependent activity maxima
Redox potential	Concentrations and ratios of electron donors/acceptors determine pathways and efficiency of degradation
Solubility, volatility, particle size, sorption, occlusion	Bioavailability is determined
Temperature	Composition of communities (all the interacting organisms living together in a specific habitat) and velocity of degradation; pollutants persist longer at lower temperature
Water content	Transport of pollutants and the degraded products; degradation of pollutants

Table 10.2 *Some commonly known soil contamination and treatment methods*

Method	Pollutants	Reference
Biosparging	Organic	Doelman 1999, EPA 2005a
Bioventing	Petroleum hydrocarbon, nonchlorinated solvents	FRTR 2005, EPA 2005a
Chemical oxidation	Toxic organic chemicals	FRTR 2005, EPA 2005a
Chemical reduction	Chlorinated solvents and metals	EPA 2005a
Electrokinetic separation	Metals, polar organics and heavy metals	FRTR 2005, EPA 2005a, EPA 2005b
Enhanced bioremediation/ biorestauration	Petroleum hydrocarbons, solvents, pesticides	Doelman, 1999; EPA, 2005a; FRTR 2005
Fracturing	Other in situ methods	FRTR, 2005; EPA 2005a
Pytoremediation	Organic and inorganic contaminants	FRTR, 2005; Adams et al., 2000
Soil Flushing	Volatile organic contaminants and semivolatiles	FRTR, 2005; EPA, 2005a
Soil vapor extraction/dual thermal treatment	Volatile organic contaminants, some semivolatile, light nonvolatiles as well as pesticides and fuels	FRTR, 2005
Solidification	Organic and inorganic contaminants	FRTR, 2005
Stabilization	Organic and inorganic contaminants	FRTR, 2005; EPA, 2005a

DDT, with the Fenton reaction (Beard, 2006). Heavy metals have also been widely used in various industries such as Pb, Ni, Cd, Cr, Hg and As. As certain heavy metals such as lead, cadmium, and mercury have been recognized to be potentially toxic within specific limiting values, a considerable potential hazard exists for human nutrition through the food chain. A number of aerobic and anaerobic microorganisms are capable of reducing such metals through bioremediation pathways (Turusov *et al.*, 2002). In the presence of oxygen, microbial reduction of heavy metals is commonly catalyzed by soluble enzymes, except in *Pseudomonas maltophilia* and *Bacillus megaterium*, which utilize membrane-associated reductases. Some of the commonly known soil contamination and treatment methods are described in Table 10.2.

Organic pollutants in sediments from different sources are a worldwide problem because sediments act as sinks for hydrophobic, recalcitrant, and hazardous compounds. Depending on biogeochemical processes these hydrocarbons are involved in adsorption, desorption and transformation processes and can be made available to benthic organisms as well as organisms in the water column through the sediment–water interface. Most of these recalcitrant hydrocarbons are toxic and carcinogenic in nature; they may enter the food chain and accumulate in biological tissues of human beings. Several approaches are being investigated and have been already used to remove organic hydrocarbons as well as other pollutants from sediments in many different countries: these are summarized in Table 10.3. Bioremediation is playing the vital role in all those cases. Some new approaches with emphasis on bioremediation, like biostimulation, bioaugmentation and phytoremediation are applied to sediments nowadays and impressive results have been found (Rittman 1985). These new techniques promise to

Table 10.3   *Some commonly known worldwide sites using bioremediation technology*

Technology	Site	Pollutants
Anaerobic oxidation	The Netherlands	Aromatic compounds (BTEX), nitrate, benzene, mineral oil
Ex situ bioremediation of soil and sludge	Southeastern Wood Preserving Superfund site, Canton, Mississippi	PAHs (slurry phase bioremediation)
In situ bioremediation	Germany	Polyaromatic hydrocarbon (PAH)
In situ bioremediation of composting	Ciba-Geigy, Dover Township, New Jersey	VOCs (volatile organic compounds)
In situ bioremediation of soil	Dover Air Force Base, Building 719 site, Delaware	TCE (Trichloroethylene), TCA (1,1,1-trichloroethane) *cis*-DCE (*cis*-dichloroethylene) (In situ cometabolic bioventing)
In situ bioremediation of water	Avco Lycoming Superfund site, Williamsport, Pennsylvania	TCE, *cis*-DCE vinyl chloride (VC), hexavalent chromium, cadmium
In situ bioremediation of water	France	Cesium and Strontium in the soil
In situ chemical oxidation (ISCO), In situ chemical Reduction (ISCR)	Different petrochemical facility in Northern Italy	1,2 dichloroethane (1,2-DCA),vinyl chloride (VC)
In situ reductive dechlorination	The Netherlands	*cis*-DCE, perchloroethylene (PCE) and trichloroethylene (TCE), HCH isomers, β-HCH
In situ source treatment	US DOE, Savannah River, River Site, SC	TCE, PCE (polychloroethylene) (nitrogen, phosphorus, methane addition)
In situ surface soil contamination	Czech Republic	Petroleum hydrocarbons
In situ bioreclamation	Asten, The Netherlands	Oily wastes and PAH in sediments
Land treatment	Burlington, Northern Superfund site, MN	PAHs, SVOCs (semi-VOCs) (treatment with lime, cow manure)
Land treatment	Scott Lumber Company, Alton, MO	PAHs, Benzo(a)pyrene
Slurry-phase bioremediation	French Limited Superfund site, Crosby, Texas	VOCs, SVOCs, PCBs (Polychlorinated biphenyls) PCP

be of lower impact and more cost efficient than traditional management strategies. Oil spills have also become a serious problem in cold environments with the ever-increasing resource exploitation, transportation, storage, and accidental leakage of oil. Several techniques, including physical, chemical, and biological methods, are used to recover spilled oil from the environment which can be seen in Table 10.3.

Bioremediation is a promising option for remediation since it is effective and economic in removing oil with less environmental damage. However, it is a relatively slow process in cold regions and the degree of success depends on a number of factors, including the properties and fate of oil spilled in cold environments, and the major microbial and environmental limitations of bioremediation. However, the challenges and constraints for bioremediation in cold environments remain large. Due to different types of bioremediation technologies as well as different parameters to study, the bioremediation technology requires the use of some experimental controls and performance indicators (Table 10.4) for both process optimization and implementation of regulatory decisions.

Each year agricultural effluents, industrial residues, and industrial accidents contaminate surface waters, soils, air, streams, and reservoirs. A new compost technology, known as

Table 10.4   *Experimental controls and performance indicators for bioremediation technologies. (Adapted from NRC, 1997)*

Experimental control
- Baseline data collection
- Conservative tracers
- Controlled contaminant injection
- Partitioning tracers
- Sequential start-and-stop testing
- Side-by-side and sequential application of technologies
- Systematic variation of technology's control parameters
- Untreated controls

Performance indicator
- Bacterial adaptation to contaminant in treatment zone
- Changes in carbon isotope ratios (or, in stable isotopes consistent with the biological process)
- Decreased electron acceptor concentration
- Decreased ratio of reactant to inert tracer (or decreased ration of transformable to non-transformable substances)
- Increased concentrations of intermediate-stage and final products
- Increased inorganic carbon concentration
- Increased number of bacteria in treatment zone
- Increased number of protozoa in treatment zone
- Increased rates of bacterial activity in treatment zone
- Increased ratio of transformation product to non-transformable substances
- Increased ratio of transformation product to reactant
- Indicators of liquid/gas flow filed consistent with technology, indicating that treatment fluids have been successfully delivered to the contaminant area
- Relative rates of transformation for different contaminants consistent with laboratory data
- Stoichiometry and mass balance between reactants and products

compost bioremediation, is currently being used to restore contaminated soils, manage storm water, control odors, and degrade volatile organic compounds (VOCs). Compost bioremediation refers to the use of a biological system of microorganisms in a mature, cured compost to sequester or break down contaminants in water or soil. Microorganisms consume contaminants in soils, ground and surface waters, and air. The contaminants are digested, metabolized, and transformed into humus and inert byproducts, such as carbon dioxide, water, and salts. Compost bioremediation has proven effective in degrading or altering many types of contaminants, such as chlorinated and nonchlorinated hydrocarbons, wood-preserving chemicals, solvents, heavy metals, pesticides, petroleum products, and explosives. Compost used in bioremediation is referred to as "tailored" or "designed" compost in that it is specially made to treat specific contaminants at specific sites. The ultimate goal in any remediation project is to return the site to its pre-contamination condition, which often includes re-vegetation to stabilize the treated soil. In addition to reducing contaminant levels, compost advances this goal by facilitating plant growth. In this role, compost provides soil conditioning and also provides nutrients to a wide variety of vegetation.

10.3 Water Management

At present, man-made activities have seriously disturbed the delicate balance existing between humans and their environment. Deep-seated technologies have emerged which are enhancing the rate of pollution in the environment leading to the crisis of ecological access to pure, clean, and potable water.

Wastewater treatment or sewage reclamation is exercised, aiming to co-operate the aerobic bacteria and microalgae (bio-based) with each other to bring about complete biodegradation of the organic matter present in it, since the bio-based way of sewage reclamation is eco-friendly, cost effective and resource generating (Markandey and Markandey, 2002).

Drastic increase in population and economic growth has resulted in a huge increase in water demand. Changes in the climatic conditions, rainfall, flood, and contamination has led to create the problem of water shortage. Remediation of wastewater from various sources is an immediate need of the day. Water consumption in 1680 was estimated at $86 \, km^3$, in 1900, this amount increased to $522 \, km^3$, but from 1980–2120 it will reach to $2700 \, km^3$ per year. It is a concerning factor that the human water consumption may triple in the next 30 years (Koltuniewicz and Drioli, 2008).

As water is one of the most important natural resources in our environment, the problem of its consumption must be solved globally. One of the most important solutions to this problem lies in water management and its reuse. Waste water which comes out of various industries, municipalities, communities (as sewage) can be treated and reused for various purposes. There are various methods available involving the treatment of the waste water; namely sewage treatment plants (STP) and effluent treatment plants (ETP), but these are very expensive methods that rely on high-cost chemicals and high input of energy. Conventional methods such as coagulation and flocculation, solid/liquid separation, precipitation, aeration, adsorption, evaporation, oxidation-reduction and so on, are being widely used. Certain limitations such as excessive chemical consumption, sludge production, loss of expensive chemicals to wastewater, strong pH sensitivity, and treatment cost of chemicals have encouraged the search for better technologies.

Emerging technologies such as membrane technology, involving ultrafiltration, micro-filtration, nanofiltration, reverse osmosis and so on, have proven beneficial for the treatment of wastewater and its further use. Though this is promising technology, there lay certain limitations, such as high pressure requirement, low water permeability, low rejection potential, severe fouling problems, high pressure consumptions and an overall higher cost. With emphasis on sustainable wastewater treatment, we are keen on pursuing a method that can be cost effective and can provide a sustainable, long-term solution for treatment of wastewater and sewage.

Thus, it is better to produce a method which can overcome all of these problems. The solution to all of this can be obtained considering a biological method. Municipalities, communities and industries the world over who are producing wastewater are keenly exploring bioremediation as an important way by which they can clean it up. Such an avenue provides an economical and environmentally sustainable treatment method.

Wastewater treatment systems rely on microbes to perform the function of the break-down of sewage influent (Metcalf and Eddy, 1991). These microbes live in the sludge of treatment plants and holding tanks. They digest the solids and breakdown various com-pounds. Some wastewater treatment systems are exceptionally efficient, others perform effectively soon after start up, but later stop performing as well. The key to under-standing how and why some wastewater treatment systems work well and others don't is the need to understand what these microbes need to function. As microbes are liv-ing organisms, they require certain nutrients and environments to survive, multiply and perform. In any wastewater treatment system there is a vast array of microbes present, that is; aerobic, anaerobic, and facultative, each performing specific functions in their respective parts of the system. Each species has a tolerance of ecological minima and maxima with regard to various conditions; pH, temperature, dissolved oxygen levels, and nutrient levels. All microbes require optimal conditions in order to proliferate and infuse the system in sufficient numbers to maximize the efficiency of the wastewater treatment plant (Radjenovi, 2008).

People must have necessary knowledge of (1) wastewater constituents, (2) impact on the environment when disposed, (3) the treatment process, transformation and long-term fate of these constituents, (4) modification or eradication of the constituents from wastew-ater treatment methods that can be used, (5) disposal methods for generated solids after treatment procedures, in order to protect public health and the environment. Microorgan-isms are utilized to accomplish dissolved and particulate carbonaceous BOD and organic matter stabilization found in wastewater (Metcalf and Eddy, 1991). The sewer lines usu-ally carry the untreated sewage whose BOD is around $700-750 \, \text{mg} \, \text{L}^{-1}$ and suspended matter of around $4000-5000 \, \text{ppm}$. The soluble and colloidal organic material contains mainly carbohydrates, lignin, proteins, lipids, and amino acids. "Temple pond alga" is the nickname given to *microcystis* microorganisms in temple ponds, rich in organic matter. Accelerated activities of the aerobic saprobes can mineralize the organic matter present in sewage. The creation of an anaerobic zone can be prevented by continuous aeration of the sewage and hence encourages the activity of the aerobic heterotrophs. This avoids septic stage creation and also brings about complete oxidation of organic matter leading to mobilization of nutrients. Microorganisms like microalgae are the most versatile tiny inhabitants of the sewage, among them a few are bioindicators (microcystis, oscillato-ria, desmids and diatoms) of the type and extent of pollution. A group of microscopic photosynthetic plants constitute an important component of the biota of water bodies. These are highly efficient scavengers of carbon dioxide due to their minute size and planktonic nature along with affiliated production of oxygen and active users of the incident light. Waste water treatment efficiency can be enhanced by increasing the population of these microalgae (Markandey and Markandey, 2002). Aerobic reaction uses oxygen as an electron acceptor and, on the other hand, anaerobic reactions involves other electron acceptors (Metcalf and Eddy, 1991). In aerobic conditions, microalgae can help to remove nitrogen, phosphorous, carbon, sulfur and pathogens (Marais, 2001). However, removal of pathogens and carbon, mainly through sedimentation, can be accelerated by increasing pH to 9 or above during rapid algal photosynthesis (Parhad, 1974). Substantial amounts of nitrogen and phosphorous can be scavenged from sewage as microalgae have a luxurious consumption of power (Kuhl, 1974) and can also exterminate heavy metals like Cu, Fe, Zn, and Cd effectively (Hutchins, 1986; Malik, 2004).

The capability of denitrification has been seen in wide range of bacteria, but it is absent in algae or fungi; bacteria, which are capable of denitrification, can be both heterotrophic and autotrophic. The heterotrophic organisms include the following genera: *Achromobacter*, *Agrobacterium*, *Bacillus*, *Chromobacterium*, *Corynebacterium*, *Flavobacterium*, *Hypomicrobium*, *Moraxella*, *Neisseria*, *Paracoccus*, *Pseudomonas*, *Rhizobium*, *Rhodopseudomonas*, *Spirilhom*, and Vibrio (Payne, 1981). Heterotrophs uses organic carbon for the production of new biomass (i.e., cell growth), while autotrophic microorganisms derive cell carbon from carbon dioxide for cell growth, which is a reductive process, resulting in lower yields of cell mass and growth rates (Markandey and Markandey, 2002). *Pseudomonas* species are frequently and widely distributed of all the denitrifiers. Simultaneous nitrification-denitrification in activated sludge systems can remove more than 90% of nitrogen for waste water treatment (Rittman, 1985). To reproduce and function properly, energy sources are also required for microbes, like phototrophic organisms are able to use light and chemotrophs derive their energy from chemical reactions. These are vital classifications for microbial synthesis and growth (Markandey and Markandey, 2002).

Much microbial biomass does not show any specific binding and acts as a broad range biosorbent, while some exhibit preferences for certain heavy metals (Horikoshi, 1981). An exhaustive list of microbes and their binding capacity with metal has been compiled (Volesky, 1995). Particular genera of yeast and fungi have shown excellent potential of metal biosorption; *Rhizopus*, *Aspergillus*, *Streptoverticillum* and *Saccharomyces* (Brady and Duncan, 1993; de Rome, 1987; Galun, 1984; Leuf, 1991; Puranik, 1997; Volesky, 1981). Successful bioremediation programs can be achieved only by close collaboration of microbiologists, chemists and engineers and are the ideal reference to accomplish the desired goal in future.

Vetiver, also known as "wonder grass", is efficient enough to act as a "biological sieve" in preventing the movement of soil (and the attached pollutants), by conserving and cleaning water, and by strengthening through its root system, soil profiles, thus preventing water induced slippage and hence preventing subsequent damage to life and property. Vetiver roots can absorb and repeatedly accumulate some of the heavy metals present in soil and water (Truong and Baker, 1998). Further studies have indicated that very little (1–5%) of As, Cd, Cr, Hg, and very moderate amounts (16–33%) of Cu, Pb, Ni and Se absorbed were transported to the shoots. Vetiver's capacity to withstand prolonged submergence in water also makes it behave as a wetland plant (Malik, 2004; Hashim 2011). It can efficiently absorb dissolved nitrogen, phosphorous, mercury, cadmium, lead and all other heavy metals from polluted streams, ponds and lakes, and its efficiency increases with age.

Constructed wetland technology (CWT) is a technology which uses aquatic and wetland plants in artificially created wetlands for municipal wastewater/storm water treatment and the purification process are also considered as a part of phytoremediation technology. Vetiver easily thrives in wetlands and can be grown in constructed wetlands for removal of nitrogen and phosphorus and heavy metals from the polluted storm water, municipal and industrial wastewater, and effluents from abattoirs, feedlots, piggeries and other intensive livestock industries (Singh and Tripathi, 2007).

Treatment of water with natural coagulants has become a common practice because awareness that agencies are burdened with the responsibility of providing potable water to the public and cannot cope with the present demand, this often leads to scarcity and supply is rather unpredictable. Thus a technique of treating waste water with latex from *Calotropis procera* has proven to be extremely useful. It has shown satisfactory clarifying properties. It reduces the turbidity, color, odor, pH, microbial load, and total coliforms of all highly turbid samples. The water treatment potential of latex of *Calotropis procera* on turbidity, pH, odor, microbial load and total coliform reduction is not only of good economic importance but could be the most cost effective alternative method to prevent pollution. This biological healing (bioremediation) is, therefore, favored over and above other treatment methods because its techniques are cost effective, and do not need extensive guidance and controls (Okonko, 2007).

10.4   Overview of Bioremediation in Wastewater Treatment and Ground Water Contamination

Bioremediation eliminates the chance of future liability associated with treatment and disposal of contaminated material. This can prove less expensive than other technologies that are used for the clean-up of hazardous waste. Bioremediations are widely used in many different applications which are summarized in Table 10.5.

Table 10.5   *Industrial sources of water pollution*

Origin of Waste	BOD kg/ton	TSS kg/ton
Food beverage	2.5–220	1.3–257
Dairy industry	5.3	2.2
Domestic sewage	0.025	0.022
Fruit and vegetables	12.5	4.3
Pulp and paper	4–130	11.5–26
Starch and glucose industry	13.4	9.7
Tannery industry	48–86	85–155
Textile industry	30–314	55–196
Yeast industry	125	18.7

Bioremediation is limited to those compounds that are biodegradable. Not all compounds are susceptible to rapid and complete degradation. There are some concerns that the products of biodegradation may be more persistent or toxic than the parent compound.

Bioremediation provides a safe and economic alternative to commonly used physicochemical strategies. However, it is not feasible at present because it is difficult to obtain such valuable microbes from numerous kinds of microorganisms for heavy metal bioremediation, and the adaptation abilities and the remediation efficiencies of reported microorganisms are not enough for practical application. As is known, mechanisms may facilitate the remediation efficiencies of valuable microbes and finally make them feasible for practical application (Thakur, 2006). Microorganisms and microbial products have been reported to efficiently remove soluble and particulate forms of metals, especially from dilute solutions through bioaccumulation and therefore microbe-based technologies provide an alternative to the conventional techniques of metal removal/recovery (Sylvia, 2005).

Metal waste is often a result of industrial activities, such as mining, refining, and electroplating. Mercury, arsenic, lead, and chromium are often prevalent at highly contaminated sites. This fact holds significant challenges for industries because these metals are difficult to remove. Therefore, researchers and industries are researching metals that undergo methylation, complexation, or changes in valence state. Remediation of metals often involves five general approaches: isolation, immobilization, mobilization, physical separation, and extraction. Immobilization and mobilization involve bioremediation processes. Industries use a combination of more than one approach to properly treat metal-contaminated sites. The combination of the approaches can be cost-effective. Immobilization is a technique used to reduce the mobility of contaminants by altering the physical or chemical characteristics of the contaminant. This remediation approach can use microorganisms to immobilize metal contaminants (Malik, 2004). Most immobilization technologies are performed in situ or ex situ. Chemical reagents and bacterial reagents assist with the immobilization of metal contaminants. Most sites contaminated with metals use the solidification and stabilization approach to immobilize metals. Microorganisms can mobilize metals through autotrophic and heterotrophic leaching, chelation by microbial metabolites and siderophores, methylation, and redox transformations. The way in which a particular metal becomes mobilized depends on the metal's chemical and physical characteristics.

Nitrate (NO_3^-) is the second most common pollutant of groundwater after pesticides. Thereby it is vital to protect groundwater resources from pollutants containing nitrates which threaten quality. Nitrate pollution mainly originates from agricultural activities and industrial effluents. It commonly accumulates in soils either due to fertilizer addition or when the rate of NO_3^- N production is much greater than a crop demand. Nitrate comes into the environment not only from anthropogenic activities but some geological formations also contributing to the process. Atmospheric precipitation washes out an appreciable quantity of N in form of nitrate and ammonia. Among the major contributors of nitrate pollution in village environment, animal wastes play a vital role. Little NO_3^- N is leached beneath well-stocked and well-managed cattle feedlots due to the hoof action of the animals. By leaching of nitrate beneath the root zones cultivation also contribute to groundwater nitrate pollution. Nitrate is a potential human health threat, especially to toddlers. Development of a treatment technology for removing nitrate from groundwater is inevitable as nitrous compounds formed from nitrate and nitrites are potentially carcinogenic.

Polycyclic aromatic hydrocarbons (PAHs) are a class of chemicals with two or more fused benzene rings in linear, angular or cluster arrangements. The aromatic compounds due to their persistence and toxicity are of major concern along with other most abundant environmental pollutants. They are produced during fossil fuel combustion, waste incineration, or as byproducts of industrial processes, such as coal gasification, and petroleum refining. The decontamination of PAH-polluted sites is mandatory because many PAH compounds are known or suspected to be toxic, mutagenic or carcinogenic. PAHs are predominantly distributed in nature as components of surface waxes of leaves, plant oils, cuticles of insects and lipids of microorganisms. Petroleum and coal provide the largest source of mononuclear and polynuclear compounds. Other major sources are anthropogenic, geochemical, and biosynthetic. Anthropogenic sources are the result of accidental spillage and intentional dumping of such materials as creosote, coal tar, and petroleum products. Biological technologies are now being explored for their potential in the remediation of contaminated sites. The biodegradation of PAHs are inconsistent and dependent not only on PAH structure, but also on the physicochemical parameters of the site, as well as the number and types of microorganisms present. A few microorganisms have been shown to utilize four-ring PAHs for their growth in the absence of co-factors or surfactants (Churchill, 1999; Singh and Tripathi, 2006). Bacteria genera, capable of degrading PAHs commonly, include species of *Pseudomonas*, *Alcaligenes*, *Rhodococcus*, *Sphingomonas*, and *Mycobacterium* (Churchill, 1999).

10.5   Membrane Separation in Bioremediation

Innovations in the area of wastewater treatment focus on the premise of membrane separation in bioremediation where wastewater is considered to be a resource. With continuing depletion of fresh water resources worldwide, focus has been shifted more towards water recovery, reuse and recycling. The prospect of such ventures requires an extension of conventional wastewater treatment technologies. In recent years it has become more and more apparent that conventional waste water treatment processes such as biological conversion, sedimentation, flocculation and so on, are often unsatisfactory when large quantities of industrial effluents, containing large amounts of highly toxic, biologically non-degradable, or high oxygen demanding constituents, have to be treated. Particularly in heavily industrialized areas with a thick populations and where huge surface water is consumed domestically, the conventional wastewater treatment methods have to be supplemented by more efficient physical or chemical procedures (Chakraborty, 2012; Strathmann, 1976).

The membrane separation process is regarded as the key exercise for advanced waste water reclamation and reuse, which becomes increasingly competitive and is considered superior to the traditional water treatment technology with proven economic performance and process (Radjenovi, 2008) The most widely applied membrane separation processes are microfiltration (MF), ultrafiltration (UF), nanofiltration (NF), reverse osmosis (RO), and electrodialysis (ED) which are utilized in areas where discharge requirements are very strict and direct reuse of the effluent is desired (Metcalf and Eddy, 1991). Firstly, application of membrane-based technologies in wastewater treatment focuses on tertiary treatment of secondary effluent, so as to obtain a high-quality final effluent that can be reused for different purposes.

Membrane bioreactor (MBR) technology provides a good alternative to the conventional treatment of municipal wastewater (Huber, 2004). Membrane Bioreactors can be broadly defined as systems integrating biological degradation of waste products with membrane filtration (Judd, 2006). They have proven quite effective in removing organic and inorganic contaminants as well as biological entities from wastewater. Advantages of the MBR include good control of biological activity, high quality effluent free of bacteria and pathogens, smaller plant size, and higher organic loading rates. Current applications include water recycling in buildings, wastewater treatment for small communities, industrial wastewater treatment, and landfill leachate treatment (Cicek, 2003). However, membrane fouling remains a major drawback, limiting the wider application of this process. Fouling leads to a significant increase in hydraulic resistance, manifested as permeate flux decline or transmembrane pressure (TMP) increase. Many anti-fouling strategies have been proposed for MBR applications including backwashing, cleaning agents (like hypochloride) and so on.

The main applications of membrane technology reported in industry are for treatments of heavily loaded wastewaters such as oily wastewaters (Zaloum, 1994), or discharges from tanneries (Reemtsma, 2002) and textile industries (Rozzi, 2000). Promising applications also exist in treating landfill leachate, chlorinated solvents in manufacturing wastewater, and for groundwater remediation.

MBR technology is currently used worldwide to treat water for reuse, and these plants are growing in size. Over 1000 MBRs are currently in operation around the world with approximately 66% in Japan, and the remainder largely in Europe and North America. About 55% of these installations use submerged membranes while the rest have external membrane modules (Cicek, 2003). A number of plants with a treatment capacity of around 5 million L per day (MLD) to 10 MLD have been in operation for several years now whilst the next generations have design capacities up to 45 MLD (SAWC, 2001).

There are two MBR configurations as shown in Fig. 10.2: (a) external/side stream, where membranes are a separate unit process requiring an intermediate pumping step,

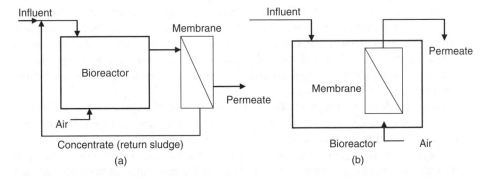

Figure 10.2 *(a) Side-stream MBR. (b) Submerged MBR*

and in other configuration (b) the membranes are immersed in and integral to the bio-logical reactor. Membrane separation is carried out either by pressure-driven filtration in side-stream MBRs or with vacuum-driven membranes immersed directly into the biore-actor, which operates in dead-end mode in submerged MBRs. The more common MBR configuration for wastewater treatment is the submerged one, with immersed membranes, although a side-stream configuration is also possible, with waste water pumped through the membrane module and then returned to the bioreactor.

The energy consumption required for filtration in submerged MBR is significantly lower. In order to prevent the membrane fouling, turbulence over the membrane surface is required by both the configurations with the constituents of mixed liquor. Side-stream MBRs and other membrane processes provide this shear through pumping, whereas immersed processes employ aeration in the bioreactor. Shear enhancement is critical in promoting permeate flux and suppressing membrane fouling, but generating shear also demands energy, which is probably the reason for submerged configuration predom-inance. Due to high permeate flux, fouling in the side-stream MBR module is more pronounced. There are five principal membrane configurations currently employed in practice (Radjenovi, 2008).

In general, MBR applications for wastewater treatment can be classified into four groups, namely (Visvanathan, 2000):

1. Extractive Membrane Bioreactors
2. Bubble-less Aeration Membrane Bioreactors
3. Recycle Membrane Bioreactors
4. Membrane Separation Bioreactors

1. *Extractive Membrane Bioreactors*: The performance capabilities of biological waste water treatments are enhanced by extractive membrane bioreactors. This enhance-ment is obtained by exploiting the membrane's ability to achieve a high degree of separation where components are transported from one phase to another. The separa-tion aids are maintained in optimal conditions within the bioreactor for the biological degradation of wastewater pollutants. For treatment of priority pollutants such as chloroethanes, chlorobenzenes, chloroanilines, toluene and so on, this technology has been successfully demonstrated.

2. *Bubble-less Aeration Membrane Bioreactors*: The membrane aeration bioreactor pro-cess utilizes gas permeable membranes to directly supply high purity oxygen without bubble formation to a biofilm. To transfer large quantity of air/oxygen into the waste water a synthetic polymer membrane is placed in between a gas phase and a liquid phase, thereby obtaining the bubble free aeration. A very high air transfer rate is attained as the gas is diffuse through the membrane. Here, the membrane also acts as a support medium for the biofilm formation, which reduces the potential for bubble formation and air transfer rate.

3. *Recycle Membrane Bioreactors*: The membrane recycle bioreactor comprises a reac-tion vessel operated as a stirred tank reactor and an externally attached membrane module. The substrate (feed wastewater) and biocatalyst are added to the reaction ves-sel in pre-determined concentrations. Thereafter the mixture is continuously pumped through the external membrane circuit. The smaller molecular compounds, the end products of the biodegradation reaction, are permeated through the membrane, while

the large molecular size biocatalysts are rejected and recycled back into the reaction tank. For the removal of aromatic pollutants and pesticides, this type of bioreactor has been tested on industrial scale for bioremediation activities.

4. *Membrane Separation Bioreactors*: Membrane separation bioreactors are a combination of suspended growth reactors for biodegradation of wastes and membrane filtration. To treat both municipal and industrial wastewater, the most widely aerobic activated sludge process is used. However, the quality of the final effluent from this treatment system is highly dependent on the hydrodynamic conditions in the sedimentation tank and the settling characteristics of the sludge. To obtain adequate solid/liquid separation, several hours of residence time in large volume sedimentation tanks are required. At the same time, close control of the biological treatment unit is necessary to avoid conditions, which could lead to poor settle ability and/or bulking of sludge. The disadvantages of the sedimentation and biological treatment steps of the conventional activated sludge process can be overcome by the application of membrane separation (MF, UF, RO, NF) techniques for biosolid separation. The membrane offers a complete barrier to suspended solids and yields higher quality effluent.

Much research has confirmed that MBR is a highly viable wastewater treatment technology regarding nitrification-denitrification and phosphorus removal as well as heavy metals. With optimized design and operating parameters it warrants high effluent quality in terms of ammonia, nitrates, and phosphates present in wastewater. Current European regulation describes restriction guidelines for total phosphorus and nitrogen in treated effluent to $1-2\,mg\,L^{-1}$ and $10-15\,mg\,L^{-1}$, respectively (Radjenovi, 2008). When mixing an activated sludge with real wastewater, a COD removal mechanism is still scarce. The microbial response to dynamic conditions in a real wastewater treatment unit can be different from a simple increase in cell number (i.e., growth of microbial population), and include other substrate-removal mechanisms like sorption, accumulation, and storage. There have been several investigations on treatment efficiencies of MBR operating under comparable conditions that have shown significantly improved performance of an MBR in terms of COD, NH_3-N and suspended solid removals (Radjenovi, 2008). The removal of higher performances heavy metals like Pb, Fe, Cu, Ni, Cd, and Al have been detected mostly in the bound matter (always $>70\%$). On the other hand, some other compounds like As, Hg, Cr, and Zn showed smaller removal percentages. Micropollutants abd toxic heavy metals are concentrated into waste sludge, which is useful for land application (Hashim, 2011).

Removal efficiencies for the most common parameters are given in Table 10.6. In a MBR all of the suspended solids are removed. As a consequence the removal of heavy metals and micro pollutants attached to the suspended solids is also improved (Fenglian, 2011).

Table 10.6   *Removal efficiencies of different contaminants*

Parameter	MBR (mg L^{-1})	Conventional Treatment (mg L^{-1})
Suspended solids	0	10
COD	30	40–50
N Total	10	10
P Total	0.15–0.5	0.5–1.0

Table 10.7   *Industrial wastes and its re-use in different areas*

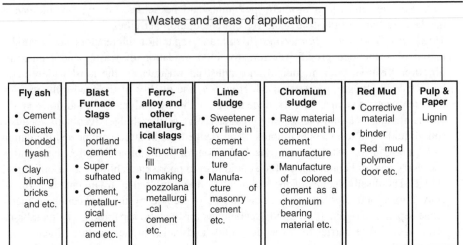

Fly ash	Blast Furnace Slags	Ferro-alloy and other metallurg-ical slags	Lime sludge	Chromium sludge	Red Mud	Pulp & Paper
• Cement • Silicate bonded flyash • Clay binding bricks and etc.	• Non-portland cement • Super sufhated • Cement, metallur-gical cement and etc.	• Structural fill • Inmaking pozzolana metallurgi-cal cement etc.	• Sweetener for lime in cement manufac-ture • Manufa-cture of masonry cement etc.	• Raw material component in cement manufacture • Manufacture of colored cement as a chromium bearing material etc.	• Corrective material • binder • Red mud polymer door etc.	Lignin

The industrial wastes can be re-applied in many areas, which are shown in Table 10.7.

Disposal of industrial wastes, which has been the least controlled, has again been brought into the limelight during the last decade in developing countries by the legislative parties aiming to curb it from society. There are three predominant and widely accepted practiced methods for disposal: (1) landfill (2) incineration and (3) composting. Land filling is still the major disposal method in many countries. And for hazardous wastes incinerating is the best option, while in the case of groundwater monitoring, composting is preferred (Shakya, 2012).

10.6   Case Studies

The following are some case studies related to environmental contamination and the way to solve it by applying bioremediation technologies.

10.6.1   Bioremediation of Heavy Metals

The permanent existence of Cd(II), Pb(II), Cu(II), and Cr(VI) in polluted ecosystems threaten the health of human beings all the time. Thus, removal of Cd(II), Pb(II), Cu(II), and Cr(VI) from synthetic solutions has been studied (Guo, 2010; Zahoor and Rehman, 2009; Hashim, 2011). Multi-metal resistant endophytic bacteria L14 (EB L14) isolated from the cadmium hyperaccumulator *Solanum nigrum L.* effectively uptake 75.78, 80.4%, and 21.25% of Cd(II), Pb(II) and Cu(II) under the initial concentration of 10 mg L^{-1}. Some recent case studies are summarized in Table 10.8.

Srivastava *et al.* (2012) have shown that the arsenic hypertolerant bacteria *Staphylococcus arlettae* strain NBRIEAG-6 isolated from an arsenic contaminated site resulted in reduction of arsenic. *S. arlettae* possesses the *arsC* gene which is responsible for arsenate

Table 10.8   *Types of heavy metal pollution and removal efficiency by microorganisms*

Type of Pollutants	Microorganisms	Removal/Uptake (%)
Arsenic	*Micrococcus roseus*	85.61 (Shakya 2012)
Cadmium	*EB L14*	75.78 (Guo, 2010)
Chromium(VI) and S. capitis	*Bacillus sp. JDM-2-1 & and S. capitis*	85 and 81 (Zahoor and Rahman, 2009)
Copper	*EB L14*	21.25 (Guo, 2010)
Lead	EB L14	80.48 (Guo, 2010)
Mercury	*Pseudomonas and Psychrobacter spp.*	99%, efficient in the presence of a culture medium plus mercury, and 96% efficient when leachate solutions were added (Pepi, 2011)

reductase activity and is able to remove arsenic from liquid media. Liao (2011) isolated, identified and cultured aerobic arsenite-oxidizing and facultative anaerobic arsenate-reducing microbial bacteria that exist in highly arsenic-contaminated groundwater in Taiwan. All of the isolates exhibited high levels of arsenic resistance with minimum inhibitory concentrations of arsenic. Without the addition of any electron donors or acceptors, Strain AR-11 was able to rapidly oxidize arsenite to arsenate at concentrations relevant to environmental groundwater samples. The results support the hypothesis that bacteria capable of either oxidizing arsenite or reducing arsenate coexist, and are ubiquitous in arsenic-contaminated groundwater. Sakya (2011) randomly isolated the culturable arsenic-resistant bacteria indigenous to surface water as well as groundwater from Rautahat District of Nepal. Nine morphologically distinct potent arsenate tolerant bacteria showed relatedness with *Micrococcus varians, Micrococcus roseus, Micrococcus luteus, Pseudomonas maltophilia, Pseudomonas sp., Vibrio parahaemolyticus, Bacillus cereus, Bacillus smithii 1*, and *Bacillus smithii 2* by the morphological and biochemical tests. The isolates were capable of tolerating more than $1000 \, mg \, L^{-1}$ of arsenate and $749 \, mg \, L^{-1}$ of arsenite. Bioaccumulation capability was lowest with *B. smithii* (47.88%) and highest with *M. roseus* (85.61%) indicating the potential of the organisms in arsenic resistance and in bioremediation.

Immobilized mercury bioaccumulating *Bacillus cereus* cells were used for the remediation of mercury from synthetic effluent (Sinha *et al.*, 2012). The maximum biosorption capacity of the immobilized cells was found to be $104.1 \, mg \, g^{-1}$. They studied the feasibility of using immobilized cells in a continuous column for effective mercury remediation. The alginate immobilized *B. cereus* cells can effectively be used in mercury contaminated aqueous environments and which constitute a prospective mercury remediation system. Considerable numbers (the magnitude of $105 \, g^{-1}$ soil) of bacteria with the combined potential for hydrocarbon-utilization and mercury-resistance from rhizospheric soils of three tested legume crops: broad beans (*Vicia faba*), beans (*Phaseolus vulgaris*) and peas (*Pisum sativum*), and two non-legume crops: cucumber (*Cucumis sativus*) and tomato, (*Lycopersicon esculentum*) were isolated (Sorkhoh *et al.*, 2010), which was affiliated to *Citrobacter freundii, Enterobacter aerogenes, Exiquobacterium aurantiacum, Pseudomonas veronii, Micrococcus luteus, Brevibacillus brevis, Arthrobacter sp.* and *Flavobacterium psychrophilum*. Mercury removal from the surrounding media was equivalent with rhizospheric bacterial consortia in the absence of the plants and rhizobacteria-free plants. They concluded that both the collector plants and their rhizospheric bacterial consortia contributed equivalently to mercury removal from soil. Clam cultivation is one of the main activities in the region's Marano and Grado lagoons which are polluted by mercury from the Isonzo River and a chlor-alkali plant. Two Gram-positive (*Staphylococcus* and *Bacillus*) and two Gram-negative (*Stenotrophomonas* and *Pseudomonas*) Hg-resistant bacterial strains were isolated. Two were able to produce Hg^0, but just one contained a *merA* gene; while other two strains did not produce Hg^0 even though they were able to grow at 5 µg ml of $HgCl_2$. (Baldi, 2012).

Bacillus sp. JDM-2-1 and *Staphylococcus capitis* were assessed to reduce hexavalent chromium into its trivalent form. *Bacillus sp. JDM-2-1* could tolerate Cr(VI) (4800 µg ml^{-1}) and *S. capitis* could tolerate Cr(VI) (2800 µg mlL^{-1}). Both organisms were able to resist Cd^{2+} (50 µg ml^{-1}), Cu^{2+} (200 µg ml^{-1}), Pb^{2+} (800 µg ml^{-1}), Hg^{2+} (50 µg ml^{-1}) and Ni^{2+} (4000 µg ml^{-1}). After 96 h it was observed that *Bacillus sp*. JDM-2-1 and *S. capitis* could reduce 85% and 81% of hexavalent chromium from the medium. They were also capable of reducing hexavalent chromium, by 86% and 89% respectively, from the industrial effluents after 144 h. The isolated bacteria have the potential to reduce the toxic hexavalent form to its nontoxic trivalent form (Zahoor and Rehman, 2009). Srivastava and Thakur (2006) isolated *Aspergillus niger* from a soil polluted with leather tanning effluent to remove chromium. *A. niger* introduced in the soil microcosm (40% moisture content) with different concentrations of chromate (250, 500, 1000, 1500, and 2000 ppm) removed more than 70% chromium in soil contaminated by 250 and 500 ppm of chromate. The results of chromate toxicity in the wheat plants revealed that the peroxidases were induced due to increase of metal stress which was reversed in a soil microcosm amended with compost.

A hybrid system was developed by Rosales *et al*. (2012) to treat polluted effluents with dye and toxic metals combining the ion-exchange properties of a NaY zeolite and the characteristics of the bacterium *Arthrobacter viscosus*. A removal of higher than 50% for chromium(VI) and higher than 99% for dye was observed after 8 days operating at the optimized conditions in the treatment of single pollutants, a mixture of dye and metal solutions (Rosales, 2012). Augustynowicz (2010), found an excellent chromium accumulator *Callitriche cophocarpa* (water-starwort) – aquatic widespread macrophyte. These plants were exposed to various chromium(VI) concentrations ranging from 50–700 µM in a hydroponic culture for three weeks. It was observed that plants grown in 50 µM Cr(VI) solution exhibited photosynthetic activity, and shoot and leaf morphology similar to the control plants. The results revealed *Callitriche cophocarpa* to be a very promising species for the chromium(VI) phytoremediation mechanisms as well as a good candidate for wastewaters remediation purpose.

10.6.2 Bioremediation of Nitrate Pollution

In recent years, there has been growing interest in bacteria which can remove nitrates from wastes. Researchers have already isolated a denitrifying bacterium, *Halomonas campisalis* (Mormile *et al*., 1999). The transformation of uranyl nitrate and other compounds in high ionic strength brines by a *halomonas sp*. under denitrifying conditions has also been demonstrated (Singh and Tripathi, 2007). Denitrifying bacteria was newly isolated from coastal sediments and waste water contaminated by marine water. All strains were in α-*Proteobacteria*. The newly isolated strain PH32, PH34, and GRP21 may be the first halophilic bacteria to degrade trimethylamine under denitrifying conditions (Kim, 2003). Hu (2003) investigated the characteristics of denitrifying phosphorus removal bacteria by using three different types of electron acceptors. Denitrifying phosphorous removal bacteria was enriched under anaerobic–anoxic (A/A) conditions and the results obtained indicated that oxygen, nitrate, and nitrite were able to act as electron acceptors successfully. Electro-bioremediation, a hybrid technology of bioremediation and electrokinetics, was employed to remove nitrate from soil. Choi (2009) collected the abundance of *Bacillus spp.* as nitrate reducing bacteria from a greenhouse from the soil sample at Jinju City of Gyengsangnamdo, South Korea and this was isolated and identified. The experiment was conducted in an electrokinetic (EK) cell by applying 20 V across the electrodes. Experimental results showed that the electro-biokinetic processes through electroosmosis and physiological activity of bacteria reduced nitrate in soil environment effectively. To improve nitrate removal efficiency, Wang (2012) isolated and identified new aerobic denitrifying bacteria from nitrate-contaminated sediments and utilized modified immobilized techniques for immobilizing isolates. Three purified aerobic bacteria (HS–N2, HS–N25 and HS–N62 identified as *Salmonella sp.*, *Bacillus cereus*, and *Pseudomonas sp.* respectively) capable of NO_3^-–N removal from sediments were obtained. The removal efficiency of NO3–N by immobilized HS–N62 pellets was enhanced with no nitrite accumulation. Immobilized pellets exhibited more desirable denitrification after 24 h under stressful batch conditions, indicating they had stable and strong denitrifying capabilities. Nitrate-contaminated soil-and-aquifer systems were investigated in a desert environment, a former uranium mill in Monument Valley, Arizona, using biological methods (Jordan, 2008). The native saltbush (*Atriplex canescens*) and black greasewood (*Sarcobatus vermiculatus*) plant community was extracting water from the plume, offering a possible means of preventing migration of the plume by controlling grazing over the site. It was concluded that biological remediation was a feasible alternative to pump/treat solutions at this type of site. Alfano (2011) studied bioremoval of nitrate and sulfate salts using nitrate and sulfate reducing bacteria from the tough stone surfaces of the twelfth-century Matera Cathedral. Previously-isolated *Pseudomonas pseudoalcaligenes* KF707 strain and *Desulfovibrio vulgaris* ATCC 29579 cells were used in the bioremoval treatment. The two strains were trapped in a Carbogel, and applied individually and together to the vertical wall. The biological procedure resulted in the efficient, homogeneous removal of 55% nitrate and 85% of the sulfate deposits, respectively.

10.6.3 Bioremediation in the Petroleum Industry

The oil industry effluents, oily sludge and oil spills on land and water are a major menace to the environment as they severely damage the surrounding ecosystems including human health (Mandal, 2011). Oil refineries generate huge quantities of hydrocarbon waste which create environmental pollution. It needs a well-planned alternative ecofriendly strategy to minimize the threat to the environment. Since, there is a limit to the area available within a refinery for the eradication of those contaminants. *In situ* bioremediation process was employed microorganisms capable of degrading toxic contaminants for the reclamation of polluted sites in the refineries. Energy and Resources Institute (TERI) were developed an efficient bacterial consortium, named Oilzapper (Lal, 2000), that degrades aliphatic, aromatic, asphaltene, and NSO (nitrogen, sulfur, and oxygen compounds) fractions of crude oil and oily sludge very fast. Feasibility study on indigenous crude oil/oily sludge shows the degradation of bacterial population of only 104 c fu/g soil. The application of both bacterial consortium and nutrients gave maximum response in 48.5% biodegradation of TPH (total petroleum hydrocarbons) in four months as compared to only 17% biodegradation of TPH in soil treated with nutrients alone. TERI and IOC's R&D centers (India) jointly developed another microbial-based products Oilivorous–S and Oilivorous–A for the application to the specific quality of oily sludge. It was found that Oilivorous–S proved to be more effective against the oily sludge with a high sulfur content whereas Oilivorous–A was more suitable specifically for oily sludge which is highly acidic in nature (Lal, 2000). Several freeze dried bacteria types like *Pseudomonas, Aeromonas, Moraxella, Beijerinckia, F lavobacteria, chrobacteria, Nocardia, Corynebacteria, Atinetobacter, Mycobactena*, Modococci, *Streptomyces, Bacilli, Arthrobacter, Aeromonas, Cyanobacteria* and so on (Thapa, 2012) are commercially available for hydrocarbon decomposition.

Biocatalysis is another new approach to improve the development of bioremediation strategies for aliphatic (n-alkanes), alicyclic, and aromatic hydrocarbons. It is an enzymatic remediation that can be easier to work than whole organisms, especially in extreme environments (Peixoto, 2011). Alkane biodegradation can be taken place by alkane-degrading enzymes like alkane hydroxylases (soluble methane monooxygenase (sMMO), particulate methane monooxygenase (pMMO), AlkB-related alkane hydroxylases, eukaryotic P450 (CYP52, class II), Bacterial P450 oxygenase system and dioxygenase (CYP153, class I) which are distributed among many different species of bacteria, yeast, fungi, and algae (van Beilen, 2007). Aromatic hydrocarbons, such as benzene, toluene, xylene, and naphthalene, can also be degraded in aerobic conditions. Formation of catehol (dioxygenase class of bacterial iron-containing enzymes) is the initial step for the degrading of these compounds. Once formed, catechol can be degraded, resulting in compounds that can be introduced into the citric acid cycle. Also, these compounds can be completely degraded to CO_2 (Cao, 2009). Polycyclic aromatic hydrocarbons (PAHs) are also hydrocarbon pollutants in crude oil. The enzymatic remediation of PAHs involves the oxidation of the aromatic ring under aerobic conditions by specific dioxygenases, and a complete biotransformation into CO_2 and water (Whiteley and Lee, 2006). Detoxification of PAHs can also be achieved by the use of laccases (Alcalde *et al.*, 2005) (enzymes capable of catalyzing the oxidation of phenols, polyphenols, and anilines). Recently, Scott (2010) successfully reported an enzyme-based product (enzyme TrzN), for efficient remediation of water bodies contaminated with herbicides.

Pseudomonas bacteria, capable of degrading naphthalene along with other PAHs, can be readily isolated from soils contaminated with PAHs (Mueller, 1996). A *Rhodococcus opacus* strain M213, capable of growing on naphthalene as a sole carbon source, was described in which salicylate does not appear to be an intermediate suggesting a different degradation pathway (Singh and Tripathi, 2006).

10.7   Conclusions

Bioremediation is a process that uses naturally occurring microorganisms to enhance normal biological breakdown. It is an effective method for treating many hazardous materials. Of all the different processes available for clean-up of sites, bioremediation is the best and most cost effective method for remediation, with respect to environmental liability. The nature and location of the contamination, the type of soil and geological conditions, determine which method of remediation is best for each individual clean-up site. In this case people are ready to pay much for the treatment of water. Bioremediation is applied heavily. In comparison of costs between conventional methods and bioremediation, it should be kept in mind that, in the case of in situ remediation, costs for transport and excavation cease. In terms of sustainability, bioremediation has priority, because it leads to real reduction of waste, not just storage or displacement of pollutants. There is a huge amount of substances that we do not know how to degrade. Study of the effluent quality produced by conventional secondary treatment processes reveals that such treatment methods do not remove many pollutants, which may create a pollution problem, or prevent reuse of the effluent. MBRs offer a proven alternative due to their ability to handle high organic loadings and wide fluctuations in flow and strength. Activated sludge scrubbing may also be able to be incorporated into these systems for odor control and air pollution management. High quality effluent produced by the MBR would provide pathogen and bacteria control and assist the facility in complying with strict environmental regulations. The industrial effluents released into the environment should have less number of pathogen and low BOD. To keep the environment safe from pathogens and BOD, constant monitoring by indicators is required for humans and wildlife. Thereby smarter solutions for bio-based wastewater management are socially and culturally appropriate for safe future. Education and awareness plays the central role in wastewater management. Therefore overall volume and harmful content of the waste water must be reduced and nutrients which remain in the sludge must be reclaimed and reused. The use of green technologies and ecosystem management should be practiced more actively in rural and urban areas with regard to both water supply and wastewater management.

List of Acronyms

BOD:	Biochemical oxygen demand
COD:	Chemical Oxygen Demand
CWT:	Constructed wetland technology
DDT:	dichlorodiphenyltrichloroethane
ED:	electrodialysis
EPA:	Environment Protection Agency
ETP:	effluent treatment plants
FRTR:	Federal Remediation Technologies Roundtable
MBR:	Membrane Bioreactor
MF:	microfiltration
MLD:	Million liters/day
NF:	nanofiltration
PAHs:	Polycyclic aromatic hydrocarbons

PCB:	Polychlorinated Biphenyls
PCE:	polychloroethylene
pMMO:	particulate methane monooxygenase
RO:	reverse osmosis
sMMO:	soluble methane monooxygenase
STP:	sewage treatment plants
sVOC:	semi-volatile organic compounds
TCE:	Trichloroethylene
TMP:	Transmembrane Pressure
TPH:	total petroleum hydrocarbons
TSS:	Total suspended solid
UF:	ultrafiltration
VOC:	volatile organic compounds

References

A.J.Y. Hu, S.L. Ong, W.J. Ng, *et al.* (2003), New method for characterizing denitrifying phosphorus removal bacteria by using three different types of electron acceptors, *Water Research*, **37**(14), 3463–3471.

N. Adams, D. Carroll, K. Madalinski, *et al.* (2000), *Introduction to Phytoremediation*. National Risk Management Research Laboratory. Office of Research and Development. US EPA Cincinnati, Ohio. EPA/600/R-99/107.

M. Alcalde, T. Bulter and M. Zumárraga, (2005), Screening mutant libraries of fungal laccases in the presence of organic solvents, *Journal of Biomolecular Screening*, **10**(6), 624–631.

G. Alfano, G. Lustrato, C. Belli, *et al.* (2011), The bioremoval of nitrate and sulfate alterations on artistic stonework: The case-study of Matera Cathedral after six years from the treatment, *International Biodeterioration & Biodegradation*, **65**(7), 1004–1011.

J. Augustynowicz, M. Grosicki, E.H. Fajerska, *et al.* (2010), Chromium (IV) bioremediation by aquatic macrophyte, Callitriche cophocarpa. *Chemosphere*, **79**(11), 1077–1083.

F. Baldi, M. Gallo, D. Marchetto, *et al.* (2012), Seasonal mercury transformation and surficial sediment detoxification by bacteria of Marano and Grado lagoons, *Estuarine, Coastal and Shelf Science*, DOI: 10.1016/j.ecss.2012.02.008.

J. Beard (2006), DDT and human health. *Sci. Total Environ.*, **355**, 78–89.

D. Brady and J.R. Duncan. (1993). Bioaccumulation of metal cations by *Saccharomyces cerevisiae*. In: *Biohydrometallurgical Technologies* (eds), A.E. Torma, M.L. Apel and C.L. Brierley. The Minerals, Metals and Materials Society, TMS Publication, Wyoming, Vol. **II**, 711–723.

B. Cao, K. Nagarajan and K.C. Loh, (2009), Biodegradation of aromatic compounds: current status and opportunities for biomolecular approaches, *Applied Microbiology and Biotechnology*, **85**(2), 207–228.

S. Chakraborty, E. Drioli and L. Giorno, (2012), Development of a two-separate phase submerged biocatalytic membrane reactor for the production of fatty acids and glycerol from residual vegetable oil streams, *Biomass and Bioenergy*, **46**, 574–583, DOI 10.1016/j.biombioe.2012.07.004.

J.-H. Choi, S. Maruthamuthu, H.-G. Lee, *et al.* (2009), Nitrate removal by electro-bioremediation technology in Korean soil, *Journal of Hazardous Materials*, **168**(2–3), 1208–1216.

S.A. Churchill, J.P. Harper and P.F. Churchill (1999), Isolation and Characterization of a *Mycobacterium* Species Capable of Degrading Three- and Four-Ring Aromatic and Aliphatic Hydrocarbons, *Appl. Environ. Microbiol.*, **65**(2), 549–552.

N. Cicek (2003). A review of membrane bioreactors and their potential application in the treatment of agricultural wastewater. *Canadian Biosystems Engineering*, Canada, **45**, 6.37–6.49.

L. de Rome (1987). Copper adsorption by *Rhizopus arrhizus*, *Cladosporium resinal* and *Penicillium italicum*. *Appl. Microbial. Biotechnol.*, **26**, 84–90.

P. Doelman and G. Breedveld. (1999), In situ versus on site practices. In: D.C. Adriano, J.-M. Bollag, W.T. Frankenberger, Jr and R.C Sims. (eds), *Bioremediation of Contaminated Soil*. Agronomy No. 37, American Society of Agronomy, Inc. Crop Science Society of America, Inc. Soil Science Society of America, Inc. Madison, Wisconsin. pp. 539–558.

E. Eljarrat and D. Barcelo (2003): Priority lists for persistent organic pollutants and emerging contaminants based on their relative toxic potency in environmental samples. *Trends in Analytical Chemistry*, **22**(10), 655–665.

EPA (2005a) *EPA REACH IT Database*. Remediation and Characterization Innovative Technologies, Available at: http://clu-in.org/vendor/ (accessed February 23, 2013).

EPA (2005b) *Innovative Remediation Technologies: Field-Scale Demonstration Projects in North America: Report and Database*. Available at: http://www.clu-in.org/products/nairt/default.cfm (accessed February 23, 2013).

F. Fenglian and Q. Wang (2011), Removal of heavy metal ions from wastewaters: A review, *Journal of Environmental Management*, **92**(3), 407–418.

FRTR (2005), *Federal remediation Technologies Roundtable Remediation Technologies Screening Matrix and Reference Guide*, Version 4.0. Available at: http://www.frtr.gov/matrix2/top_page.html (accessed February 23, 2013).

M. Galun, P. Keller, D. Malki, *et al.* (1984). *Water, Air, Soil Pollut.*, **21**, 411–414.

H. Guo, S. Luo, L. Chen, *et al.* (2010). *Bioremediation Bioresource Technology*, **101**(22), 8599–8605.

M.A. Hashim, S. Mukhopadhyay, J.N. Sahu and B. Sengupta (2011). Remediation technologies for heavy metal contaminated groundwater, *Journal of Environmental Management*, **92**, 2355–2388.

T. Horikoshi, A. Nakajima and T. Sakaguchi (1981). Studies on the accumulation of uranium by microorganisms. *Eur. J. Appl. Microbiol. Biotechnology*, **12**, 90–96.

Huber VRM Membrane Bioreactor, (2004). Available at: http://www.huber.de/products /membrane-bioreactor-mbr/huber-vrmr-bioreactor.html (accessed February 23, 2013).

S.R. Hutchins, M.S. Davidson, J.A. Brierley and C.L. Brierley (1986). Microorganisms in reclamation of metals. *Ann. Rev. Microbiol.*, **40**, 311–316.

F. Jordan, W.J. Waugh, E.P. Glenn, *et al.* (2008), Natural bioremediation of a nitrate-contaminated soil-and-aquifer system in a desert environment, *Journal of Arid Environments*, **72**(5), 748–763.

S. Judd (2006). *The MBR Book*. Elsevier, Oxford.

S.G. Kim, H.S. Bae, H.M. Oh and S.T. Lee (2003), Isolation and characterization of novel halotolerant and/or halophilic denitrifying bacteria with versatile metabolic pathways for the degradation of trimethylamine. *FEMS Microbiology Letters*, **225**(2), 263–269.

A.B. Koltuniewicz and E. Drioli (2008). *Membranes in Clean Technologies*, Vol. **2** Wiley-VCH Verlag GmbH & Co. KGaA, Weinheim.

A. Kuhl (1974). Phosphorous. In: *Algal Physiology and Biochemistry*. W.D.P. Stewart (ed.). Blackwell Scientific Publications, Oxford, pp. 636–654.

B. Lal, S. Mishra, D. Bhattacharya and P.M. Sarma (2000), Oily sludge management by Oilzapper: hydrocarbon degrading bacterial consortium, In *Proceedings of the 41st Association of Microbiologists of India.*, 10.

E. Leuf, T. Prey and C.P. Kubicek (1991), Biosorption of zinc by fungal mycelial wastes. *Appl. Microbiol. Biotechnol.*, **34**, 688–692.

V.H.C. Liao, C. Yu-Ju, S. Yu-Chen, *et al.* (2011), Arsenite-oxidizing and arsenate-reducing bacteria associated with arsenic-rich groundwater in Taiwan, *Journal of Contaminant Hydrology*, **123**(1–2), 20–29.

A. Malik (2004), Metal bioremediation through growing cells, *Environment International*, **30**, 261–278.

A.K. Mandal, P.M. Sarma, B. Singh, *et al.* (2011), Bioremediation: A sustainable eco-friendly biotechnological solution for environmental pollution in oil industries, *Journal of Sustainable Development and Environmental Protection*, **1**(3), 5–23.

G.V.R Marais. (2001), Dynamic behaviour of oxidation ponds. *Proc. 2nd Symp. Waste treatment Lagoons, Kansas City, USA*. Missouri Basic Engineering Health Council and Federal Water Quality Administration.

D.K. Markandey and N.R. Markandey (2002), *Microorganisms in Bioremediation*, Capital Publishing Company.

M.R. Mormile, M.F. Romine, M. Teresa Garcia, *et al.* (1994), *Halomonas campisalis* sp. nov., a denitrifying, moderately haloalkaliphilic bacterium, *Systematic and Applied Microbiology*, **22**(4) 551–558.

Metcalf and Eddy Inc. (1991). *Wastewater Engineering: Treatment, Disposal and Re-Use*, 4th edition revised by G. Tchobanoglous and F.L. Burton, McGraw Hill Inc. International.

J. Michels, T. Track, U. Gerhrke and D. Sell (2000), *Biologische Verfahren zur* Bodensanierung. Bonn: Grun-weise Reihe des BMBF.

J.G. Mueller, C.E. Cerniglia and P.H. Pritchard (1996), Bioremediation of Environments Contaminated by Polycyclic Aromatic Hydrocarbons. In *Bioremediation: Principles and Applications*, 125–194, Cambridge University Press, Cambridge.

National Research Council (NRC) (1997), *Innovations in Ground Water and Soil Cleanup: From Concept to Commercialization*. Washington, DC.

I. Okonko and B. Shittu (2007), Bioremediation of waste water and municipal water treatment using latex exudate from *Calotropis procera* (Sodom apple). *EJEAF Chem.*, **6**(3), 1890–1904.

N.M. Parhad and N.V. Rao (1974), Effect of pH on survival of *Escherichia coli*. i, **46**, 980–986.

W.J. Payne (1981), *Denitrification*, John Wiley & Sons, Inc., New York.

R.S. Peixoto, A.B. Vermelho and A.S. Rosado (2011), *Petroleum-Degrading Enzymes: Bioremediation and New Prospects*, Federal University of Rio de Janeiro, Brazil.

M. Pepi, C. Gaggi, E. Bernardini, *et al.* (2011), Mercury-resistant bacterial strains *Pseudomonas* and *Psychrobacter spp.* isolated from sediments of Orbetello Lagoon (Italy) and their possible use in bioremediation processes, *International Biodeterioration & Biodegradation*, **65**(1), 85–91.

P.R. Puranik and K.M. Paknikar (1997). *J. Biotechnol.*, **55**, 113–124.

A.S. Purnomo, T. Mori, I. Kamei, *et al.* (2010), Application of mushroom waste medium from *Pleurotus ostreatus* for bioremediation of DDT-contaminated soil. *International Biodeterioration & Biodegradation*, **64**, 397–402.

A.S. Purnomo, I. Kamei and R. Kondo (2008), Degradation of 1,1,1-trichlro-2,2-bis (4-chlorophenyl) ethane (DDT) by brown-rot fungi. *Journal of Bioscience and Bioengineering*, **105**, 614–621.

J. Radjenovi, M. Matosi, I. Mijatovi, Petrovi', *et al.* (2008), Membrane Bioreactor (MBR) as an Advanced Wastewater Treatment Technology, *Env. Chem.*, **5**, 37–101.

T. Reemtsma, B. Zywicki, M. Stueber, *et al.* (2002), *Environ. Sci. Technol.*, **36**,1102.

B.E. Rittman and W.E. Langeland (1985), Simultaneous Denitrification with nitrification in single-channel oxidation ditches, *Journal of the Water Pollution Control Federation*, **57**, 300.

E. Rosales, M. Pazos, M.A. Sanromán and T. Tavares (2012), Application of zeolite-*Arthrobacter viscosus* system for the removal of heavy metal and dye: chromium and Azure B. *Desalination*, **284**(4), 150–156.

A. Rozzi, F. Malpei, R. Bianchi and D. Mattioli, (2000), *Water Sci. Technol.*, **41**, 189.

C. Scott, S.E. Lewis and R. Milla (2010), A free-enzyme catalyst for the bioremediation of environmental atrazine contamination, *Journal of Environmental Management*, **91**(10), 2075–2078.

S. Shakya, B. Pradhan, L. Smith, *et al.* (2011), Isolation and characterization of aerobic culturable arsenic-resistant bacteria from surface water and groundwater of Rautahat District, *Nepal. Journal of Environmental Management*, **95**, S250–S255.

S. Shakya, B. Pradhan, L. Smith, *et al.* (2012), Isolation and characterization of aerobic culturable arsenic-resistant bacteria from surface water and groundwater of Rautahat District, Nepal, *Journal of Environmental Management*, **95**, Supplement S250–255.

S.N. Singh and R.D. Tripathi (2006), *Environmental Bioremediation Technologies*, Springer, New York, ISBN 3-540-34790-9.

A. Sinha, K.K. Pant and S.K. Khare (2012), Studies on mercury bioremediation by alginate immobilized mercury tolerant *Bacillus cereus* cells. *International Biodeterioration & Biodegradation*, **71**, 1–8.

N.A. Sorkhoh, N. Ali, H. Al-Awadhi, *et al.* (2010), Phytoremediation of mercury in pristine and crude oil contaminated soils: Contributions of *rhizobacteria* and their host plants to mercury removal, *Ecotoxicology and Environmental Safety*, **73**(8), 1998–2003.

South Australia Water Corporation (SAWC) (2001), *Survey of Immersed Membrane Bioreactor Technology*.

S. Srivastava and I.S. Thakur (2006), Evaluation of bioremediation and detoxification potentiality of *Aspergillus niger* for removal of hexavalent chromium in soil microcosm, *Soil Biology and Biochemistry*, **38**(7), 1904–1911.

S. Srivastava, P.C. Verma, V. Chaudhry., *et al.* (2012), Influence of inoculation of arsenic-resistant Staphylococcus arlettae on growth and arsenic uptake in *Brassica juncea(L.) Czern. Var.* R-46, *Journal of Hazardous Materials*, DOI: http://dx.doi.org/10.1016/j.jhazmat.2012.08.019.

H. Strathmann (1976), Membrane separation processes in advanced waste water treatment, *Pure & Appl. Chem.*, **46**, 213–220.

D.M. Sylvia, J.F. Fuhrmann, P.G. Hartel and D.A. Zuberer (2005), *Principles and Applications of Soil Microbiology*. Pearson Education Inc., New Jersey.

I.S. Thakur (2006), *Environmental Biotechnology*, I. K. International Pvt. Ltd, New Delhi, India.

B. Thapa, A.K.C. Kumar and A. Ghimire (2012), A review on bioremediation of petroleum hydrocarbon contaminants in soil, *Journal of Science, Engineering and Technology*, **8**(1), 164–170.

P.N. Truong and D. Baker. (1998). Vetiver grass system for environmental protection. *Technical Bulletin No. 1998/1*. Pacific Rim Vetiver Network. Office of the Royal Development Projects Board, Bangkok, Thailand.

V. Turusov, V. Rakitsky and L. Tomatis. (2002), Dichlorodiphenyltrichloroethane (DDT): Ubiquity, persistence and risks. *Environ. Health Perspect.*, **110**, 125–128.

J.B. van Beilen and E.G. Funhoff, (2007), "Alkane hydroxylases involved in microbial alkane degradation," *Applied Microbiology and Biotechnology*, **74**(1), 13–21.

C. Visvanathan, R. Ben Aim and K. Parameshwaran (2000), Membrane Separation Bioreactors for Wastewater Treatment, *Environmental Science and Technology*, **30**(1), 1–48.

B. Volesky and N. Kuyucak (1995). Bio-sorption of heavy metals. *Biotechnol. Prog.*, **11**, 235–250.

B. Volesky and Z.R Holan. (1981). Separation of Uranium by biosorption. US patent 4/ 1981, 320, 093.

P. Wang, Y. Yuan, Q. Li, *et al.* (2012), Isolation and immobilization of new aerobic denitrifying bacteria, *International Biodeterioration & Biodegradation*, DOI: 10.1016/j.ibiod.2012.06.008.

C.G. Whiteley and D.J. Lee, (2006). Enzyme technology and biological remediation, *Enzyme and Microbial Technology*, **38**(3–4), 291–316.

A. Zahoor and A. Rehman. (2009), Isolation of Cr(VI) reducing bacteria from industrial effluents and their potential use in bioremediation of chromium containing wastewater, *Journal of Environmental Sciences*, **21**(6), 814–820.

R. Zaloum, S. Lessard, D. Mourato and J. Carriere (1994), *Water Sci. Technol.*, **30**, 21.

11

Effective Remediation of Contaminated Soils by Eco-Compatible Physical, Biological, and Chemical Practices

Filomena Sannino[1] and Alessandro Piccolo[2]
[1] Dipartimento di Scienze del Suolo, [2] Università degli Studi di Napoli "Federico II",
Portici, Italy

11.1 Introduction

The large increase of industrial development in the last century, population growth and urbanization led to the release of hazardous chemicals in the environment and general global pollution. Environmental pollutants are characterized by different structures and toxicity levels, and have origins in both natural (biogenic and geochemical) and anthropogenic (agricultural and industrial) sources. Several chemicals, including heavy metals and radionuclides, as well as organic compounds such as pesticides, dyes, and Polycyclic Aromatic Hydrocarbons (PAHs), may persistently accumulate in soils and sediments, thus potentially menacing human health and environment quality due to their carcinogenic and mutagenic effects, and ability to bioconcentrate throughout the trophic chain (Semple *et al.*, 2001).

The concern of toxicity risk and environmental pollution associated with chemical contaminants has called for the development and application of remediation techniques. In fact, large effort has been devoted to find ways to remove contaminants from ecosystems.

Sustainable Development in Chemical Engineering – Innovative Technologies, First Edition.
Edited by Vincenzo Piemonte, Marcello De Falco and Angelo Basile.
© 2013 John Wiley & Sons, Ltd. Published 2013 by John Wiley & Sons, Ltd.

In particular, several strategies were devised to remediate and restored polluted soils, based on physical, chemical, and biological methods. These techniques may be applied *in situ*, that is, in the very contaminated soil, thus offering numerous advantages over *ex situ* technologies, whereby the soil is removed to be treated elsewhere. Thus, in situ remediation techniques do not require soil transportation costs and can be applied to diluted and widely diffused contaminations, thus minimizing dangerous intensive environmental manipulation. Conversely, ex situ processes imply the excavation of polluted soil and their decontamination to be conducted in a separate processing plant (Iwamoto *et al.*, 2001). Table 11.1 summarizes the main technologies for cleaning up polluted soils and the estimated costs for each treatment (Semple *et al.*, 2001).

The efficacy of remediation methods is governed by several factors, including soil characteristics as well as structure and physico-chemical properties of pollutants.

Table 11.1 *Main technologies for cleaning up of polluted soils and the estimated costs of each treatment. Reproduced from* Environmental Pollution, *112/2, Semple, Reid and Fermor, Impact of composting strategies on the treatment of soils contaminated with organic pollutants, 269–283, April 2001, with permission from Elsevier.*

Treatment	Approximate remediation cost (£/ton)
Removal to landfill	Up to 100
Solidification	
Cement and Pozzolan based	25–175
Lime based	25–50
Vitrification	50–525
Physical processes	
Soil washing	25–150
Physicochemical washing	50–175
Vapor extraction	75
Chemical Processes	
Solvent extraction	50–600
Chemical dehalogenation	175–450
In situ flushing	25–80
Surface amendments	10–25
Thermal treatment	
Thermal desorption	25–225
Incineration	50–1200
Biological treatment	
Windrow turning	10–50
Land farming	10–90
Bioventing	15–75
Bioslurry	50–85
Biopiles	15–35
In situ bioremediation	175

The bioavailability of pollutants is also an essential feature that affects remediation processes. There are several definitions of bioavailability (Madsen, 2003). In its simplest form, bioavailability can be defined as: "the amount of contaminant that can be readily taken up by living organisms, e.g., microbial cells" (Maier, 2000). Another definition: "a measure of the potential of a chemical to entry into ecological or human receptors", takes into account the assessment of an inherent risk. Although the determination of bioavailability for risk assessment is an undisputed requirement, numerous factors influence bioavailability and its quantification is extremely complex.

Bioavailability of pollutants in soils declines with time (the ageing phenomenon) and several abiotic processes concur to make contaminants less available to biological uptake and, consequently, less subjected to bioremediation processes in environmental compartments. This "ageing effect" may result from the formation of chemical bonds between pollutants and soil organic components, or simply by their physical entrapment in soil organic matter (SOM) or mineral lattices. Depending on the nature of established interactions, sequestered pollutants may later be released back into solution. The formation of either covalent bonds between SOM and pollutants or their entrapment into SOM by multiple hydrophobic interactions (Piccolo *et al.*, 1998; Celano *et al.*, 2008) increases their residence times in soil and leads to the so-called bound residues, which are considered non-bioavailable in the time scale of years or longer.

The mechanisms of interaction of organic pollutants in soil and the relative importance of soil constituents to ageing have been extensively investigated (Nam and Alexander, 1998; Piatt and Brusseau, 1998). Soil-xenobiotic interactions were shown to be influenced by a number of factors including the amount and nature of SOM, the pore size and structure of soil inorganic constituents, activity of microorganisms, and pollutant concentration. Depending on contaminants characteristics and soil properties, different soil remediation technologies can be applied with variable success. However, effective eco-friendly biological, physical, and chemical remediation practices are preferred today over the techniques which have larger biotic and abiotic environmental impacts.

11.2   Biological Methods (Microorganisms, Plants, Compost, and Biochar)

Bioremediation, either as a spontaneous or as a managed strategy, involves the application of biological agents to clean-up environmental compartments polluted by hazardous chemicals. Plants, microorganisms, and plant-microorganism associations, either naturally occurring or tailor-made for the specific purpose, represent the main bioremediation active factors.

11.2.1   Microorganisms

Degrading microorganisms are ubiquitously distributed in natural environments, such as in soils (bacteria and non-ligninolytic fungi) and wood materials (ligninolytic fungi) (Roesch *et al.*, 2007). Microbes have developed enzymatic systems to degrade compounds emanating from natural processes. The diversity of soil microbial communities is enormous. In fact, it has been proposed that soil may contain $10^9 - 10^{10}$ microbial cells per soil cubic cm and an estimated number of $10^4 - 10^6$ distinct genomes per gram of soil (Roesch *et al.*, 2007).

Anthropogenic Organic Pollutants (AOPs) have been produced and discharged in the environment for only about 100 years. Most of these manmade compounds have a molecular structure very different from naturally occurring chemicals. Anthropogenic compounds are thus foreign and unfamiliar to microbes and degrading enzymes, which may exert a possible lower biodegradation activity. AOPs may be biodegraded when: (1) they are compatible with the catabolic enzymatic apparatus of a degrader microbe (Alexander, 1999; Leung *et al.*, 2007), (2) the microbial enzymatic apparatus has a wide specificity (Baldrian, 2006), and (3) genetic microbial adaptation has led to new catabolic pathways for pollutants degradation (Janssen *et al.*, 2005). In contaminated soils, aromatic AOPs can be degraded by bacteria or fungi via an aerobic or anaerobic metabolism, or both. In aerobic metabolism, molecular oxygen is incorporated into the aromatic ring prior to dehydrogenation and subsequent aromatic ring cleavage. In anaerobic metabolic processes molecular oxygen is absent, and alternative electron acceptors such as nitrate, ferrous iron, and sulfate, are necessary to oxidize aromatic compounds.

The effective agents in the transformation of organic pollutants are the microbial enzymatic system that, as powerful catalysts, extensively modify the structure and toxicological properties of contaminants or completely mineralize the organic molecule into innocuous inorganic end products. However, in order to be biodegraded, contaminants must interact with the enzymatic system within the biodegrading organisms. If soluble, they can easily enter cells, but, if insoluble, they must be transformed into soluble or more easily cell-available products. The first step for cell-transformation of insoluble pollutants into bioavailable substances involves a reaction catalyzed by ecto- and extra-cellular enzymes, which are deliberately released by microbial cells into their nearby environment. This transformation process can be quite rapid for some natural compounds like cellulose, but very slow for many xenobiotics.

Extra cellular enzymes include a large range of oxidoreductases and hydrolases. *In vivo* microbial oxidoreductases are periplasmic enzymes associated with the cell surface of viable cells. Their main sources of these enzymes are fungi, such as wood-degrading basidiomycetes, terricolous basidiomycetes, ectomycorrizal fungi, soil-borne microfungi, and actinomycetes. In fact, a great interest is placed in the application of fungi as bioremediating agents (Reddy *et al.*, 1995; Cameron *et al.*, 1999, 2000; Pointing *et al.*, 2001).

Most fungi are robust organisms and may tolerate larger concentrations of pollutants than bacteria. In particular, white-rot fungi appear unique and attractive organisms for the bioremediation of polluted sites. Various combinations of the two glycosylated heme-containing peroxidases, lignin peroxidase (LiP) and Mn-dependent peroxidase (MnP), and of one copper-containing phenoloxidase, laccase (L), constitute the lignin-degrading enzyme systems of white-rot fungi (Bumpus, 1993). Their importance as biodegrading microorganisms is due to several factors. First of all, white-rot fungi are ubiquitous organisms in natural environments. They are unique among eukaryotic or prokaryotic microorganisms, because they possess a very powerful extra-cellular oxidative enzymatic system, the lignin-degrading enzyme system (LDS), which reveals a broad substrate specificity, being able to oxidize several environmental pollutants (Reddy *et al.*, 1995; Cameron *et al.*, 1999, 2000; Pointing *et al.*, 2001). However, other non-ligninolytic enzymes, like cellobiose dehydrogenases, may concomitantly participate in the transformation of pollutants. Cellobiose dehydrogenases are usually secreted under non-ligninolytic conditions when cellulose is the nutrient carbon and either directly or indirectly may oxidize several contaminants. As a consequence, a vast range of toxic environmental pollutants, including low soluble compounds, can be mineralized or degraded by white-rot fungi. Moreover, white-rot fungi may grow in soil exploiting low-value substrates such as agricultural crop wastes that can be easily added to a contaminated site as nutrition substrates. Since white-rot fungi are filamentous fungi, they grow by hyphal extension and elongate in soil, thus enabling a contact with soil pollutants to an extent prohibited to bacteria (Bumpus, 1993; Singh, 2006).

Another class of microbial enzymes involved in pollutants transformation is the hydrolases, including proteases, carbohydrases, esterases, phosphatases, and phytases. These enzymes are physiologically necessary to living organisms. Some of them (e.g., proteases and carbohydrases) catalyze the hydrolysis of large molecules, such as proteins and carbohydrates, to smaller molecules for subsequent absorption by cells. Other enzymes, like phosphatases and phytases, contribute to the nutrition of plants and microbes by transforming organic forms of phosphorous into inorganic phosphorous, thus becoming available for plants and microbial uptake. Due to their intrinsic low substrate specificity, hydrolases may play a basic role in the bioremediation of several pollutants (Metcalfe *et al.*, 2002; Singh *et al.*, 2002).

A possible alternative to the bioremediation of polluted sites by microbial activity may be the direct application of cell-free enzymes after their isolation from microbial cultures. The properties, potentials and restrictions of isolated enzymes in the transformation and detoxification of organic pollutants have been already reviewed (Gianfreda and Bollag, 2002). The most significant features of cell-free enzymes are manifold. They exert unique substrate-specificity and catalytic power. Moreover, they maintain activity in the presence of many toxic, even recalcitrant substances, and/or under a wide range of environmental conditions, often unfavorable to active microbial cells (i.e., relatively wide temperature, pH and salinity ranges, high and low concentrations of contaminants). Additionally, extracellular enzymes apparently reveal low sensitivity or susceptibility to predators, inhibitors of microbial metabolism, and drastic changes in contaminant concentrations. Pesticides of different chemical nature, very recalcitrant compounds like PCBs, polychlorophenols, PAHs and others toxic pollutants were reported to be transformed by oxidoreductases and hydrolases isolated from fungi, bacteria and plant cells (Gianfreda and Bollag, 2002).

However, most results have been obtained under laboratory rather than real field conditions (Filazzola *et al.*, 1999; Sannino *et al.*, 1999; Sannino and Gianfreda, 2000). For example, it has been shown that the decrease of 2,4-dichlorophenol (2,4-DCP) in soil induced by laccases from various origins (i.e., *Cerrena unicolor, Trametes villosa*) may be differently affected (Table 11.2) by the co-presence of either one or two other chlorophenols such as 4-chlorophenol or 2,4,6 trichlorophenol (Bollag *et al.*, 2003). The decrease of 2,4-DCP mediated by laccase can also be affected by another phenol such as catechol and by the presence of its parent precursor, 2,4-D, or of a triazine, such as simazine, that is usually supplied to soil in combination with 2,4-D (Filazzola *et al.*, 1999). In another study, Sannino and Gianfreda (2000 and references herein) studied the enhancing or depressing effects of a laccase from *Rhus vernicifera* on the transformation

Table 11.2 *Substrate removal (%) and residual enzymatic activity (%) after laccase action. Reproduced from Enzyme and Microbial Technology, 35/4, Gianfreda and Rao, Potential of extra cellular enzymes in remediation of polluted soils: a review, 339–354, September 2004, with permission from Elsevier.*

Substrate	Substrate decrease	Laccase activity
Single Phenols		*Cerrena unicolor*[a]
2,4-DCP	66	34
Tyrosol	11	88
Resorcinol	40	76
Methylcatechol	76	24
Hydroxytyrosol	86	18
Pyrogallol	100	89
Gallic acid	98	83
Phenol mixtures		*Trametes villosa*[b]
2,4-DCP	66	34
+4-CP	56	56
+2,4,6-TCP	58	58
+2,4-D	82	20
		Cerrena unicolor[c]
+Catechol	77	20
+Simazine	46	30
+4-CP + 2,4,6-TCP	58	58
+Catechol + Simazine	39	60
+2,4-D + Catechol	100	0
		Rhus vernicifera[d]
Catechol	58	70
+Methylcatechol (M)	38	11
+Tyrosol (T)	100	66
+ Hydroxytyrosol (H)	99	27
+ M + T	16	9
+T + H	95	29
+M + H	56	10
+M +T +H	63	11

The substrate decrease values have been normalized by laccase units.
From ([a]Gianfreda et al., 1998; [b]Bollag et al., 2003; [c]Filazzola et al., 1999; [d]Gianfreda et al., 2003)

of catechol, when in model mixture of four phenols (catechol, methylcatechol, tyrosol and hydroxytytrosol).

While several extra-cellular enzymes, either as cell-associated or cell-free enzymes, may behave as effective catalysts in the biodegradation of harmful pollutants, their large-scale application for remediation of polluted soils is greatly limited. This may be due to several drawbacks which depend on both pollutants and enzymes (Alcade *et al.*, 2006). The simultaneous presence of several polluting substances in a contaminated site may result in synergistic, often negative, effects on the enzyme efficiency. The high costs associated with the isolation and purification of free enzymes discourages their application in massive soil remediation processes. The low stability of enzymes in the multiphase and bioactive conditions of soil severely limits a wide use of enzymes as remediating agents of polluted soils. Although immobilized enzymes may present a greater stability under soil conditions, they are not yet widely applied in the remediation of polluted soils (Khan and Husain, 2007).

Bioremediation of contaminants can be more rapidly accomplished by two methods, bioaugmentation and/or biostimulation (Cunningham and Philp, 2000). The process of bioaugmentation, as it applies to remediation of petroleum hydrocarbon contaminated soils, involves the introduction in a contaminated system of microorganisms that have been exogenously cultured with the aim to degrade specific chains of hydrocarbons (Mrozik and Piotrowska-Seget, 2010). These microbial cultures may be derived from the very same contaminated soil or obtained from a stock of microbes that have been previously proven to degrade hydrocarbons. On the other hand, the biostimulation process implies the addition to polluted soils of nutrients in the form of organic and/or inorganic fertilizers, in order to stimulate the activity and proliferation of indigenous microbes (Pankrantz, 2001). These may or may not be proved to aim the polluting hydrocarbons as a primary food source. However, the hydrocarbons are assumed to be degraded more rapidly in comparison to natural attenuation processes, probably because of the increased number of microorganisms induced by the greater amount of nutrients provided to the contaminated soil.

11.2.2 Plants

Phytoremediation is the use of plants for decontaminating polluted sites and is presently considered to be a practically useful, low cost, and green technology. Several plants are being identified for their efficient use in phytoremediation trials (Singh and Tripathi, 2007). The most versatile plant species for phytoremediation applications, both terrestrial and aquatic, are listed in Table 11.3. These plants are highly tolerant to salinity and pollutant toxicity, possibly because of their extensive root binding systems (Sinha *et al.*,

Table 11.3 *The most versatile terrestrial and aquatic plant species identified to be useful for phytoremediation. Adapted from R.K. Sinha et al., 2007*

1) Vetiver grass (*Vetiveria zizanioides*)	2) Barmuda grass (*Cynodon dactylon*)
3) Bahia grass (*Paspalaum notatum*)	4) Sunflower oil plant (*Helianthus annus*)
5) Poplar tree (*Populus spp.*)	6) Mustard oil plant (*Brassica Juncea*)
7) Periwinkle (*Catheranthus roseus*)	8) Cumbungi (*Typha angustifolia*)
9) Water hyacinth (*Eichornia cressipes*)	10) Duck weed (*Lemna minor*)
11) Red Mulberry (*Morus rubra*)	12) Kochia (*Kochia scoparia*)
13) Foxtail barley (*Hordeum Jubatum*)	14) Switch grass (*Panicum variegatum*)
15) Musk Thistle (*Carduus nutans*)	16) White radish (*Raphanus sativus*)
17) Catnip (*Nepeta cataria*)	18) Big bluestem (*Andropogan gerardii*)
19) Indian grass (*Sorghastrum nutans*)	20) Canada wild rye (*Elymus candensis*)
21) Nightshade (*Solanum nigrum*)	22) Wheat grass (*Agopyron Cristatum*)
23) Alfa-alfa (*Medicago sativa*)	24) Tall Fescue (*Festuca anundinacea*)
25) Lambsquarters (*Chenopodium berlandieri*)	26) Reed grass (*Phragmites australis*)

2007 and references herein). The most ideal species for phytoremediation are members of the *Gramineae* and *Cyperaceae* grass family and members of the *Brassicaceae* and *Salicaceae* families. For example, the sunflower plant (*Helianthus annus*) is reported to adsorb radionuclides from soil to an extent that allows its effective decontamination (Prasad and Freitas, 2003). Large scale plantations of sunflower have been set around Chernobyl, where the 1986 nuclear disaster spread a great amount of radioactive material, thus badly contaminating the soil in the accident area (Sinha *et al.*, 2007). The phytoremediation of soil radioisotopes by sunflower was rather inexpensive as it cost about $US 2 per ha, while other techniques for similar soil decontamination may have cost millions of US dollars.

Phytoremediation of organic and inorganic contaminants involves either a physical removal of pollutants or their bioconversion (biodegradation or biotransformation) into biologically inactive forms. The conversion of metals into inactive forms can be enhanced by external conditioning of soils: enhancement of soil pH (e.g., through liming), addition of organic matter (e.g., sewage sludge, compost, etc.), inorganic anions (e.g., phosphates) and metal oxides and hydroxides (e.g., iron oxides) (Varun *et al.*, 2012). Concomitantly, plants can play a role here in transforming contaminants in inactive forms by releasing different anionic species in soil and altering soil redox conditions (Peng *et al.*, 2009).

In general, plants affect the fate of soil contaminants by: (1) stabilizing and degrading pollutants in the rhizosphere, (2) uptaking pollutants into plant tissues, (3) subjecting the uptaken pollutants to transport, accumulation, and degradation within the vegetal tissues, and (4) volatizing pollutants in to the atmosphere. These processes are generally grouped under the definition of phytostabilization, phytodegradation, phytoextraction, phytostimulation, and phytovolatilization (Pilon-Smits, 2005). Phytostabilization is a process by which plants stabilize pollutants in soil either by either simply preventing their erosion, leaching, and/or run-off, or converting pollutants to less bioavailable forms. In the phytostimulation or rhizodegradation processes, plants facilitate microbial degradation of organic pollutants by exuding from roots simple organic molecules, which promote the rhizospheric growth of degrading microbes (McCutcheon and Schnoor, 2003). Phytodegradation occurs when plants directly degrade organic pollutants by releasing degrading enzymes such as dehydrogenases, oxygenases and reductases, either inside the plant issues or into the soil. The phytovolatilization involves the transformation of pollutants, which have been uptaken in plant tissues, in volatile forms which can then be released from plant to the atmosphere. Finally, phytoextraction is the general process by which pollutants are uptaken by plants and accumulated in plant tissues.

The uptake of AOPs by plants occurs through two pathways (Collins *et al.*, 2006). One pathway is the soil-water-plant cycle in which pollutants are uptaken from the soil solution and then transported up plant shoots within the xylem transpiration system. A second pathway involves the soil-air-plant cycle, in which AOPs are uptaken by aerial parts of plants either from soil particles adsorbed on plant leaves or directly as gaseous forms of AOPs after their volatilization from soil. Following plant uptake, AOPs are further translocated, sequestered, and degraded in plant tissues by other processes. The key parameters which influence the translocation of contaminants from soil to plant include the content of contaminants in soil (or water), their physicochemical properties, the plant species, the soil types, and the time of exposure to plant (Chiou, 2002).

Phytoremediation becomes largely more effective in enhancing degradation of organic pollutants in the rhizosphere as a result of greater densities and activities of microorganisms in the surrounding soil. In fact, progressive plant root growth and extension in soil stimulate contaminant bioavailability by exposing the adsorbed or entrapped contaminants to enhanced solubilization by root exudates and increased microbial activity (Parrish *et al.*, 2005). The effect of plant root exudates in phytoremediation of soil contaminated by PAHs was shown to be enhanced by amendments with a recycled biomass such as compost (Wang *et al.*, 2012). The advantages of phytoremediation over other approaches is due to the inherent preservation of soil natural structure and to the free sunlight energy involved in the process that enhances the content of degrading microbial biomass in soil.

11.2.3   Plant-Microorganism Associations: Mycorrhizal Fungi

Mycorrhizas are symbiotic associations between some soil fungi and plant roots and are ubiquitous in natural environments. Their role in the nutrient transport throughout ecosystems and in the protection of plants against environmental and cropping stresses has long been known. The majority of mycorrhizas are obligated symbionts because of their poor or low ability for independent growth. Smith and Read (1997) defined a mycorrhiza as "a symbiosis in which an external mycelium of a fungus supplies soil-derived nutrients to a plant root".

The vesicular-arbuscular mycorrhiza (VAM) is the most ancient type of association and colonizes plants due to their capacity of phosphate scavenging. About two-thirds of plant species are known to be associated with VAM. The second common type in the ecosystem is the ectomycorrhizal (ECM) fungi, which mostly specialize in nutrient capture. Both fungi are also called "extremophiles" due to their occurrence in extreme habitats, including high or low temperature, pH, salt and metal concentration, and drought conditions. The mycorrhiza's activity has been also observed in conditions of low nutrient availability where they function as activators of plant growth. However, VAM and ECM differ in their ability to sustain plant biomass production.

The role of mycorrhizal fungi in regulating metal uptake by plants and in favoring degradation of persistent organic compounds in soil is of great importance. Moreover, the symbiotic relationship with plants allows mycorrhiza to overcome the harsh conditions of toxic contamination. In fact, the rhizosphere is nutrient-depleted and the microbially active layer immediately surrounding the largest plant roots is where the metabolic activity that sustains fungal growth and survival in soil ecosystems becomes distinguished from the more biologically inert bulk soil (Pinton, 2007).

The qualitative and quantitative differences in soil characteristics between rhizosphere and bulk soil significantly affect the degradation of organic contaminants and the uptake of heavy metals (Vetterlein, 2004). Several different mechanisms have been proposed for the interactions between mycorrhizas colonization and accumulation of heavy metals (Smith and Read, 1997). These include the interactions with the phosphorous nutrition processes induced by the mycorrhizal fungus, which implies a dilution of toxic elements in tissues and consequent tolerance of toxicity. A protective effect of mycorrhiza against plant Zn uptake was reported (Li and Christie, 2001) and it was correlated either with variations in Zn solubility, brought about by changes in soil solution pH or immobilization of Zn in extra-radical mycelium. The ability of *Paxillus involutus* to accumulate Cd in different plant compartments was also shown (Blaudez *et al.*, 2000) as occurring by two mechanisms: a binding of Cd in cell walls and an accumulation of Cd in the vacuolar compartment.

Several studies have been conducted on the degradation of PAHs and PolyChlorinated Biphenyls (PCBs) by arbuscular mycorrhiza (AM). The PAHs in soils may either be degraded in the rhizosphere, or uptaken by plants (Watts *et al.*, 2006). The effects of AM fungi on the uptake by maize of phenanthrene (PHE) and pyrene (PYR) and on their dissipation in soil were investigated using three compartmentalized rhizoboxes (Wu *et al.*, 2011a). Inoculation of *Glomus mosseae* significantly increased PHE and PYR concentrations in maize roots and enhanced PYR translocation from roots to stems in PHE+PYR spiked-soils. Teng *et al.* (2011) studied the effects of inoculation with *Rhizobium meliloti* on the phytoremediation capacity of alfalfa grown for 90 days in an agricultural soil contaminated with PAHs. In comparison with an uncropped control soil, the planting with uninoculated alfalfa (P) and alfalfa inoculated with *R. meliloti* (PR) significantly lowered the initial soil PAH concentrations by 37.2% and 51.4%, respectively. Moreover, the inoculation with *R. meliloti* increased the presence of PAH-degrading bacteria, the overall soil microbial activity, and the ultimate C utilization capacity by the soil microbial community. This suggests that the symbiotic association between alfalfa and *Rhizobium meliloti* can stimulate the activity of a PAH degrading microflora in the rhizosphere and it may thus become an effective bioremediation strategy.

11.2.4 Compost and Biochar

The composting process is the biological decomposition of organic wastes under controlled aerobic conditions. In contrast to uncontrolled natural decomposition of organic compounds, the temperature in composting waste heaps can increase by self-heating to the ranges which are typical of mesophilic (25–40 °C) and termophilic microorganisms (50–70 °C). The end product of composting is a biologically stable humus-like product that can be employed in several applications, for example: soil conditioner, fertilizer, biofiltering material, fuel, and so on. The composting process can concomitantly reach different objectives, such as the volume and mass reduction of biomasses, their stabilization and drying, and the elimination of phytotoxic substances and pathogens (Mena *et al.*, 2003).

Composting is also a method to be employed in the decontamination of polluted soils, because compost is capable of sustaining various microbial populations potentially hydrocarbon degraders, such as bacteria, including bacilli, pseudomonas, mesophilic and thermophilic actinomycetes, and lignin-degrading fungi (Adekunle, 2011). Compost can also improve the chemical and physical properties of soil to be decontaminated, since it affects soil pH, nutrients and moisture content, soil structure, and microbial biomass population (Tejada, 2009). The application of compost for the remediation of soils contaminated by organic anthropogenic compounds has become progressively popular and successful (Purnomo *et al.*, 2011; Castagnero, 2011; Quilty and Cattle, 2011).

A so-called biochar recalcitrant pyrogenic material has been recently pushed to the forefront as a general solution to raise the quality of agricultural soil, including carbon storage (Atkinson *et al.*, 2010). While it appears that agricultural wastes may be usefully disposed by "baking" biomass under low oxygen, the potential of the resulting pyrolyzed biochar to store carbon and reduce greenhouse gases emissions from soil remains still highly controversial (Rogovska *et al.*, 2011; Jones *et al.*, 2011; Shackley *et al.*, 2011; Zimmerman *et al.*, 2011). Unless coupled with more bioactive compost materials, the possible use of biochar in the remediation of contaminated soil appears limited by its inherent biological recalcitrance that depresses the activity of pollutant microbial degraders (Beesley *et al.*, 2011).

11.3   Physicochemical Methods

The most common physical and physicochemical methods for treating contaminated soils, such as landfill disposal, incineration, thermal desorption and vapor extraction have proven to be highly expensive. Soil washing may be a valid and relatively inexpensive alternative in soil remediation (Kuhlman and Greenfield, 1999). It is based on desorption of pollutants from contaminated soils through the action of micellar solutions capable to removing pollutants form solid surfaces. Although water is still employed with poor results (Griffiths, 1995), synthetic organic surfactants are widely and effectively applied to decontamination by washing soils (Chun *et al.*, 2002; Garon *et al.*, 2002). The liquid phases containing the desorbed pollutants are then disposed or subjected to further chemical or biochemical treatments for complete detoxification.

Surfactants are amphiphilic compounds which reduce the free energy of the system by replacing the bulk molecules of greater energy at an interface. They contain both a polar ionic hydrophilic moiety (referred to as the head group) and a non-polar hydrophobic moiety (the tail group). According to the charge present in the hydrophilic portion of the molecule, surfactants are classified as anionic, cationic, nonionic, and zwitterionic (with both negative and positive charges) surfactants. A phenomenon unique to surfactants is the self-assembly of molecules above a certain concentration in water into dynamic clusters called micelles. The concentration of a surfactant to form micelles is referred to as *Critical Micelle Concentration* (CMC). CMC is different for each surfactant, typically ranging between 0.1–10 mM (West and Harwell, 1992). At concentrations below CMC, the dissolved surfactant is all in monomeric form, while the surfactant molecules in water associate to themselves above CMC and organized micelles are formed. In a micelle, the individual monomers are oriented with their hydrophilic moieties in contact with the aqueous phase, while their hydrophobic moieties are tucked into the interior of the aggregate. Thus the main characteristic of surfactants resides in their chemical structure to reach the most adequate balance of hydrophilic and hydrophobic interface interactions, in order to achieve a largest reduction of the interfacial tension.

Surfactants are not only industrially synthesized (SDS, triton, etc.) but can be also biologically produced by yeast and bacteria and are known as biosurfactants (Lima *et al.*, 2011). Biosurfactants are glycolipids, lipopeptides, phospholipids, fatty acids, and neutral lipids. Most of biosurfactants are either anionic or neutral, while only few are cationic such as those containing amine groups. The hydrophobic part of biosurfactants consists of either long-chain fatty acids, or hydroxy- or α-alkyl-β-hydroxy fatty acids, while the hydrophilic part is generally represented by carbohydrates, amino acids, cyclic peptides, phosphates, carboxylic acids or alcohols.

A wide variety of microorganisms can produce biosurfactants (Table 11.4). A class of mostly nonionic surfactants, the alkyl polyglucosides, are produced from renewable resources such as fatty alcohols and sugars (Fukuda *et al.*, 1993). Due to their low toxicity and excellent biodegradability, these compounds have been studied extensively and used for many applications (Park *et al.*, 2007). Cyclodextrins represent another class of biosurfactants, as cyclic macromolecules consisting of a taurus of $\alpha-1$-4-linked glucose units. They show high aqueous solubility although their inner hydrophobic cavity can form 1:1 host-guest complexes with hydrophobic organic molecules (Martin Del Valle, 2003). Although contaminants in complexes with cyclodextrins are supposed to be less available to microorganisms, it has been shown that formation of inclusion complexes enhances the degradation of organic pollutants, such as phenantrene (Wang *et al.*, 1998).

Different classes of surfactants can then be employed for the washing of contaminated soils depending on the nature of contaminants to be removed. For example, pesticides are removed by non-ionic synthetic surfactants such as Triton X-100 and biosurfactants like rhamnolipids, whereas PAH can be washed away by alkylphenol ethoxylate, hydroxypropyl-β-cyclodextrins and other anionic surfactants (Conte *et al.*, 2005; Joshi and Desai, 2010; Pacwa-Płociniczak *et al.*, 2011).

11.3.1   Humic Substances as Natural Surfactants

Soil organic matter (SOM) has a multi-functional and important significance in environment and agriculture, since it is related to the productivity and sustainability of agroecosystems and to the conservation and protection of the environment. It is composed of humic substances (70–80%), which are supramolecular self-associations of heterogeneous molecules of biological origin with a mass lower than 1000 Da and held together, in apparently large sizes, by dispersive forces such as $\pi-\pi$, CH-π, and van der Waals interactions (Piccolo, 2002). Due to the coexistence of both hydrophilic and hydrophobic domains in humic substances, they behave in aqueous phase as natural surfactants. Piccolo and coworkers thus proposed that humic substances may exert micellar capacities to emulsify a wide variety of organic molecules, favor their solubilization in water, and, thus, enhance their bioavailability (Fava and Piccolo, 2002; Conte *et al.*, 2005).

Humic Substances (HS) were found to significantly enhance mobility and biodegradability of PCBs in slurry-phase microcosms of a spiked model soil (Fava and Piccolo, 2002). In particular, large PCB dechlorination and biodegradation were obtained in microcosms added with 1.5% HS rate in the slurry phase. The HS favored PCB biodegradation by both increasing PCBs bioavailability and sustaining the co-metabolism PCB biodegrading bacteria in a soil microcosm. Apolar PCBs must have been repartitioned from the model soil into the humic hydrophobic domains, thereby increasing PCB accessibility to the degrading bacteria which easily accommodate in the proxy hydrophilic domains of HS. These results showed that natural HS enhance the PCB bioavailability and aerobic biodegradation to a larger extent than that obtained by synthetic surfactants.

The effects of both HS and another biosurfactant, such as Soya Lecithin (SL), on the aerobic bioremediation of a soil historically contaminated by PAHs were studied in aerobic solid-phase and slurry-phase reactors (Fava *et al.*, 2004). The overall removal of PAHs in the presence of SL or HS was fast and accompanied by a large soil detoxification, especially under slurry-phase conditions. Both HS and SL could be metabolized by

Table 11.4 *Type and microbial origin of biosurfactants. Reproduced from* Environmental Pollution, *133/2, Mulligan, Environmental applications for biosurfactants, 183–198, January 2005, with permission from Elsevier.*

Type of biosurfactants	Microorganism
Trehalose lipids	*Arthrobacter paraffineus,* *Corynebacterium* spp. *Mycobacterium* spp. *Rhodococus erythropolis, Nacardia* spp.
Rhamnolipids	*Pseudomonas aeruginosa,* *Pseudomonas* sp., *Serratia rubidea*
Sophorose lipids	*Candida apicola, Candida bombicolica,* *Candida lipolytica,* *Candida bogoriensis*
Glycolipids	*Alcanivorax borkumensis,* *Arthrobacter* sp., *Corynebacterium* sp. *R. erythropolis, Serratia marcescens,* *Tsukamurella* sp.
Cellobiose lipids	*Ustilago maydis*
Polyol lipids	*Rhodotorula glutinus,* *Rhodotorula graminus*
Diglycosyl diglycerides	*Lactobacillus fermentii*
Lipopolysaccharides	*Acinetobacter calcoaceticus* (RAGI) *Pseudomonas* sp., *Candida lipolytica*
Arthrofactin	*Arthrobacter* sp.,
Lichenysin A, Lichenysin b	*Bacillus licheniformis*
Surfactin	*Bacillus subtilis, Bacillus pumilus*
Viscosin	*Pseudomonas fluorescens*
Ornithine, lysine peptides	*Thiobacillus thiooxidans,* *Streptomyces sioyaensis,* *Glucanobacter cerinus*
Phospholipids	*Acinetobacter* sp.
Sulfonylipids	*T. thiooxidans, Corynebacterium alkanolyticum*
Fatty acids (corynomycolic acids, Spiculisporic acids, etc.)	*Capnocytophaga* sp., *Pennicilium spiculisporum,* *Corynebacterium lepus,* *Arthrobacter paraffineus,* *Talaramyces trachyspermus,* *Nocardia erythropolis*
Alasan	*Acinetobacter radioresistens*
Streptofactin	*Streptomyces tendae*
Particulate surfactant (PM)	*Pseudomonas marginalis*
Biosur (PM)	*Pseudomonas maltophilla*

soil aerobic microorganisms, thereby enhancing the content and activity of indigenous aerobic PAH-degrading bacteria in the reactor water phase. These results indicate that the presence of biodegradable HS and SL efficiently enhance PAH bioavailability in soil.

Moreover, biogenic agents, such as cyclodextrins, humic substances, and rhamno-lipids, which may be used in soil washing processes, were shown to enhance pollutants biodegradation and dechlorination by also stimulating the growth of specialized bacteria in the reactors (Berselli *et al.*, 2004). Thus, these efficient and nontoxic surfactants appear to be promising in integrating the physicochemical soil washing method with a post-washing progressive biological attenuation of pollutants until a complete soil restoration.

A lignite humic acid (a North-Dakota leonardite) has been used as natural surfactant in aqueous solutions for the washing of highly contaminated soils (Conte *et al.*, 2005). In fact, soil samples, differing in texture and content of organic hydrophobic contaminants, collected from the heavily polluted site of ACNA (Aziende Chimiche Nazionali Associate), a former color-making industrial area near Savona in northern Italy, and included in the list of national top priorities for soil remediation, were subjected to washing with lignite humic solutions. The efficiency of pollutant removal was compared to that obtained with washings with either water or aqueous solutions of two synthetic surfactants (sodium dodecylsulfate, and Triton X-100). Soil clean-up by washing with water was unable to exhaustively remove contaminants from both soils, whereas the pollutants removing efficiency of humic natural surfactants was up to 90% and similar to that obtained by synthetic surfactants. Hence, this study showed for the first time that solutions of natural humic acids are a better choice for soil washings than synthetic surfactants (Figure 11.1). In fact, not only the removal capacity of the two surfactant types are comparable, but humic acids leave in soil a metabolic substrate capable to promote microbial activity, in contrast to the assessed biological toxicity of synthetic surfactants when applied at the CMC (Sandbacka *et al.*, 2000).

Humic acids with molecular characteristics very similar to those from soil may be obtained in good yields from composted biomasses (Spaccini and Piccolo, 2009). In fact, humic acids from compost can be generally isolated in 12% yield from food and green waste compost (Quagliotto *et al.*, 2006). The surfactant properties of these compost humic acids in aqueous solution were also found to enhance the water solubility of hydrophobic substances like phenantrene (Montoneri *et al.*, 2009). These results indicate that not only humic acids of geochemical origin but also the humic materials extracted from recycled biomasses can be applied to remediation contaminated soil by micellar washing (Salati *et al.*, 2011).

11.4   Chemical Methods

Polluted soils can be remediated by using different chemical processes such as organic solvent extraction (Jonsson *et al.*, 2010; Wu *et al.*, 2011b), chemical dehalogenation (Henry and Warner, 2003; Aristov and Habekost, 2010), in situ flushing (Navarro and Martinez, 2010; Pasha *et al.*, 2011) and surface amendments (Miretzky and Cirelli, 2010; Wang *et al.*, 2011). These methods have all been proven effective, although expensive and generally toxic to the environment.

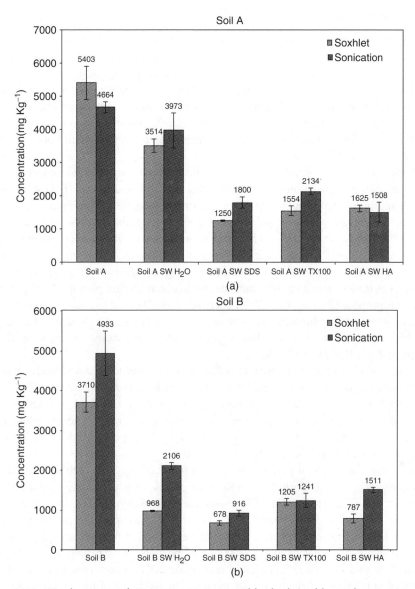

Figure 11.1   *Total amount of contaminants extracted by both soxhlet and sonication from soil A and soil B after soil washing (SW) with H₂O, SDS, TX100, and HA. Reproduced from* Environmental Pollution, *135/3, Conte et al.,* Soil remediation: humic acids as natural surfactants in the washing of highly contaminated soils, 515–522, *June 2005, with permission from Elsevier.*

11.4.1   Metal-Porphyrins

Ecologically compatible technologies to breakdown and transform xenobiotics may rely on the use of metal-porphyrins as catalysts (Buschmann *et al.*, 1999; Baeseman and Novak, 2001). Recent studies have shown that metal-porphyrins used as biomimetic catalytic systems may be an ecofriendly method of controlling the mass of natural organic matter (NOM) and have useful applications in environmental chemistry (Smejkalova and Piccolo, 2006; Fontaine and Piccolo, 2011). In fact, the transformation of environmental contaminants is often mediated by HS, the major component of natural organic matter (Piccolo *et al.*, 2002).

HS are recognized to facilitate the free-radical mechanism of redox reactions which favor the abiotic degradation of xenobiotics via either photocatalysis (Fukushima *et al.*, 2000; Fukushima and Tatsumi, 2001), or enzymatic oxidation (Morimoto *et al.*, 2000), or other abiotic and/or biotic processes (Curtis and Reinhard, 1994; Schwarzenbach *et al.*, 1997). In particular, HS are believed to enhance the formation of reactive oxygen species which become strong oxidants of organic molecules. For example, the photoexcitation of humic chromophores can produce oxygen species such as singlet oxygen, hydrogen peroxide, and hydroxyl radicals, which increase the photolysis of pesticides and chlorophenols (Prosen *et al.* 2005; Richard *et al.*, 2004). Moreover, the presence of redox couples as the quinone/hydroquinone systems in HS are involved in enhancing the reductive dehalogenation of trichloroethylene, hexachloroethane, carbon tetrachloride, and bromoform (Curtis *et al.*, 1994).

The role of HS in removing organic pollutants was found to be significantly enhanced in the presence of biomimetic catalytic systems such as metal-porphyrins (Fukushima *et al.*, 2003; Hahn *et al.*, 2007). Metal-porphyrins can catalyze oxidation reactions similarly to those of enzymes such as peroxidase, ligninase, and monoxygenase (Song *et al.*, 1997). For example, chlorophenols are oxidized by hydrogen peroxide (H_2O_2) in the presence of metal-porphyrins (Labat *et al.*, 1990).

The active site of metal-porphyrins is a tetra-pyrrole ring to which a metal atom is coordinated. The lateral constituents of the tetra-pyrrole ring, depending on the particular porphyrin, are added to avoid the intermolecular formation of μ-oxo complexes among porphyrin molecules, which would lead to their inactivation. In the presence of H_2O_2 as oxygen donor, metal-porphyrins can form very reactive metal-oxo components that are involved in oxidative coupling reactions of phenols and other organic compounds (Smejkalova and Piccolo, 2006; Fukushima *et al.*, 2007). Other investigations showed that metal-porphyrins efficiently induce the catalytic oxidation of hydrocarbons by using UV light as oxidizing agent (Maldotti *et al.*, 2002).

HS were shown to positively influence the oxidative degradation of pentachlorophenol under iron-porphyrin (FeP) biomimetic catalysis (Fukushima *et al.*, 2003). Phenolic compounds, such as catechol, caffeic acid, and *p*-coumaric acid, which are common constituents of natural HS, were found to undergo oxidative oligomerization catalyzed by metal-porphyrins either in the presence of an oxygen donor or dissolved O_2 under daylight (Smejkalova and Piccolo, 2006; Smejkalova *et al.*, 2006, 2007).

A direct polymerization of HS was found to occur in the presence of hydrogen peroxide as oxygen donor and under the catalysis of a water-soluble FeP (Piccolo *et al.*,

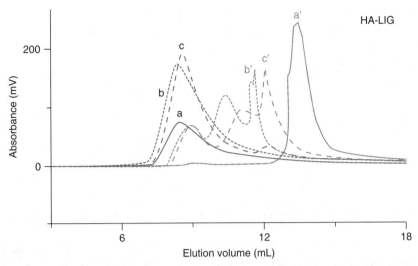

Figure 11.2   *HPSEC chromatograms of HA-LIG (Humic Acids extracted from a North Dakota Lignite) solutions under different treatments: (a) HA control solution before and (a') after acetic acid addition; (b) HA + FeP reaction solution without PCP and (b') after acetic acid addition; (c) HA + FeP + PCP reaction solution and (c') after acetic acid addition. Adapted from Fontaine and Piccolo, 2011.*

2005). Such polymerization of humic molecules took also place under solar electromagnetic radiation, thereby suggesting that dissolved oxygen alone may act as oxidant specie in the photooxidative coupling of humic components (Smejkalova and Piccolo, 2005). These results agree with the photocatalytic oxidation of hydrocarbons, alkanes and alkenes, obtained with synthetic metal-porphyrins in the only presence of dioxygen (Maldotti, 2002).

Moreover, Fontaine and Piccolo (2011) recently demonstrated that the removal of pentachlorophenol by oxidative FeP catalysis and in the presence of HS, was mainly due to a co-polymerization of the pollutant with humic molecules (Figure 11.2 and Table 11.5). In particular, the effects of the catalyst-assisted oxidative co-polymerization of PCP into the humic molecular structure were assessed by using High Performance Size Exclusion Chromatography (HPSEC) to measure the apparent weight-average molecular weight (Mw) of reaction products of two different humic acids (Figure 11.2).

The co-polymerized PCP-humic solutions were injected into the HPSEC system either before or after addition of acetic acid that lowered the mixture pH to 3.5. It has been shown that humic supramolecular conformations, stabilized mainly by hydrophobic interactions at pH 7, are altered when they are added with organic acids to lower the pH to 3.5, as new and stronger intermolecular hydrogen bonds are formed between complementary humic functions (Piccolo *et al.*, 2001). However, if humic conformations are more firmly stabilized by intermolecular covalent bonds induced by the catalytic oxidative polymerization, their alteration by pH lowering is significantly reduced. The limited capacity of acetic acid to modify the HPSEC chromatographic profile of both HS and HS+PCP solutions after their treatment with FeP suggested that intermolecular covalent bonds must have been formed among humic molecule and/or between these and PCP

Table 11.5 *Added and recovered (mg L⁻¹) pentachlorophenol (PCP), as detected by GC/ECD, when present alone in solution (blank), in addition with HA (PCP + HA), and in the co-polymerization reaction with iron-porphyrin (FeP) as oxidative catalyst (PCP + HA + FeP). Adapted from Fontaine and Piccolo, 2011*

Added PCP	Recovered PCP		
	Blank	PCP + HA	PCP + HA +FeP
HA–LIG			
10	9.68±0.20	9.61±0.42	0.02±0.00
5	4.66±0.36	3.84±0.24	0.21±0.02
1	1.28±0.05	0.21±0.01	0.04±0.01
0.5	0.39±0.03	0.42±0.04	0.22±0.01
0.1	0.09±0.00	0.10±0.01	0.05±0.01

(Fontaine and Piccolo, 2011). This result implied that the oxidative biomimetic catalysis was successful in stabilizing the loose humic conformation, thus proving the occurrence of both the HS polymerization and PCP-HS co-polymerization. These findings have highlighted that the catalyzed co-polymerization reaction between soil pollutants and HS may become a useful technology to remediate soils and waters contaminated by persistent aromatic compounds.

11.4.2   Nanocatalysts

Nanoporous materials are a large family of solid-state materials. According to the pore size definition by the International Union of Pure and Applied Chemistry (IUPAC), porous materials can be divided into three classes: microporous (<2 nm), mesoporous (2 ~ 50 nm) and macroporous materials (>50 nm). Most or parts of these three classes of materials can be included in the so-called nanoporous materials (1 ~ 100 nm), which have found important applications in the field of adsorption, separation, catalysis, optoelectric devices, petroleum and chemical industries, owing to their well-defined pore structures, large surface areas, high thermal stability, and tunable pore surface characters (Shi *et al.*, 2007).

To meet the ever increasing need for environmental protections and environmental friendly processes, nanoporous materials are expected to play important roles in the environmental field due to their high catalytic, adsorbing, and sensing performances. For example, metal-loaded composite catalysts show significantly lower ignition temperature for catalysis, higher stability, and lower noble metal consumption, as compared to the common noble metal-loaded zirconia/ceria powders (Shi *et al.*, 2007). The reason is the large dispersion of noble metal particles in the pore channels, which provide abundant active sites for the exhaust conversion and prevent aggregation of metal particles. The nanoporous materials, especially the conductive mesoporous carbons, are promising as electrode reaction catalysts for polymer electrolyte membrane cells (Antolini, 2009). Moreover, thanks to the very great surface area of the pore channels and their tunable wall and pore surface chemical properties, these materials show very promising applications for the detection and removal of harmful substances such as heavy metals, toxic gases and organic pollutants (Savage and Diallo, 2005).

An active research field is the photocatalysis for the decomposition of different organic pollutants (Marin *et al.*, 2012). In particular, the heterogeneous photocatalysis is a process that involves the absorption of radiation of appropriate wavelength by a semiconductor, which is present in a suspension containing the contaminant to be degraded. Because the process gradually breaks down the contaminant molecule, no residue of the original material is left and hence no sludge requiring landfill disposal is produced. The catalyst itself is unchanged during the process and no consumable chemicals are required. Such process results in considerable savings, simpler operation, and less equipment. Additionally, because the contaminant is attracted strongly to the surface of the catalyst, the process will continue to work at very low concentrations allowing sub part-per-million consents to be achieved (Marin *et al.*, 2012). All together, these advantages also mean lower water production cost and greater environmental protection.

The photocatalysis over a metal oxide semiconductors, such as TiO_2, is initiated by the absorption of a photon with energy equal or greater than the band gap of the semiconductor producing electron-hole (*e-/h*+) pairs, as in the following equation:

$$TiO_2 \xrightarrow{hv} e^-_{cb}(TiO_2) + h^+_{vb}(TiO_2)$$

where cb is conduction band and vb is the valence band. Consequently, following irradiation, the TiO_2 particle can act as either an electron donor or acceptor for molecules in the surrounding medium. The electron and the hole can recombine, releasing the absorbed light energy as heat, with no chemical effect. Otherwise, the charges can move to "trap" sites at slightly lower energies. The charges can still recombine, or they participate in redox reactions towards adsorbed species. A simplified mechanism for the photo-activation of a semiconductor catalyst is presented in Figure 11.3. The valence band hole is strongly oxidizing, and the conduction band electron is strongly reducing. At the external surface, the excited electron and the hole can take part in redox reactions with adsorbed species such as water, hydroxide ion (OH^-), organic compounds, or oxygen. The charges can react directly with adsorbed pollutants, but reactions with water are far more likely since the water molecules are far more frequent than contaminant molecules. Oxidation of water or OH^- by the hole produces the hydroxyl radical (•OH), an extremely powerful and unspecific oxidant.

Alvarez *et al.* (2007) demonstrated that 2,4-dichlorophenoxyacetic acid (2,4-D) was photodegraded by zirconium oxide doped with transition metals (Mn, Fe, Co, Ni and Cu). Moreover, Phanikrishna Sharma *et al.* (2008) showed that TiO_2 supported over SBA-15

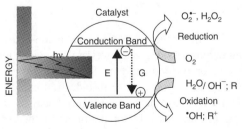

Figure 11.3 *Simplified mechanism for the photo-activation of a semiconductor catalyst. Adapted from Mills and Le Hunte, 1997.*

was an efficient photocatalyst for different pesticides, such as isoproturon, imidacloprid and phosphamidon. Occulti *et al.* (2008) were first to propose an integrated physicochemical approach to remediate PCB-polluted soil through a soil washing with a biosurfactant (soya lecithin) followed by a photocatalytic treatment (TiO_2).

Organic-inorganic hybrids can be applied in many areas of materials chemistry because they are simple to process and easy to be designed on the molecular scale. The most important advantage offered by these materials stems from the possibility of combining the mechanical, thermal, and structural stability of a rigid inorganic framework together with the high reactivity of the organic component.

In a special issue devoted to hybrid materials, Sanchez *et al.* (2011) reviewed the preparation methods, properties and applications of hybrid organic-inorganic materials. In particular, a growing interest has been shown for bio-hybrid materials formed by entrapping and/or encapsulating biomolecules within inorganic matrices, and for their applications in fields such as green energy production, smart cell-therapy and environmental remediation.

In the realm of environmental remediation, a class II hybrid sol-gel material was recently prepared starting from zirconium (IV) propoxide and 2,4-pentanedione (Aronne *et al.*, 2012). This material showed a catalytic activity in the transformation of the herbicide 4-chloro-2-methylphenoxyacetic acid (MCPA) that was found to be removed by 98% from aqueous solutions. A thermal and structural characterization, obtained by thermogravimetry, differential thermal analysis and DRIFT spectroscopy, showed the hybrid nature of the prepared material (Figure 11.4 and Figure 11.5). The material structure was described as a polymeric network of zirconium oxo clusters, on whose surface large part of Zr^{4+} ions are involved in strong complexation equilibria with acetylacetonate (*acac*) ligands. A two-step mechanism was proposed for the MCPA removal, in which a reversible first-order adsorption of the herbicide is followed by its catalytic degradation. Since neither the products of the MCPA catalytic degradation nor the adopted reaction conditions supported the typical oxidation pathways involving radicals, the existence of a different mechanism was suggested. This implied that the MCPA degradation was achieved by a Zr^{4+}:*acac* enol-type complex acting as Lewis acid catalyst (Aronne *et al.*, 2012). Moreover, the catalyst did not undergo significant structural modifications under the reaction conditions, thereby suggesting the possibility of its reuse. The high efficiency of this hybrid material, its stability, as well as the simplicity of its preparation makes it as a promising catalyst for the herbicides removal from soil.

11.5   Conclusions

Soil is an important ecosystem for the planet's life and a non-renewable natural resource. A vital soil provides numerous services to the welfare of society, and it is of fundamental importance for the production of food and feed. There is a growing necessity to recover soils which have been degraded after long term industrial exploitations, accidental disasters or intensive chemical amendments, and return them to sustainable agricultural production. Considering the current world population trend, the expected enhancement in

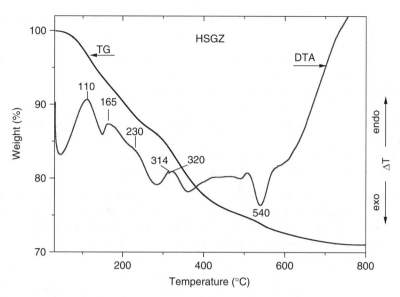

Figure 11.4 *TG/DTA curves of HSGZ matrix recorded in nitrogen at 10°C min⁻¹. Adapted from Aronne et al., 2012.*

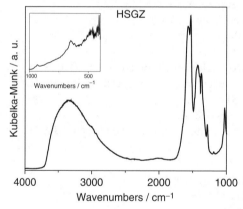

Figure 11.5 *DRIFT spectra recorded at room temperature of HSGZ matrix. Adapted from Aronne et al., 2012.*

feed and food demand, and the increasingly limited availability of productive agricultural land, it becomes essential to develop ecologically friendly technologies to remediate and restore polluted soils to their original agricultural quality.

In particular, new sustainable strategies ranging from biological to physical or advanced chemical processes can be applied for soil decontamination. However, it should be emphasized that benefits and limits must be taken into account for each technique. The useful criteria to select the best possible method for soil remediation can be grouped in the following technical, environmental and economic aspects:

- *Technical aspects* include information on soil physical and chemical properties, qualitative and quantitative analysis of soil pollutants, on general site climatic conditions (temperature, moisture, radiation, wind).
- *Environmental aspects* concern the occurrence of possible hazardous by-products arising from remediation processes and their impact on the quality of waters, soils, and sediments.
- *Economic aspects* consist in outlining and quantifying all costs inherent to the fulfillment of the technical and environmental criteria.

Consequently, the selection of the best method for soil remediation is always a complex matter, having to take into account a variety of factors, which can be altogether limited by their economic cost. In conclusion, research and development are still needed to improve emerging eco-friendly technologies and bring them at the market level for full-scale implementation.

List of Symbols and Acronyms

Anthropogenic Organic Pollutants (AOPs)
Aziende Chimiche Nazionali Associate (ACNA)
Critical Micelle Concentration (CMC)
DiChloroPhenol (DCP)
Ectomycorrhizal (ECM)
High Performance Size Exclusion Chromatography (HPSEC)
Humic Acids (HA) extracted from a North Dakota lignite (HA-LIG)
Humic Substances (HS)
Hybrid Sol-Gel Zirconia (HSGZ)
International Union of Pure and Applied Chemistry (IUPAC)
Laccase (L)
Lignin Peroxidase (LiP)
Lignin-degrading Enzyme System (LDS)
Natural Organic Matter (NOM)
Pentachlorophenol (PCP)
Phenanthrene (PHE)
PolyChlorinated Biphenyls (PCBs)
Polycyclic Aromatic Hydrocarbons (PAHs)
Pyrene (PYR)
Sodium dodecyl sulfate (SDS)
Soil Organic Matter (SOM)
Soya Lecithin (SL)
TriChloroPhenol (TCP)
Vescicular-arbuscular mycorrhiza (VAM)
cb conduction band
vb valence band

Acknowledgments

The authors gratefully acknowledge the collaboration of Dr Barbara Fontaine and Dr Anna Agretto in providing some of the experimental data.

References

E. Antolini, Carbon supports for low-temperature fuel cell catalysts, *Appl. Catal. B-Environ.*, **88**, 1–24 (2009).

A. Aronne, F. Sannino, S.R. Bonavolontà, *et al.*, Use of a new hybrid sol-gel zirconia matrix in the removal of the herbicide MCPA: a sorption/degradation process, *Environ. Sci. Technol.*, **46**, 1755–1763 (2012).

I.M. Adekunle, Bioremediation of soils contaminated with Nigerian petroleum products using composted municipal wastes, *Bioremediation Journal*, **15**, 230–241 (2011).

M. Alcade, M. Ferrer, F.J. Plou, and A. Ballesteros, Environmental biocatalysis: from remediation with enzymes to novel green processes, *Trends Biotechnol.*, **24**, 281–287 (2006).

M. Alexander, *Biodegradation and Bioremediation*, Academic Press, San Diego, (1999).

M. Alvarez, T. Lopez, J.A. Odriozola, *et al.*, 2,4-Dichlorophenoxyacetic acid (2,4-D) photodegradation using an M^{n+}/ZrO_2 photocatalyst: XPS, UV-vis, XRD characterization, *Appl. Catal. B-Environ.*, **73**, 34–41 (2007).

N. Aristov and A. Habekost, Heterogeneous dehalogenation of PCBs with iron/toluene or iron/quicklime, *Chemosphere*, **80**, 113–115 (2010).

C.J. Atkinson, J.D. Fitzgerald, and N.A. Hipps, Potential mechanisms for achieving agricultural benefits from biochar application to temperate soils: a review, *Plant Soil*, **337**, 1–18 (2010).

J.L. Baeseman and P.J. Novak, Effects of various environmental conditions on the transformation of chlorinated solvents by *Methanosarcina thermophila* cell exudates, *Biotechnol. Bioeng.*, **75**, 634–641 (2001).

P. Baldrian, Fungal laccases-occurrence and properties, *FEMS Microbiol. Rev.*, **30**, 215–242 (2006).

L. Beesley, E. Moreno-Jiménez, J.L. Gomez-Eyles, *et al.*, A review of biochars' potential role in the remediation, revegetation and restoration of contaminated soils, *Environ. Pollut.*, **159**, 3269–3282 (2011).

S. Berselli, G. Milone, P. Canepa, *et al.*, Effects of cyclodextrins, humic substances, and rhamnolipids on the washing of a historically contaminated soil and on the aerobic bioremediation of the resulting effluents, *Biotechnol. Bioeng.*, **88**, 111–120 (2004).

D. Blaudez, B. Botton, and M. Chalet, Cadmium uptake and subcellular compartmentation in the ectomycorrhizal fungus *Paxillus involutus*, *Microbiology*, **146**, 1109–1117 (2000).

J.M. Bollag, R. Chu, M.A. Rao, and L. Gianfreda, Enzymatic oxidative transformation of chlorophenol mixtures, *J. Environ. Qual.*, **32**, 62–71 (2003).

J. Buschmann, W. Angst, and R.P. Schwarzenbach, Iron porphyrin and cysteine mediated reduction of ten polyhalogenated methanes in homogeneous aqueous solution: product analyses and mechanistic considerations, *Environ. Sci. Technol.*, **33**, 1015–1020 (1999).

J.A. Bumpus, White-rot fungi and their potential use in soil bioremediation processes, in *Soil Biochemistry*, J.M. Bollag and G. Stotzky (eds), Marcel Dekker, New York, (1993).

M.D. Cameron and S.D. Aust, Degradation of chemicals by reactive radicals produced by cellobiose dehydrogenase from *Phanerochaete chrysosporium*, *Arch. Biochem. Biophys.*, **367**, 115–121 (1999).

M.D. Cameron, S. Timofeevski, and S.D. Aust, Enzymology of *Phanerochaete chrysosporium* with respect to the degradation of recalcitrant compounds and xenobiotics, *Appl. Microbiol. Biotechnol.*, **54**, 751–758 (2000).

C. Castagnero, Economic realities of using compostable products, *BioCycle*, **52**, 45–47 (2011).

G. Celano, R. Spaccini, D. Smejkalova, and A. Piccolo, Interactions of three s-triazines with humic acids of different structure, *J. Agr. Food. Chem.*, **56**, 7360–7366 (2008).

C.T. Chiou, Partition and adsorption of organic contaminants in environmental systems, John Wiley & Sons, Inc., New York (2002).

C.L. Chun, J.J. Lee, and J.W. Park, Solubilization of PAH mixtures by three different anionic surfactants, *Environ. Pollut.*, **118**, 307–313 (2002).

C. Collins, M. Fryer, and A. Grosso, Plant uptake of non-ionic organic chemicals, *Environ. Sci. Technol.*, **40**, 45–52 (2006).

P. Conte, A. Agretto, R. Spaccini, and A. Piccolo, Soil remediation: humic acids as natural surfactants in the washings of highly contaminated soils, *Environ. Pollut.*, **135**, 515–522 (2005).

C.J. Cunningham and J.C. Philp, Comparison of bioaugmentation and biostimulation in ex situ treatment of diesel contaminated soil, *Land Contamination & Reclamation*, **8**, 261–269 (2000).

G.P. Curtis and M. Reinhard, Reductive dechlorination of hexachloroethane, carbon tetrachloride, and bromoform by anthrahydroquinone disulfonate and humic acid, *Environ. Sci. Technol.*, **28**, 2393–2401 (1994).

F. Fava, S. Berselli, P. Conte, *et al.*, Effects of humic substances and soya lecithin on the aerobic bioremediation of a soil historically contaminated by Polycyclic Aromatic Hydrocarbons (PAHs), *Biotechnol. Bioeng.*, **88**, 214–223 (2004).

F. Fava and A. Piccolo, Effects of humic substances on the bioavailability and aerobic biodegradation of polychlorinated biphenyls in a model soil, *Biotechnol. Bioeng.*, **77**, 204–211 (2002).

M.T. Filazzola, F. Sannino, M.A. Rao, and L. Gianfreda, Effect of various pollutants and soil-like constituents on laccase from *Cerrena unicolor*, *J. Environ. Qual.*, **28**, 1929–1938 (1999).

B. Fontaine and A. Piccolo, Co-polymerization of penta-halogenated phenols in humic substances by catalytic oxidation using biomimetic catalysis, *Environ. Sci. Pollut. Res.*, DOI 10.1007/s11356-011-0626-x (2011).

K. Fukuda, O. Soderman, B. Lindman, and K. Shinoda, Microemulsions formed by alkyl polyglucosides and an alkyl glycerol ether, *Langmuir*, **9**, 2921–2925 (1993).

M. Fukushima and K. Tatsumi, Degradation pathways of pentachlorophenol by photo-Fenton systems in the presence of iron(III), humic acid and hydrogen peroxide, *Environ. Sci. Technol.*, **35**, 1771–1778 (2001).

M. Fukushima, A. Sawada, M. Kawasaki, *et al.*, M. Aoyama, Influence of humic substances on the removal of pentachlorophenol by a biomimetic catalytic system with a water-soluble iron(III)porphyrin complex, *Environ. Sci. Technol.*, **37**, 1031–1036. (2003).

M. Fukushima, K. Tatsumi, and K. Morimoto, Influence of iron(III) and humic acid on the photodegradation of pentachlorophenol, *Environ. Toxicol. Chem.*, **19**, 1711–1716 (2000).

M. Fukushima, Y. Tanabe, K. Morimoto, and K. Tatsumi, Role of humic acid fraction with higher aromaticità in enhancing the activity of a biomimetic catalyst, tetra(*p*-sulfonatophenyl)porphyneiron(III), *Biomacromolecules*, **8**, 386–391 (2007).

D. Garon, S. Krivobok, D. Wouessidjewe, and F. Seigle-Murandi, Influence of surfactants on solubilization and fungal degradation of fluorine, *Chemosphere*, **47**, 303–309 (2002).

L. Gianfreda and J.M. Bollag, Isolated enzymes for the transformation and detoxification of organic pollutants, in *Enzymes in the Environment: Activity, Ecology and Applications*, R.G. Burns and R. Dick (eds), Marcel Dekker, New York (2002).

L. Gianfreda and M.A. Rao, Potential of extracellular enzymes in remediation of polluted soils: a review, *Enzyme Microb. Tech.*, **35**, 339–354 (2004).

L. Gianfreda, F. Sannino, M.T. Filazzola, and A. Leonowicz, Catalytic behavior and detoxifying ability of a laccase from the fungal strain *Cerrena unicolor*, *J. Mol. Catal. B: Enzymatic*, **14**, 13–23 (1998).

L. Gianfreda, F. Sannino, M.A. Rao, and J.M. Bollag, Oxidative transformation of phenols in aqueous mixtures, *Water. Res.*, **37**, 3205–3215 (2003).

R.A. Griffiths, Soil washing technology and practice, *J. Hazard. Mater.*, **40**, 175–189 (1995).

D. Hahn, A. Cozzolino, A. Piccolo, and P.M. Arenante, Reduction of 2,4-dichlorophenol toxicity to *Pseudomonas putida* after oxidative incubation with humic substances and a biomimetic catalyst, *Ecotox. Environ. Saf.*, **66**, 335–342 (2007).

S.M. Henry and S.D. Warner, Chlorinated solvent and DNAPL remediation: An overview of physical, chemical, and biological processes. In *Chlorinated Solvent and DNAPL Remediation: Innovative Strategies for Subsurface Cleanup*, S.M. Henry, C.H. Hardcastle, S.D. Warner (eds) ACS Symposium Series, **837**, 1–20 (2003).

T. Iwamoto and M. Nasu, Current bioremediation practice and perspective, *J. Biosci. Bioeng.*, **92**, 1–8 (2001).

D.B. Janssen, I.J.T. Dinkla., G.J. Poelarends, and P. Terpstra, Bacterial degradation of xenobiotic compounds: evolution and distribution of novel enzyme activities, *Environ. Microbiol.*, **7**, 1868–1882 (2005).

D.L. Jones, D.V. Murphy, M. Khalid, *et al.*, Short-term biochar-induced increase in soil CO_2 release is both biotically and abiotically mediated, *Soil Biol. Biochem.*, **43**, 1723–1731 (2011).

S. Jonsson, H. Lind, S. Lundstedt, *et al.*, Dioxin removal from contaminated soils by ethanol, *J. Hazard. Mat.*, **179**, 393–399 (2010).

S.J. Joshi and A.J. Desai, Biosurfactant's role in bioremediation of NAPL and fermentative production, *Adv. Exp. Med. Biol.*, **672**, 222–235 (2010).

A.A. Khan and Q. Husain, Decolorization and removal of textile and non-textile dyes from polluted wastewater and dyeing effluent by using potato (*Solanum tuberosum*) soluble and immobilized polyphenol oxidase, *Biores. Technol.*, **98**, 1012–1019 (2007).

M.I. Kuhlman and T.M. Greenfield, Simplified soil washing processes for a variety of soils. *J. Hazard. Mater.*, **66**, 31–45 (1999).

G. Labat, J.L. Seris, and B. Meunier, Oxidative degradation of aromatic pollutants by chemical models of ligninase based on porphyrin complexes, *Angew. Chem. Int. Ed. Engl.*, **29**, 1471–1473 (1990).

K.T. Leung, K. Nandakumar, K. Sreekumari, *et al.*, Biodegradation and bioremediation of organic pollutants in soil, in *Modern Soil Microbiology*, J.D. Van Elsas, J.K. Jansson, and J.T. Trevors (eds), CRC Press, Boca Raton, FL (2007).

X. Li and P. Christie, Changes in soil solution Zn and pH and uptake of Zn by arbuscular mycorrhizal red clover in Zn-contaminated soil, *Chemosphere*, **42**, 201–207 (2001).

T.M.S. Lima, L.C. Procópio, F.D. Brandão, *et al.*, Biodegradability of bacterial surfactants, *Biodegradation*, **22**, 585–592 (2011).

E. Madsen, *Report on Bioavailability of Chemical Wastes with Respect to the Potential for Soil Bioremediation*. U.S. Environmental Protection Agency report EPA/600/R-03/076. U.S. Environmental Agency, Washington, D.C (2003).

R. Maier, Bioavailability and its importance to bioremediation, in *International Society for Environmental Biotechnology: Environmental Monitoring and Biodiagnostics*, J.J. Valdes (ed), Kluwer, Dordrecht, The Netherlands (2000).

A. Maldotti, A. Molinari, and R. Amadelli, Photocatalysis with organized systems for the oxofunctionalization of hydrocarbons by O_2, *Chem. Rev.*, **102**, 3811–3836 (2002).

M.L. Marin, L. Santos-Juanes, A. Arques, *et al.*, Organic photocatalysts for the oxidation of pollutants and model compounds, *Chem. Rev.*, **112**, 1710–1750 (2012).

E.M. Martin Del Valle, Cyclodextrins and their uses: a review, *Process Biochem.*, **39**, 1033–1046 (2003).

S.C. McCutcheon and J.L. Schnoor, Overview of phytotransformation and control of wastes, p. 3–58. In *Phytoremediation: Transformation and Control of Contaminants*, S.C. McCutcheon and J.L. Schnoor (eds) John Wiley & Sons, Inc., New York (2003).

E. Mena, A. Garrido, T. Hernández, and C. García, Bioremediation of sewage sludge by composting, *Commun. Soil Sci. Plan.*, **34**, 957–971 (2003).

A.C. Metcalfe, M. Krsek, G.W. Gooday, *et al.*, Molecular analysis of a bacterial chitinolytic community in an upland pasture, *Appl. Environ. Microbiol.*, **68**, 5042–5050 (2002).

A. Mills and S. Le Hunte, An overview of semiconductor photocatalysis, *J. Photoch. Photobio. A*, **108**, 1–35 (1997).

P. Miretzkyand and A.F. Cirelli, Remediation of arsenic-contaminated soils by iron amendments: a review, *Crit. Rev. Env. Sci. Tech.*, **40**, 93–115 (2010).

E. Montoneri, V. Boffa, P. Savarino, *et al.*, Use of biosurfactants from urban wastes compost in textile dyeing and soil remediation, *Waste Manage.*, **29**, 283–289 (2009).

K. Morimoto, K. Tatsumi, and K. Kuroda, Peroxidase catalyzed co-polymerization of pentachlorophenol and a potential humic precursor, *Soil Biol. Biochem.*, **32**, 1071–1077 (2000).

A. Mrozik and Z. Piotrowska-Seget, Bioaugmentation as a strategy for cleaning up of soils contaminated with aromatic compounds, *Microbiol. Res.*, **165**, 363–375 (2010).

C.N. Mulligan, Environmental applications for biosurfactants, *Environ. Pollut.*, **133**, 183–198 (2005).

K. Nam and M. Alexander, Role of nanoporosity and hydrophobicicty in sequestration and bioavailability: test with model soils, *Environ. Sci. Technol.*, **32**, 71–74 (1998).

A. Navarro and F. Martinez, The use of soil-flushing to remediate metal contamination in a smelting slag dumping area: column and pilot-scale experiments, *Eng. Geol.*, **115**, 16–27 (2010).

F. Occulti, G.C. Roda, S. Berselli, and F. Fava, Sustainable decontamination of an actual-site aged PCB-polluted soil through a biosurfactant-based washing followed by a photocatalytic treatment, *Biotechnol. Bioeng.*, **99**, 1525–1534 (2008).

M. Pacwa-Płociniczak, G.A. Płaza, Z. Piotrowska-Seget, and S.S. Cameotra, Environmental applications of biosurfactants: recent advances, *Int. J. Mol. Sci.*, **12**, 633–654 (2011).

T.M. Pankrantz, *Environmental Engineering Dictionary and Directory*, CRC Press Boca Raton, FL (2001).

J.Y. Park, H.H. Lee, S.J. Kim, *et al.*, Surfactant-enhanced electrokinetic removal of phenantrene from kaolinite, *J. Hazard. Mater.*, **140**, 230–236 (2007).

Z.D. Parrish, M.K. Banks, and A.P. Schwab, Assessment of contaminant lability during phytoremediation of polycyclic aromatic hydrocarbon impacted soil, *Environ. Pollut.*, **137**, 187–197 (2005).

A.Y. Pasha, L.M. Hu, J.N. Meegoda, *et al.*, Centrifuge modeling of in situ surfactant enhanced flushing of diesel contaminated soil. *Geotech. Test. J.*, **34**, 623–633 (2011).

J. Peng, Y. Song, P. Yuan, *et al.*, The remediation of heavy metals contaminated sediment, *J. Hazard. Mater.*, **161**, 633–640 (2009).

M.V. Phanikrishna Sharma, V. Durga Kumari, and M. Subrahmanyam TiO_2 supported over SBA-15: an efficient photocatalyst for the pesticide degradation using solar light, *Chemosphere*, **73**, 1562–1569 (2008).

J.J. Piatt and M.L. Brusseau, Rate limiting sorption of hydrophobic organic compounds by soils with well characterised organic matter, *Environ. Sci. Technol.*, **32**, 1604–1608 (1998).

A. Piccolo, P. Conte, and A. Cozzolino, Chromatographic and spectrophotometric properties of dissolved humic substances compared with macromolecular polymer, *Soil Sci.*, **166**, 174–185 (2001).

A. Piccolo, P. Conte, I. Scheunert, and M. Paci, Atrazine interactions with soil humic substances of different molecular structure, *J. Environ. Q.*, **27**, 1324–1333 (1998).

A. Piccolo, P. Conte, and P. Tagliatesta, Increased conformational rigidity of humic substances by oxidative biomimetic catalysis, *Biomacromolecules*, **6**, 351–358 (2005).

A. Piccolo, The supramolecular structure of humic substances: a novel understanding of humus chemistry and implications in soil science, *Adv. Agron.*, **75**, 57–134 (2002).

E. Pilon-Smits, Phytoremediation. *Annu. Rev. Plant Biol.*, **56**, 15–39 (2005).

R. Pinton, Z. Varanini, and P. Nannipieri, *The Rhizosphere: Biochemistry and Organic Substances at the Soil-Plant Interface*, CRC/Taylor & Francis, London (2007).

S.B. Pointing, Feasibility of bioremediation by white-rot fungi, *Appl. Microbiol. Biotechnol.*, **57**, 20–32 (2001).

M.N.V. Prasad and M. Freitas, Metal hyperaccumulation in plants – biodiversity prospecting for phytoremediation technology, *J. Biotechnol.*, **6**, 276–313 (2003).

H. Prosen and L. Zupancic-Kralj, Evaluation of photolysis and hydrolysis of atrazine and its first degradation products in the presence of humic acids, *Environ. Pollut.*, **133**, 517–529 (2005).

A.S. Purnomo, T. Mori, I. Kamei, and R. Kondo, Basic studies and applications on bioremediation of DDT: a review, *Int. Biodeter. Biodegr.*, **65**, 921–930 (2011).

P.L. Quagliotto, E. Montoneri, F. Tambone, *et al.*, Chemicals from wastes: compost-derived humic acid like matter as surfactant, *Environ. Sci. Technol.*, **40**, 1686–1692 (2006).

J.R. Quilty and S.R. Cattle, Use and understanding of organic amendments in Australian agriculture: A review, *Soil Research*, **49**, 1–26 (2011).

C.A. Reddy, The potential for white-rot fungi in the treatment of pollutants, *Curr. Opin. Biotechnol.*, **6**, 320–328 (1995).

C. Richard, O. Trubetskaya, O. Trubetskoj, *et al.*, Key Role of the low molecular size fraction of soil humic acids for fluorescence and photoinductive activity, *Environ. Sci. Technol.*, **38**, 2052–2057 (2004).

L.F.W. Roesch, R.R. Fulthorpe, A. Riva, *et al.*, Pyrosequencing enumerates and contrasts soil microbial diversity, *ISME J.*, **1**, 283–290 (2007).

N. Rogovska, D. Laird, R. Cruse, *et al.*, Impact of biochar on manure carbon stabilization and greenhouse gas emissions, *Soil Sci. Soc. Am. J.*, **75**, 871–879 (2011).

S. Salati, G. Papa, and F. Adani, Perspective on the use of humic acids from biomass as natural surfactants for industrial applications, *Biotechnol. Adv.*, **29**, 913–922 (2011).

C. Sanchez, K.J. Shea, and S. Kitagawa, Recent progress in hybrid materials science, *Chem. Soc. Rev.*, **40**, 471–472 (2011).

M. Sandbacka, I. Christianson, and B. Isomaa, The acute toxicity of surfactants on fish cells, *Daphnia magna* and fish – a comparative study, *Toxicol. in Vitro*, **14**, 61–68 (2000).

F. Sannino and L. Gianfreda, An oxidative enzyme for the detoxification of polluted systems, in *Research Advances in Water Research*, A. Wagener, G. Perin, J.L. Alonso and R.M. Mohan (eds), (2000).

F. Sannino, M.T. Filazzola, and L. Gianfreda, Fate of herbicides influenced by biotic and abiotic interactions, *Chemosphere*, **39**, 333–341 (1999).

N. Savage and M. Diallo, Nanomaterials and water purification: opportunities and challenges, *J. Nanopart. Res.*, **7**, 331–342 (2005).

R.P. Schwarzenbach, W. Angst, C. Holliger, *et al.*, Reductive transformations of anthropogenic chemicals in natural and technical systems, *Chimia*, **51**, 908–914 (1997).

K.T. Semple, B.J. Reid, and T.R. Fermor, Impact of composting strategies on the treatments of soils contaminated with organic pollutants, *Environ. Pollut.*, **112**, 269–283 (2001).

S. Shackley, S. Carter, K. Sims, and S. Sohi, Expert perceptions of the role of biochar as a carbon abatement option with ancillary agronomic and soil-related benefits, *Energ. Environ. Sci.*, **22**, 167–188 (2011).

J. Shi, H. Chen, Z. Hua, and L. Zhang, Synthesis and properties of mesoporous-based materials for environmental applications, in *Environmental Applications of Nanomaterials, Synthesis, Sorbents and Sensors*, E.G. Fryxell and G. Cao (eds), Imperial College Press, London (2007).

C.J. Singh, Optimization of an extra cellular protease of *Chrysosporium keratinophilum* and its potential in bioremediation of keratinic wastes. *Mycopathologia*, **156**, 151–156 (2002).

S.N. Singh and R.D. Tripathi, *Environmental Bioremediation Technologies*, Springer, New York (2007).

H. Singh, Mycoremediation: Fungal Bioremediation, John Wiley & Sons, Inc., Hoboken (2006).

R.K. Sinha, S. Herat, and P.K. Tandon, Phytoremediation: role of plants in contaminated site management, in *Environmental Bioremediation Technologies*, S.N. Singh and R.D. Tripathi (eds), Springer, New York (2007).

D. Smejkalova and A. Piccolo, Rates of oxidative coupling of humic phenolic monomers catalyzed by a biomimetic iron-porphyrin, *Environ. Sci. Technol.*, **40**, 1644–1649 (2006).

D. Smejkalova, A. Piccolo, and M. Spiteller, Oligomerization of humic phenolic monomers by oxidative coupling under biomimetic catalysis, *Environ. Sci. Technol.*, **40**, 6955–6962 (2006).

D. Smejkalova, P. Conte, and A. Piccolo, Structural characterization of isomeric dimers from the oxidative oligomerization of catechol with a biomimetic catalyst. *Biomacromolecules*, **8**, 737–743 (2007).

D. Smejklova and A. Piccolo, Enhanced molecular dimensions of a humic acid induced by photooxidation catalyzed by biomimetic metalporphyrins, *Biomacromolecules*, **6**, 2120–2125 (2005).

S.E. Smith and J.D. Read, *Mycorrhizal Symbiosis*, 2nd edn. Academic Press, London (1997).

R. Song, A. Sorokin, J. Bernardou, and B. Meunier, Metal porphyrin-catalyzed oxidation of 2-methylnaphthalene to vitamin k3 and 6-methyl-1,4-naphtoquinone by potassium monopersulfate in aqueous solution, *J. Org. Chem.*, **62**, 673–678 (1997).

R. Spaccini and A. Piccolo, Molecular characteristics of humic acids extracted from compost at increasing maturity stages, *Soil Biol. Biochem.*, **41**, 1164–1172 (2009).

M. Tejada, Application of different organic wastes in a soil polluted by cadmium: effects on soil biological properties, *Geoderma*, **153**, 254–268 (2009).

Y. Teng, Y. Shen, Y. Luo, *et al.*, Influence of *Rhizobium meliloti* on phytoremediation of polycyclic aromatic hydrocarbons by alfalfa in an aged contaminated soil, *J. Hazard. Mater.*, **186**, 1271–1276 (2011).

M. Varun, R. D'Souza, J. Pratas, and M.S. Paul, Metal contamination of soils and plants associated with the glass industry in North Central India: prospects of phytoremediation, *Environ. Sci. Pollution R.*, **19**, 269–281 (2012).

D. Vetterlein and R. Jahn. Gradients in soil solution composition between bulk soil and rhizosphere-in situ measurement with changing soil water content, *Plant Soil*, **258**, 307–327 (2004).

J.M. Wang, E.M. Marlowe, R.M. Miller-Maier, and M.L. Brusseau, Cyclodextrin-enhanced biodegradation of phenanthrene, *Environ. Sci. Technol.*, **32**, 1907–1912 (1998).

M.C. Wang, Y.T. Chen, S.H. Chen, *et al.*, Phytoremediation of pyrene contaminated soils amended with compost and planted with ryegrass and alfalfa, *Chemosphere*, **87**, 217–225 (2012).

Z.Y. Wang, Y. Xu, J. Zhao *et al.*, Remediation of petroleum contaminated soils through composting and rhizosphere degradation, *J. Hazard. Mat.*, **190**, 677–685 (2011).

A.W. Watts, T.P. Ballestero, and H.K. Gardner, Uptake of polycyclic aromatic hydrocarbons (PAHs) in salt marsh plants *Spartina alterniflora* grown in contaminated sediments, *Chemosphere*, **62**, 1253–1260 (2006).

C.C. West and J.H. Harwel, Surfactants and subsurface remediaton, *Environ. Sci. Technol.* **26**, 234–2330 (1992).

F.Y. Wu, X.Z. Yu, S.C. Wu, *et al.*, Phenanthrene and pyrene uptake by arbuscular mycorrhizal maize and their dissipation in soil, *J. Hazard. Mater.*, **187**, 341–347 (2011a).

G.Z. Wu, F. Coulon, H. Li, *et al.*, Recycling of solvent used in a solvent extraction of petroleum hydrocarbons contaminated soil, *J. Hazard. Mat.*, **186**, 533–539 (2011b).

A.R. Zimmerman, B. Gao, and M.Y. Ahn, Positive and negative carbon mineralization priming effects among a variety of biochar-amended soils, *Soil Biol. Biochem.*, **43**, 1169–1179 (2011).

12

Nanoparticles as a Smart Technology for Remediation

Giuseppe Chidichimo[1], Daniela Cupelli[2], Giovanni De Filpo[1], Patrizia Formoso[1], and Fiore Pasquale Nicoletta[1]

[1]*University of Calabria, Department of Chemistry and Chemical Technologies, University of Calabria, Rende, Italy*
[2]*University of Calabria, Department of Pharmacy, Health and Nutritional Sciences, University of Calabria, Rende, Italy*

12.1 Introduction

Environmental remediation deals with the removal of contaminants from environment for maintaining and restoring the quality of soil, water, and air and, more in general, for protecting human health. Remediation is based on many technologies and can be divided in two main categories: *ex situ*, where contaminated media are "excavated or pumped" to the surface and purified, and *in situ*, where pollutants are removed without removing media.

Remediation with engineered nanomaterials and, in particular, with nanoparticles (NPs) represents a smart and cheap method to clean air and decontaminate soils and waters from pollutants; this is, of course, a great challenge. Nanotechnology deals with materials, one dimension of which is less than 100 nm. The very low dimensions of nanomaterials confer upon them physico-chemical properties which can differ from those of the same materials in the bulk form: for example, bulky gold looks yellow, whereas a solution of gold NPs reflects red light. Due to their small size NPs are characterized by a high surface area to volume ratio showing larger activities in surface related phenomena (e.g., adsorption, reaction rates, etc.) than bulky systems with same mass.

Sustainable Development in Chemical Engineering – Innovative Technologies, First Edition.
Edited by Vincenzo Piemonte, Marcello De Falco and Angelo Basile.
© 2013 John Wiley & Sons, Ltd. Published 2013 by John Wiley & Sons, Ltd.

Consequently, there has been a very rapid spread of nanotechnologies in many fields of daily life including fillers, cosmetics, catalysts, pharmaceuticals, lubricants, electronic devices, advanced materials, biomedicine, and imaging tools [1].

Environmental remediation is based on adsorptive or reactive methods applied both in situ and ex situ [2]. Both methods allow one to state that NPs for remediation can solve big problems with little particles [3].

Some environmental applications include the use of nanoscale zero-valent iron for soil and water decontamination based on redox reactions, NPs for adsorption of heavy metals and organic compounds from wastewaters, and titanium dioxide NPs for photo-catalytic degradation of organic pollutants. This chapter will focus on silica and titania NPs for their adsorption, and photo-catalysis-assisted remediation of soils and groundwater. NPs can easily adsorb pollutants from contaminated areas: then, they can be removed from soils and wastewaters. An easier recovery of NPs can be gained if they are made with magnetic materials: a wide overview is dedicated to magnetic NP synthesis and characterization. Environmental remediation by NPs requires addition of nanomaterials to soil and groundwater exposing environment to a possible source of secondary pollution. In fact, it is known that NPs can be easily ingested, inhaled or permeated through the skin of the human body and taken up by animal cells [4]. Nevertheless the toxicity of nanomaterials to humans is still unknown. Some factors influencing the toxicity of NPs will be outlined.

12.2   Silica Nanoparticles for Wastewater Treatment

12.2.1   Silica Nanoparticles: An Overview

Silica NPs can be considered, without a doubt, to be the most investigated class of nanoparticles for environmental remediation.

Silica NPs are generally irregular shaped and can exist in fused, aggregated, or agglomerated forms [5]. Naturally occurring silicas, such as quartz sands, rocks, and clays, are used as raw materials in nanotechnology. In order to produce industrial silica products, such as silica sol (colloidal silica), silica gel, fumed silica (pyrogenic silica), and precipitated silica, primary raw materials are chemically treated to produce direct-silica sources, such as sodium silicate, silicon tetrachloride, and alkoxysilane. At present, a wide range of silica products is manufactured for applications in optical communications, thin-film technology, microelectronics, and so on. [6–9]. Nanosilica is also used in pharmacy as a booster agent [10] and enterosorbent [11], in medicine as carrier for drug delivery [12, 13]. Gold-coated silica has been used in treatments for benign and malignant tumours [14–16].

A more ambitious use of nanosilicas is remediation of contaminated environments and, recently, their application in wastewater treatment has emerged as a fast-developing and attractive area of interest. Because of high surface area, silica NPs show a very good adsorption capacity and a large specific reactivity [6, 17, 18]. Silica NPs have a hydrophilic surface due to the presence of silanol groups, which are weakly acidic and, therefore, are very reactive. Moreover, these groups allow chemical modifications of particle surface, which can be transformed into a hydrophobic or organophilic form [19].

Nanosilica clusters, $(SiO_2)_n$, are characterized by a chain-like geometry, if $n > 12$, and some Si atoms do not have a full fourfold coordination [5]. Hydrophilic nanosilica (e.g.

A-300 HDK® by Wacker Silicones) is made up of $SiO_{4/2}$ tetrahedrons joined together by siloxane bridges (Si–O–Si bonds) and contain around two silanol groups ($\equiv$Si–OH) per nm^2 [20]. Silanol groups act as reactive centres and form hydrogen bonds with polar substances.

If Si–OH groups are allowed to react with organosilicon compounds, hydrophobic properties can be induced on NP surface. HDK® hydrophobic nanosilicas can have 0.5–1.0 Si–OH groups per nm^2.

Chemical addition of hydrophobic agents to silica surfaces is achieved by means of hydrolysis-resistant Si–O–Si bonds. Frequently, hydrophobic surface modifications are obtained by means of modifiers such as organochlorosilanes, polydimethylsiloxanes, hexamethyldisilazane, and long chain alkylsilanes, such as octylsilanes [21–24].

Recent studies have shown that hydrophilic and hydrophobic silica NPs are very efficienct in removing pollutants such as dyes, metallic species, and pesticides from water, and wastewaters. Another interesting benefit of silica NPs as adsorbents is that they are considered non-toxic and not harmful to the environment. Moreover, nanosilica-based technology shows an attractive cost-effectiveness ratio as it requires relatively cheap processing of raw materials.

12.2.2 Preparation of Nanosilica

Among all methods existing to produce NPs [25–27, 18] the sol–gel process and flame synthesis are the main routes. In a sol–gel process silica clusters are allowed to condense to colloidal particles, which form aggregates. The sol–gel technique produces high purity products and also offers a control in composition and structure at a molecular level [26, 28, 29].

Pyrogenic or fumed silica NPs are formed from the gas phase at high temperature. Usually, highly dispersed silica NPs can be prepared from volatile silicon tetrachloride by flame hydrolysis. This is continually vaporized, mixed with dry air and hydrogen, fed to a burner, and finally, hydrolyzed [20].

SiO_2 primary particles are characterized by nanometer diameters (5–50 nm) and non-porous surfaces. In the oxyhydrogen flame primary particles fuse together permanently to form larger units (100–1000 nm in size) with a planar or angular structure. On cooling, aggregates mechanically entangle to form agglomerates, known as tertiary structures, of about 1–250 µm in size, which contain primary particles fused to form a face-to-face sintered structure. These agglomerates are mesoporous due to their open-structure and have a very high specific surface area because of small diameters of primary particles.

12.2.3 Removal of Dyes by Silica Nanoparticles

Dyes are the first pollutants to be recognized in wastewaters [30]. The presence of very small amounts of dyes in water (sometimes less than 1 ppm for some dyes) is highly visible and undesirable [31, 32]. Dye contamination exists in wastewaters of many industries (dyestuffs, textiles, papers, printings, leathers, cosmetics, plastics, paints, tanneries, pharmaceuticals, petrochemicals, etc.) [33]. At present, large amounts of dyes produced annually, are discharged directly in industrial aqueous effluent [34]. Due to their good solubility, the release of some dyes in water streams has serious environmental impact. Many dyes, especially those that are not easily biodegradable, are toxic and cause health problems such as skin irritation, allergic dermatitis, cancer, and mutation in humans. In addition, different dyes pose a serious hazard to aquatic living organisms [35, 36] because they adsorb and reflect sunlight within water media and, therefore, hinder photosynthesis of aquatic plants.

In the last 40 years, several physical, chemical, and biological decolorization methods have been developed for wastewater treatments [37]. Adsorption is the most frequently used procedure and gives the best results with different types of non-degradable dyes. In particular, it is possible to remove dyes which are resistant to aerobic digestion and/or oxidizing agents or are very stable to light and/or heat [32, 33, 38, 39]. If an adsorption system is designed correctly it will produce a high-quality treated effluent and can provide effective and economical removal of pollutants.

Recently, silica particles have been suitably designed to control removal of anionic dyes [40]. Silica based adsorbents are of particular interest because of their thermal and chemical stability, large surface area, high mechanical resistance, possible reuse, and relative rapidity in reaching equilibrium [41–44]. Silica is a surface modifiable material [45–48] and opportune functional groups can improve adsorption capacity and selectivity to various organic compounds [17].

Silica gels functionalized with monochloro-triazinyl β-cyclodextrin were used for removal of Acid Blue 25 and gave adsorption values up to $45.80\,\mathrm{mg\,g^{-1}}$ [49].

Asouhidou *et al.* have compared the performance of aminopropyl- and monochloro-triazinyl β-cyclodextrin-modified hexagonal mesoporous silicas, HMS-NH$_2$ and HMS-CD respectively, for adsorption of Remazol Red 3BS, a dye often used in textile industries. HMS-CD particles (10–100 μm in size) were characterized by significantly higher adsorption capacity than HMS–NH$_2$ material (0.01–0.1 μm in size), while unmodified silica showed a negligible efficiency [50]. The effect of several parameters, such as pH, dye concentration, and contact time, on adsorption performance was also studied. The surface of all silicas is positively charged at pH 2.0. HMS–NH$_2$ and HMS–CD retain their positively charged surface up to pH ~6.5, due to the presence of –NH$_2$ and –NH groups, which are protonated at low/moderate pH (Figure 12.1). Adsorption of the Remazol Red 3BS was favored in acidic conditions ($2.0 < \mathrm{pH} < 4.0$), while it was gradually reduced with increasing pH, as expected for anionic dyes. The maximum adsorption values obtained at pH 2 were $0.28\,\mathrm{mmol\,g^{-1}}$ for HMS-CD and $0.14\,\mathrm{mmol\,g^{-1}}$ for HMS-NH$_2$ particles. The sorption behavior of HMS–CD was not attributed to the formation of inclusion complexes in the interior of cyclodextrin cavities, but was rather related with specific electrostatic interactions, hydrogen bonding, and $\pi-\pi$ interactions between dyes and adsorbent.

In addition, HMS-CD keeps its properties after repeated adsorption–regeneration cycles with sodium dodecyl sulfate. Adsorbent reuse is a very important factor when developing a new adsorption methodology [32].

Another interesting, low cost adsorbent with selective affinity to targeted dyes was synthesized by Joo *et al.* by functionalization of silica surface with a polyelectrolyte [51]. They impregnated silica adsorbent with poly(diallydimethylammonium chloride), PDDA, and investigated the adsorption of negative, positive, and neutral dyes. Due to the positively charged surface of PDDA/silica, adsorption for negative or acidic dyes was greatly enhanced (more than 10 times) with respect to non-functionalized silica, with values of $360\,\mathrm{mg\,g^{-1}}$ for Rifazol Yellow GR, $281\,\mathrm{mg\,g^{-1}}$ for Rifazol Red BB 150, and $138\,\mathrm{mg\,g^{-1}}$

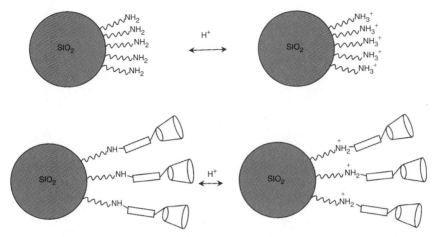

Figure 12.1   *HMS–NH₂ and HMS–CD NPs. Reproduced from* Colloids and Surfaces A: Physicochemical and Engineering Aspects, *346/1–3, Asouhidou et al., Adsorption of Remazol Red 3BS from aqueous solutions using APTES- and cyclodextrin-modified HMS-type mesoporous silicas, August 2009, with permission from Elsevier.*

for Rifafix Yellow 3RN 150H. However, modified silica had low efficiency towards positively charged and neutral dyes.

Cestari and co-workers have studied the removal of Blue and Red Remazol dyes by using chemically modified aminopropyl-silica. Adsorption capacities were $4.32\,\mathrm{mg\,g^{-1}}$ and $1.54\,\mathrm{mg\,g^{-1}}$ for Red and Blue Remazol, respectively, in aqueous solutions at pH 3.0. A pronounced decreasing of Blue dye adsorption was observed in presence of sulfonated sodium dodecylbenzene, whereas the presence of Hg(II) negatively interferes with red dye adsorption [42]. Temperature changes affect differently adsorption on particle surface in relation to dye chemical structure and surfactant addition [43]. The formation of dye–surfactant aggregates increases dye adsorption onto particles. In particular, Reactive Yellow adsorption decreases by changing temperature from 25–55 °C, whereas Reactive Red adsorption increases within the same temperature range [44].

Mohmoodi *et al.* have investigated the effect of adsorbent amount, pH, and salt concentration on efficiency of amine-modified silica NPs towards three textile dyes: Acid Red 14 (AR14), Acid Black 1 (AB1), and Acid Blue 25 (AB25). At pH 2.0 maximum adsorption by unmodified silica NPs was $0.031\,\mathrm{mg\,g^{-1}}$, $0.034\,\mathrm{mg\,g^{-1}}$, and $0.170\,\mathrm{mg\,g^{-1}}$ for AR14, AB1 and AB25, respectively, whereas the maximum adsorption capacity shown by amine-functionalized silica increased by a 10^4 factor for AR14 and AB1, but it was unchanged for AB25 [40].

Jesionowski and co-workers have removed Basic Blue 9, Acid Orange 52, and Mordant Red 3 dyes from aqueous solutions by using N-2-(aminoethyl)-3-amino-propyltrimethoxysilane modified silica. Adsorption capacities were $1.2\,\mathrm{mol\,m^{-2}}$, $0.261\,\mathrm{mol\,m^{-2}}$ for Basic Blue 9 and Acid Orange 52, respectively, [52, 53] and $5.49\,\mathrm{mol\,m^{-2}}$ for Mordant Red 3 [54].

Monoamine modified silica particles were prepared by reacting silica gel with 3-aminopropyltriethoxysilane as silylating agent and used for removal of Acid Orange

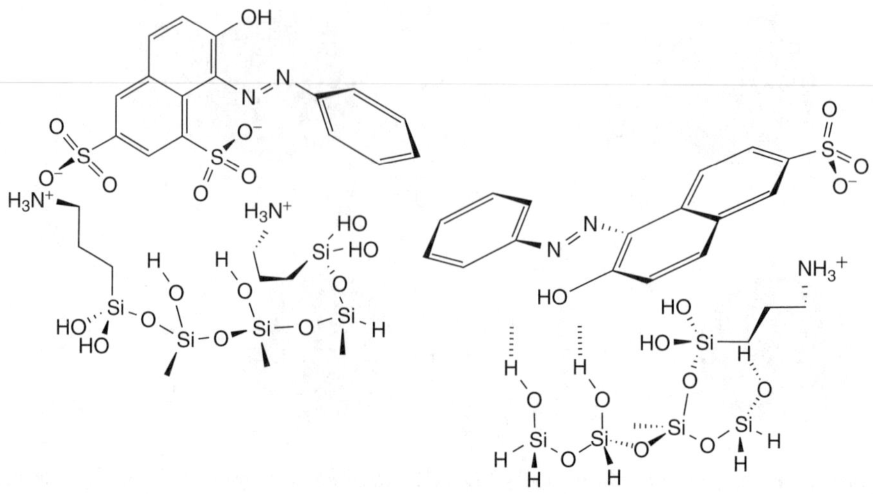

Figure 12.2 *Possible interactions between monoamine modified silica and Acid Orange 10 (a), and Acid Orange 12 (b). Reproduced from Journal of Hazardous Materials, 161/2–3, Donia et al., Effect of structural properties of acid dyes on their adsorption behaviour from acqueous solutions by amine modified silica, January 2009, with permission from Elsevier*

10 and Acid Orange 12 [47]. Different experimental conditions of pH, contact time, dye concentration, temperature, and salt amount were tested. Acid Orange 10 and Acid Orange 12 adsorptions reached a maximum value at pH 3 and 1 respectively, because of their different interactions with modified silica surface. Dye adsorption was explained by electrostatic attractions between positively charged protonated amino groups, $-NH_3^+$, on silica surface and negatively charged sulfonate groups, SO_3^-, of dyes:

$$silica\text{-}NH_2 + HCl \rightarrow silica\text{-}NH_3^+Cl^-$$

$$silica\text{-}NH_3^+Cl^- + dye\text{-}SO_3^-Na^+ \rightarrow silica\text{-}NH_3^{+-}O_3S\text{-}dye + NaCl$$

The observed efficiencies were explained in terms of dye spatial structure and different dissociation of SO_3H groups (Figure 12.2). Dye adsorption increased as temperature ranged from 25–40 °C.

The maximum adsorption of Acid Orange 10 on monoamine modified silica ($48.98\,mg\,g^{-1}$) increased if particles were prepared in presence of suspended Fe_3O_4 fine particles ($61.33\,mg\,g^{-1}$) due to a more porous structure (Figure 12.3a) [55].

Magnetic silica NPs, MNPs, are an interesting novelty in the treatment of wastewaters because of their specific magnetic separation. In fact, it is possible to avoid processes such as high speed centrifugation or filtration, which are needed to remove NPs from large volumes of wastewaters after adsorption of dyes and other pollutants [56]. Iron oxide NPs are generally used as a magnetic core and then, covered by a nonmagnetic protective shell, such as silica, which prevents agglomeration and afford an easy surface functionalization.

Recently, Fu and co-workers have synthesized carboxylic functionalized super-paramagnetic mesoporous silica for removing basic dyes from water. Microspheres had a super-paramagnetic iron oxide core coated by a mesoporous silica shell (60 nm

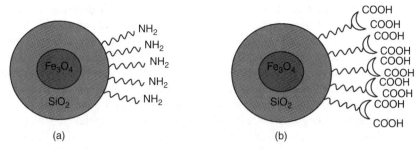

Figure 12.3 *Magnetic monoamine modified silica particle (a), and magnetic silica microsphere (b). (a) Reproduced from* Chemical Engineering Journal, *150/1, Atia, Donia, Al-Amrani, Adsoption/desorption behavior of acid orange 10 on magnetic silica modified with amine groups, July 2009, with permission from Elsevier. (b) Reproduced from* Microporous and Mesoporous Materials, *139/1–3, Fu, Chen, Wang, Liu, Fabrication of carboxylic functionalized superparmagnetic mesoporous silica microspheres and their application for removal basic dye pollutants from water, March 2011, with permission from Elsevier.*

thick). They used a sol–gel method with stearyltrimethyl ammonium bromide as surfactant template (Figure 12.3b). Methylene Blue and Acridine Orange adsorption by microspheres were $101.9 \, \text{mg g}^{-1}$ and $109.6 \, \text{mg g}^{-1}$, respectively. High adsorption values were kept within a wide range of pH (2.0–12.0, best pH value 10). These magnetic microspheres exhibited a large adsorption rate, mainly attributed to their mesoporous structure and carboxylic groups. Presence in solution of other basic dyes reduced efficiencies, whereas addition of Congo Red as acidic dye had not significant effects. Furthermore, these magnetic silica could be regenerated by simple wash with acid solutions and efficiently reused: adsorption efficiency was still above 88% after seven cycles of use/regeneration [57].

Chang *et al.* prepared hexadecyl functionalized magnetic silica NPs, which were tested to remove Rhodamine 6G dye from water. Due to cooperative effects played by hexadecyl chains (hydrophobic interactions) and Si–OH groups (electrostatic forces), hexadecyl functionalized magnetic silica had a Rhodamine 6G sorption value, $35.6 \, \text{mg g}^{-1}$ (at pH 11.0), larger than those shown by hexadecyl non-functionalized and silica free NPs [58].

12.2.4   Removal of Metallic Pollutants by Silica Nanoparticles

Human exposure to heavy metals, even at low concentrations, causes various adverse health effects by impairing mental and neurological functions [59]. Furthermore, their toxicity on flora and fauna is well documented. Heavy metals are non-biodegradable and long term persistent pollutants: they tend to accumulate in living organisms and throughout the food chain (bioaccumulation and biomagnification) [60, 61].

Heavy metals can be introduced into surface waters as well groundwater by untreated aqueous waste streams of several industrial processes such as alloy preparation, metal plating, mining operations, ceramic and battery manufacturing, dyeing, and tanneries.

Great efforts are made to limit heavy metal contamination of water resources as indicated by national and international regulations. The World Health Organisation recommends well defined maximum levels for heavy metals in drinking water. Thus, industrial effluents and agricultural wastewater must be treated prior their discharge into water ecosystem and soil. Different technologies including chemical precipitation and electrolytic separation, oxidation/reduction reactions, mechanical filtration, ion exchange, membrane separation, reverse osmosis, selective liquid–liquid extraction, photochemical process, flocculation, flotation and elimination by adsorption on activated carbon, have been reported for removing metals from industrial wastewaters [62].

Some of these technologies present disadvantages such as incomplete removal, generation of toxic residual sludge, high operational and maintenance costs, expensive facilities, and high energy consuming plants. Heavy metal separation by adsorption processes on NPs offers interesting advantages as lower environmental impact, high efficiency, cost-effectiveness, and easy handling [18]. In addition, metal recovery and regeneration of adsorbents are relatively inexpensive allowing NP reuse [63].

Amino functionalized MCM-41 mesoporous silica, prepared by a sol–gel method, was used for metal removal from aqueous solutions [64]. MCM-41 mesoporous silica was obtained by hydrothermal crystallization of fumed silica in presence of hexadecyl trimethyl ammonium bromide as template. The effects of pH, metal ion concentration, adsorbent amount, and contact time were studied: it was found that maximum adsorption capacity of amino functionalized MCM-41 silica for Ni^{2+}, Cd^{2+}, and Pb^{2+} was $12.36\,mg\,g^{-1}$, $18.25\,mg\,g^{-1}$, and $57.74\,mg\,g^{-1}$ respectively. In addition, it was confirmed that amino functionalized MCM-41 silica was characterized by a higher metal uptake than pure MCM-41 silica [65].

Pacheco and co-workers reported that silica-alumina NPs were effective in cadmium removal from model wastewaters. At pH 6.5 and room temperature cadmium concentration was reduced from 140 ppm to 5 ppb by Si–Al particles (silica core surrounded by an alumina shell) whereas Al–Si particles (alumina core surrounded by a silica shell) decreased Cd^{2+} concentration from 125 ppm to below 90 ppb [66]. Similar silica-alumina NPs were used by Medina *et al.* for lead removal from industrial streams. Pb^{2+} concentration was reduced by Al–Si NPs, with previously reported operational conditions, from 140 ppm to <100 ppb, well below legislation requirements [67]. Both studies revealed that Cd^{2+} and Pb^{2+} adsorptions depended on NP morphology, concentration of functional (hydroxyl, alkoxy, and oxy) groups on adsorbent surface, and reactivity of alumina or silica shell to link cations.

Recently, Wang *et al.* designed hierarchical structured $SiO_2@\gamma$-$AlOOH$ particles (500–700 nm in size) prepared in one step by using silica colloidal spheres as template. These nanospheres showed good efficiency in removing Cr(VI) from model wastewater. At pH 3.0 chromium maximum adsorption was $4.5\,mg\,g^{-1}$ [68].

Zong and co-workers developed dual-emission fluorescent NPs for ultrasensitive and rapid detection of Cu^{2+}. Polyethyleneimine was grafted on the surface of fluorescein isothiocyanate-doped silica, and, successively, linked to fluorescent Rhodamine B isothiocyanate (Figure 12.4). This chemosensor showed a high sensitivity (less than 10 nM) and effective specificity toward copper ions, due to the strong chelating affinity of polyethyleneimine to Cu^{2+} and selective fluorescence quenching of Rhodamine B isothiocyanate by polyethyleneimine/Cu^{2+} complexes [69].

Li *et al.* reported the preparation of thiol-functionalized magnetic mesoporous silica particles and their adsorption efficiency for Pb^{2+} and Hg^{2+}. The core-shell structure

Figure 12.4 *Silica NPs-based chemosensor for the detection of Cu²⁺. Reproduced from Analytical Chemistry, Zong, Ai, Zhang Li and Lu, Dual-Emission Fluorescent Silica Nanoparticle-Based Probe for Ultrasensitive Detection of Cu²⁺,* © *2011, American Chemical Society*

made these NPs resistant to different water matrices, even in strong acidic and alkaline conditions. NP magnetic properties made faster and easier their separation from aqueous solutions by application of a moderate magnetic field immediately after metal adsorption. Removal efficiency was kept above 90% after six subsequent cycles of adsorption/desorption. Hg^{2+} adsorption was faster than Pb^{2+} one and was explained by a monolayer adsorption process for both metal ions. At pH 6.0 maximum adsorption capacity was of 1.3 mmol g^{-1} and 0.44 mmol g^{-1}, respectively. The large efficiency in Hg^{2+} removal was explained by a large binding affinity of SH groups with mercury ions [70]. A reverse adsorption capacity for the two metal ions was observed in absence of thiol functionalization onto NP surface [71].

Recently, amino-functionalized magnetic silica NPs were tested for adsorption of Cu^{2+}, Cd^{2+}, and Pb^{2+} [56]. Metal adsorption ranked in the order Cu^{2+} > Pb^{2+} > Cd^{2+}, resulting from a higher affinity of amino groups with copper ions. Moreover, results were not much influenced by the presence of humic acid or alkali/earth metal ions (Na^+, K^+, Mg^{2+}).

N-[3-(trimethoxysilyl)propyl]-ethylenediamine was used for the surface modification of magnetite particles. The adsorption capacity of Cu^{2+} and Reactive Black 5 dye were 10.41 mg g^{-1} and 217 mg g^{-1} respectively, in pH range 3.0–5.5 confirming the high capability of these magnetic systems in removing both dyes and heavy metals [72].

12.3 Magnetic Nanoparticles: Synthesis, Characterization and Applications

12.3.1 Magnetic Nanoparticles: An Overview

A problem in the use of silica NPs in environmental remediation is their recovery after pollutant adsorption. A way to overcome this drawback is the synthesis of NPs with magnetic properties. Obviously, biomedical applications of magnetic NPs are very intriguing. Examples of the exciting and broad field of magnetic NPs applications include drug delivery, contrast agents, magnetic hyperthermia, therapeutic *in vivo* applications of magnetic carriers, *in vitro* magnetic separation and purification, molecular biology investigations, immunomagnetic methods in cell biology and in pure medical applications.

Detailed information about the physical properties, magnetic behavior, chemistry, or biomedical applications of magnetic NPs is referred to specific reviews [73].

Magnetic NPs are a class of engineered particulate materials of less than 100 nm in size that can be manipulated and recovered after use under the influence of an external magnetic field. MNPs are commonly composed of magnetic elements, such as iron, nickel, cobalt, and their oxides like magnetite (Fe_3O_4), maghemite ($\gamma\text{-}Fe_2O_3$), cobalt ferrite (Fe_2CoO_4), and chromium dioxide (CrO_2).

The classification of magnetic properties is based on material magnetic susceptibility, which is defined by the ratio of the induced magnetization to the applied magnetic field.

In ferri- and ferro-magnetic materials, magnetic moments align parallel to the external magnetic field: coupling interactions between electrons result in ordered magnetic states. The susceptibilities of these materials depend on external field strength, atomic structures, and temperature. At small sizes (in the order of tens of nanometers) ferri- or ferro-magnetic materials and MNPs maintain large magnetic moments. At sufficiently high temperatures thermal energy is sufficient to induce free rotation of particles resulting in a loss of net magnetization in the absence of an external field [74]. Lack of remnant magnetization after removal of external fields enables the particles to maintain their colloidal stability and avoids aggregation.

The magnetic properties of NPs are determined by many factors, including chemical composition, particle size and shape, morphology, type, and degree of defectiveness in the crystal lattice, interactions of particles with surrounding matrix and neighboring particles. By changing NP size, shape, composition, and structure, one can control to an extent the NP magnetic characteristics.

12.3.2   Synthesis of Magnetic Nanoparticles

Most of general methods for NP synthesis can also be used for the preparation of magnetic particles [75–77]. In fact, MNPs have to show the same characteristic of size and shape homogeneity found in conventional NPs, but with additional magnetic properties.

MNPs have been synthesized with a number of different compositions and phases, including iron oxides, such as Fe_3O_4 and $\gamma\text{-}Fe_2O_3$ [78–80], pure metals, such as Fe and Co [81, 82], spinal-type ferromagnets, such as $MgFe_2O_4$, $MnFe_2O_4$, and $CoFe_2O_4$ [83, 84], as well as alloys, such as $CoPt_3$ and FePt [85, 86].

During the last few years, many publications have described efficient synthetic routes to shape-controlled, highly stable, and monodisperse magnetic NPs. Several popular methods including co-precipitation, thermal decomposition, hydrothermal treatments, microemulsions, sonochemical and microwave-assisted procedures, chemical vapor deposition, combustion synthesis, carbon arc and laser pyrolysis, can all be directed at the synthesis of high-quality magnetic NPs.

12.3.2.1   Co-Precipitation

Co-precipitation is an easy and convenient way to synthesize MNPs (metal oxides and ferrites) from aqueous salt solutions by the addition of a base under inert atmosphere at room temperature or at elevated temperature. Iron oxide NPs (either magnetite, Fe_3O_4, or maghemite, $\gamma\text{-}Fe_2O_3$) and ferrites are usually prepared in an aqueous medium according to the following chemical reaction:

$$M^{2+} + 2Fe^{3+} + 8OH^- \rightarrow MFe_2O_4 + 4H_2O$$

where M can be Fe^{2+}, Mn^{2+}, Co^{2+}, Cu^{2+}, Mg^{2+}, Zn^{2+}, and Ni^{2+}.

Complete precipitation should be expected at pHs between 8–14 with a stoichiometric ratio of 2:1 (Fe^{3+}/M^{2+}) in a non-oxidizing oxygen environment [87].

The shape, size, and composition of magnetic NPs very much depend on Fe^{2+}/Fe^{3+} ratio, type of used salts (e.g. chlorides, sulfates, nitrates), pH value, reaction temperature, and media ionic strength. Once synthetic conditions are fixed, the quality of the MNPs is fully reproducible. Particles prepared by co-precipitation show generally a large polydispersity. It is well known that a short burst of nucleation and subsequent slow controlled growth is crucial to produce particles at a controlled size and, in particular, in the production of iron oxide magnetic NPs.

Significant advances in preparing MNPs have been made by using organic additives (e.g. polyvinyl alcohol) as stabilizers and/or reducing agents [88]. Surfactant selection is an important issue for NP stabilization. Size-tunable maghemite NPs were prepared by initial formation of magnetite in presence of trisodium salt of citric acid, in an alkaline medium, and subsequent oxidation by iron(III) nitrate [89]. The effects of several organic anions, such as carboxylate and hydroxy carboxylate ions, on the formation of iron oxides or oxyhydroxides have been studied extensively [90–92].

Recent studies showed that oleic acid can efficiently stabilize Fe_3O_4 NPs [93, 94]. The effect of organic ions on the formation of metal oxides or oxyhydroxides can be rationalized by two competing mechanisms: chelation and adsorption. Chelation of metal ions can prevent nucleation and lead to the formation of larger particles. On the other hand, adsorption of additives on nuclei and growing crystals may inhibit particle growth favoring the formation of small units.

More recently, nanostructured iron oxides were synthesized by a simple co-precipitation technique at room temperature and tested as anode materials for lithium-ion batteries [95]. The iron salt precursor has a significant effect on the morphology evolution of iron oxide. Nanosheet and NP samples were obtained by using ferrous ammonium sulfate and ferric chloride precursors, respectively. Both samples could be identified as α-Fe_2O_3 after annealing at 400 °C. The improved performance of iron oxide nanosheets toward lithium could be attributed to the high electrical conductivity and small grain size facilitating the transport of electrons and lithium ions.

12.3.2.2   Thermal Decomposition

Monodisperse magnetic nanocrystals with smaller size can be synthesized through thermal decomposition of organometallic compounds in high-boiling organic solvents containing stabilizing surfactants [84, 96, 97]. Organometallic precursors include metal acetylacetonates, $M(acac)_n$, metal cupferronates, $M_x cup_x$ (cup = N-nitrosophenylhydroxylamine) [98], or carbonyls (such as $Fe(CO)_5$) [99].

Fatty acids [100], oleic acids [101], and hexadecylamine [102] are often used as surfactants.

Thermal decomposition of organometallic precursors, the metal of which is initially zero-valent (such as $Fe(CO)_5$), leads to the formation of metal NPs. If the reaction is followed by oxidation, it can lead to high in quality monodisperse metal oxides. On the other hand, decomposition of precursors with cationic metal centers (such as $Fe(acac)_3$) leads directly to metal oxide NPs.

The reaction temperature and time, as well as the aging period, may also be crucial for the precise control of size and morphology [76]. The effect of reaction temperature and reaction time on size, morphology, and magnetic properties of NPs are schematically shown in Figure 12.5.

The main advantage of metallic NPs obtained by thermal-decomposition is magnetization values larger than metal oxide ones.

Metallic iron NPs were synthesized by thermal decomposition of $[Fe(CO)_5]$ and in presence of polyisobutene in decalin [103]. Dumestre and co-workers reported a synthesis of iron nanocubes by decomposition of $[Fe\{N\ [Si(CH_3)_3]_2\}_2]$ with H_2 in presence of hexadecylamine and oleic acid or hexadecylammonium chloride at 150 °C [104].

The synthesis of cobalt NPs by thermal decomposition allows the control of both particle shape and size [105]. Puntes and co-workers investigated the synthesis of cobalt nanodisks by thermal decomposition of a cobalt carbonyl precursor [82, 106]. The synthesis of cobalt [107, 108] and nickel nanorods [109] from high-temperature reduction of non-carbonyl organometallic complexes was also reported.

Chen *et al.* prepared nickel NPs from thermal decomposition of nickel(II) acetylacetonate in alkylamines [110]. In their study reaction temperature, heating rate and solvent type were important for the resulting crystalline phase (fcc or hcp). It is important to recall that the magnetic properties of hcp nickel NPs are quite different from those of fcc ones. Monodisperse nickel NPs were also obtained by introducing surfactants.

Metal oxides MNPs can also be synthesized by thermolysis of oxygen-rich precursors [100]. The paper by Chen *et al.* offers an alternative explanation of the mechanisms occurring in the noninjection-type systems [111].

A high-temperature reaction of iron(III) acetylacetonate, $[Fe(acac)_3]$, with 1,2-exadecanediol, oleic acid, and oleylamine in high-boiling ether solution allowed the

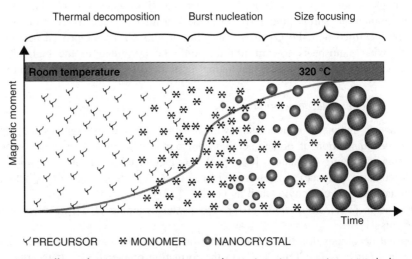

Figure 12.5 *Effect of reaction temperature and reaction time on size, morphology, and magnetic properties of MNPs. Reproduced from* Journal of the American Chemical Society, *Kwon et al.,* Kinetics of Monodisperse Iron Oxide Nanocrystal Formation by "Heating-Up" Process *© 2007, American Chemical Society*

preparation of Fe_3O_4 NPs according to the following reaction:

$$Fe(acac)_3 + C_{14}H_{29}CH(OH)CH_2OH + C_{17}H_{33}CO_2H + C_{18}H_{35}NH_2$$

$$\rightarrow nano\text{-}Fe_3O_4(L)_n$$

where nano-$Fe_3O_4(L)_n$ are magnetite NPs covered by long hydrocarbon chains: they act as a barrier that restricts the growth of particles and gives rise to a well-controlled size and size distribution [80, 97].

The size and shape of nanocrystals can be controlled by varying the reactivity and concentration of precursors.

Jana and co-workers reported a general decomposition approach for the synthesis of size- and shape-controlled magnetic oxide nanocrystals based on pyrolysis of metal fatty acid salts in non-aqueous solutions [100]. The reaction system was generally composed by metal fatty acid salts, corresponding fatty acids (decanoic, lauric, myristic, palmitic, oleic, stearic acid), a hydrocarbon solvent, and activation reagents. Nearly monodisperse Fe_3O_4 nanocrystals were obtained with controlled shapes and sizes adjustable over a wide range (3–50 nm). This method was successfully generalized for the synthesis of other magnetic nanocrystals, such as Cr_2O_3, MnO, Co_3O_4, and NiO.

Park and co-workers have also used a similar thermal decomposition approach for the preparation of monodisperse iron oxide NPs [113]. They used nontoxic and inexpensive iron(III) chloride and sodium oleate to generate an iron oleate complex in situ, which was then decomposed at temperatures between 240–320 °C in different solvents, such as 1-hexadecene, octyl ether, 1-octadecene, 1-eicosene, or trioctylamine. NPs were easily dispersible in organic solvents.

In general, one of major disadvantages of thermal decomposition methods is the production of organic soluble NPs, which limits the extent of applications and their use in biological fields, if a surface treatment is not performed after synthesis. In addition, thermal decomposition methods usually require complicated processes or relatively high temperatures.

Water soluble magnetic NPs are more desirable for applications in biotechnology. For this purpose, a very simple synthesis of water-soluble magnetite NPs was reported. Using $FeCl_3 \cdot 6H_2O$ as iron source and 2-pyrrolidone as coordinating solvent, water soluble Fe_3O_4 nanocrystals were prepared under reflux (245 °C) [114]. The same group developed a one-pot synthesis of water-soluble magnetite NPs prepared under similar reaction conditions by the addition of α,ω-dicarboxyl- terminated poly(ethylene glycol) as a surface capping agent [115]. These NPs can potentially be used as magnetic resonance imaging contrast agents for cancer diagnosis.

Recently, magnetite (Fe_3O_4) NPs with controllable size and shape were synthesized by thermal decomposition [116]. In contrast to previously reported thermal decomposition methods, their synthesis utilized a much cheaper and less toxic iron precursor, iron acetylacetonate, and environmentally non-toxic polyethylene oxide as solvent and surfactant.

More recently, water-soluble super-paramagnetic Fe_3O_4 NPs with an average diameter of 9.5 ± 1.7 nm were synthesized by thermal decomposition of $Fe(acac)_3$ in methoxy poly(ethylene glycol) used as solvent, reducing agent, and modifying agent. An obvious advantage of this approach is that no further reducing agent and surfactant are required [117]. Polyethylene oxide has been widely used as a green solvent for various organic syntheses due to its low toxicity and high boiling point [118–120].

Magnetic alloy NPs are interesting materials because of their magnetic properties and chemical stability. They have many advantages, such as high magnetic anisotropy, enhanced magnetic susceptibility, and large coercivity [121]. However, controlling the composition of magnetic alloy NPs can be difficult when they are produced from two or more precursors. This could be overcome by using a single precursor or bimetallic carbonyl cluster in a thermal decomposition process. Recently, Robinson *et al.* have used this novel synthesis method to produce $FeCo_3$, $FeNi_4$, FePt, and Fe_4Pt alloy magnetic NPs with average diameters of 7.0, 4.4, 2.6, and 3.2 nm, respectively [122].

Hexagonal iron phosphide and related materials have been intensively studied for their magneto-resistance, magneto-caloric effects, and ferromagnetism [123, 124]. Brock and co-workers have synthesized FeP and MnP NPs from the reaction of iron(III) acetylacetonate and manganese carbonyl, respectively, with tris(trimethylsilyl)phosphane at high temperatures [125, 126].

Antiferromagnetic FeP nanorods were prepared by the thermal decomposition of a precursor/surfactant mixture solution [127]. In addition, synthesis of discrete iron phosphide (Fe_2P) nanorods from thermal decomposition of continuously supplied iron pentacarbonyl in trioctylphosphane using a syringe pump was reported [113].

12.3.2.3 Hydrothermal Procedures

Hydrothermal technique is defined as any heterogeneous reaction in presence of aqueous solvents or mineralizers under high pressure and temperature conditions. These reactions are performed in reactors or autoclaves where pressure and temperature can be higher than 2000 psi and 200 °C, respectively. Under hydrothermal conditions a broad range of nanostructured materials can be formed [128–132].

Wang *et al.* reported a generalized hydrothermal method for synthesizing a variety of different nanocrystals by a liquid–solid–solution reaction [133]. The system consists of metal linoleate (solid), an ethanol–linoleic acid liquid phase, and a water–ethanol solution at different reaction temperatures under hydrothermal conditions. This strategy is based on a general phase transfer and separation mechanism occurring at the interfaces of the liquid, solid and solution phases present during the synthesis (Figure 12.6).

Deng *et al.* also reported a synthesis of monodisperse, hydrophilic, single crystalline ferrite microspheres by hydrothermal reduction. They skillfully used a multicomponent reaction mixture including ethylene glycol, sodium acetate, and polyethylene glycol to direct synthesis.

Ethylene glycol was used as a high-boiling-point reducing agent, which was known from polyol processes to produce monodisperse metal or metal oxide NPs, sodium acetate played the role of electrostatic stabilizer to prevent particle agglomeration, and polyethylene glycol acted as a surfactant against particle agglomeration [134].

Liang and co-workers synthesized iron oxides using hydrothermal and solvothermal processes [135]. In hydrothermal preparation α-Fe_2O_3 was obtained in sub-critical water via dissolution and precipitation processes. The α-Fe_2O_3 particles synthesized at 420°C had a size of about 40 nm, which was much smaller than that found in particles produced at 350 °C (60 nm). Crystalline Fe_3O_4 was formed in a solvothermal preparation due to

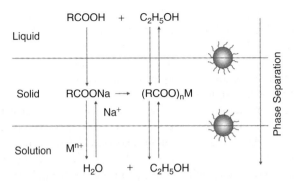

Figure 12.6 *The liquid–solid-solution phase transfer synthetic strategy. Reproduced from Nature, Wang, Zhuang, Peng and Li, A general strategy for nanocrystal synthesis © 2005, Rights Managed by Nature Publishing Group*

the reducing atmosphere resulting from oxidation of isopropanol into acetone. Acetone molecules adsorbed on Fe_3O_4 surfaces were suggested to suppress crystallite growth.

Zheng *et al.* reported the synthesis and magnetic properties of hydrothermally prepared Fe_3O_4 [136].

The hydrothermal synthesis of Fe_3O_4 NPs was carried out by Mizutani *et al.* using a starting solution containing lactate and sulfate ions at various concentrations in order to control particle size [137]. Depending on lactate and sulfate ions concentration particle size could be varied from 9.5–38.6 nm. Size-controlled magnetite NPs coated with glucose and gluconic acid were synthesized by Sun *et al.* via a simple and facile hydrothermal reduction route using $FeCl_3$ as a single iron precursor [138]. The particle size could be easily controlled in the range of 4–16 nm. Sucrose was used in order to obtain magnetite nanomaterial rather than hematite large particles. In fact, sucrose decomposes into reducing species, causing partial reduction of Fe^{3+} ions to Fe^{2+} ions, and generates a capping agent, enabling the formation of nanoscale particles.

Selective synthesis of α-FeOOH and α-Fe_2O_3 nanorods was developed using ferrous sulfate and hydrogen peroxide as raw materials [139]. α-FeOOH and α-Fe_2O_3 nanorods were obtained by adjusting the reaction temperature in hydrothermal synthesis at 150 °C and 200 °C, respectively. A facile one-step hydrothermal approach was recently made for the synthesis of Fe_3O_4 NPs with controllable diameters, narrow size distribution, and tunable magnetic properties [140]. The iron oxide NPs were synthesized by oxidation of ferrous chloride in basic aqueous solution under elevated temperature and pressure conditions.

An efficient route for simultaneous synthesis and self-assembly of 3D layered β-FeOOH nanorods was reported [141]. The nature of particles depended on a pH-induced strategy, in which the continuous change of pH was achieved by hydrolysis of $FeCl_3 \cdot 6H_2O$ in the presence of urea under hydrothermal conditions.

12.3.2.4 Microemulsions as Nanoreactors

Microemulsion systems are formed by mixtures of water, oil, and surfactant. They are transparent solutions consisting of small droplets of an immiscible phase (non-polar or polar) dispersed in a continuous phase. Surfactants are added to lower interfacial tension between the immiscible dispersed and continuous phases to stabilize droplets [142].

Microemulsions may be of oil-in-water or water-in-oil type, depending on the concentration of the different components. By varying the concentration of dispersed phase and surfactant, it is possible to tailor droplet size in the range 1–100 nm, approximately.

In water-in-oil microemulsions the aqueous phase is dispersed as microdroplets (typically 1–50 nm in diameter) surrounded by a monolayer of surfactant molecules in the continuous hydrocarbon phase. The size of reverse micelles is determined by the molar ratio of water to surfactant [143]. By mixing two identical water-in-oil microemulsions containing desired reactants, microdroplets will continuously collide, coalesce, break again, and finally, a precipitate will be formed in the micelles [144]. By addition of a solvent, such as acetone or ethanol, to microemulsions, the precipitate can be extracted by filtering or centrifuging the mixture. In this sense, a microemulsion can be used as a nanoreactor for the formation of NPs.

Using microemulsion technique metallic cobalt, cobalt/platinum alloys, and gold-coated cobalt/platinum NPs have been synthesized in reverse micelles of cetyltrimethylammonium bromide, 1-butanol, and octane [145].

MFe_2O_4 (M: Mn, Co, Ni, Cu, Zn, Mg, or Cd, etc.) are among the most important magnetic materials and have been widely used for electronic applications. Spinel ferrites can be synthesized in microemulsions and reverse micelles. For instance, $MnFe_2O_4$ NPs with controllable sizes of about 4–15 nm are synthesized through the formation of water-in-toluene reverse micelles with sodium dodecylbenzenesulfonate as the surfactant [146].

Recently, nanosized particles of cobalt ferrite, $CoFe_2O_4$, have been synthesized from single water-in-oil microemulsion technique consisting n-hexanol as an oil phase, cetyltrimethylammonium bromide as surfactant and an aqueous solution of metal salts [147]. The particles synthesized by the microemulsion route were characterized by finer sizes and better magnetic properties than those shown by particles obtained from conventional synthetic routes.

Precipitation in reverse microemulsions is a well-established technique for preparation of magnetic particles with diameters smaller than 10 nm and narrow particle size distribution [148–158]. The basis of this technique is the use of nanodroplets (in aqueous phase with diameters usually smaller than 10 nm), that contain precursor ions of magnetic NPs to carry out precipitation reactions. Further growth of particles occurs by recruiting precursor ions and precipitate molecules from droplets without particles. The increase in yield results from the differing capacities of aqueous phase dissolution between reverse and bicontinuous microemulsions. In fact, reverse microemulsions can usually accept water up to 15 wt.%, while bicontinuous microemulsions are able to accept water contents up to near 45 wt.%. Use of bicontinuous microemulsions provides a useful tool to overcome the drawback of low yields preserving magnetic NP characteristics.

Bicontinuous microemulsions are formed by interconnected aqueous channels, with diameters usually smaller than 10 nm, immersed in an oil continuous phase. Consequently, in bicontinuous microemulsions nucleation and nuclei growth are expected to occur inside channels [159].

Loo *et al.* observed that precipitation reactions, carried out in bicontinuous microemulsions with different surfactant concentrations, allow to obtain magnetic (magnetite or a mixture of magnetite-maghemite) NPs characterized by an inverse dependence of particle size on surfactant concentration [160]. This effect was probably due to a decrease in the channel diameter of microemulsions related to the increase in surfactant concentration.

Other self-assembled aggregates that can be formed by mixtures of oil, water and surfactants are, for example, reverse spherical micelles, disk-like and cylindrical micelles, hexagonal and lamellar phases [161]. Reverse micelles route can be employed for obtaining shape- and size-controlled iron oxide NPs. Geng *et al.* synthesized α-FeOOH nanorods at room temperature by using Pluronic triblock copolymer P123, poly (ethyleneoxide)-*block*-poly(propyleneoxide)-*block*-poly(ethyleneoxide), and ferric nitrate [162].

Chin and Yaacob reported the synthesis of magnetic iron oxide NPs via water in oil microemulsion [163]. Their particles were characterized by smaller size (less than 10 nm) and higher saturation magnetization values when compared with particles produced by Massart's procedure [164]. Zhang *et al.* fabricated hollow magnetite nanospheres in a microemulsion with a diameter ranging from 200–400 nm, but, unfortunately, these NPs could not be useful for drug delivery purposes [165]. Lee *et al.* reported the large-scale synthesis of uniform and crystalline magnetite NPs with well-defined nanometer sizes at high temperature using iron(III) acetyl acetonate as precursor and reverse micelles as nanoreactors [142]. Sun and co-workers reported size controlled synthesis of ultra-small magnetite (12–16 nm) NPs using $Fe(acac)_3$ as iron source in a microemulsion route. Despite the presence of surfactants, NPs needed several washing processes and further stabilization treatments to avoid aggregation [80].

The synthesis and characterization of β-FeOOH NPs obtained from a microemulsion system with a non-ionic surfactant, polyoxyethylene(4)nonylphenyl ether, has been reported [166]. The size and shape of β-FeOOH NPs could be modified by non-ionic surfactant amount.

Chitosan is a linear polysaccharide produced by deacetylation of chitin. It is currently the focus of many researches for its possible health benefits and medical applications. In-situ preparation of magnetic chitosan/Fe_3O_4 composite NPs in tiny pools of water in oil microemulsions has been reported [167]. The chitosan particle size varied from 10–80 nm as a function of chitosan molecular weight. Aslam obtained magnetically responsive microgels that consist of small iron oxide magnetic NPs ($\sim$15 nm in diameter) embedded in biocompatible microgels, the dimensions of which vary from $\sim$65 nm to $\sim$110 nm [168]. Chitosan-coated magnetic NPs have been prepared as carriers of 5-Fluorouracil, a well-known chemotherapy agent, through a reverse microemulsion method. NPs were spherical in shape, with an average size of 100 nm $\pm$ 20 nm, and showed low aggregation and good magnetic response [169].

12.3.2.5 Other Synthesis Methods

Reduction of metal salts is the most common solution-phase chemistry route developed for the preparation of metal NPs. Reducing agents, such as $NaBH_4$, are commonly employed as reported in the following reaction [170–173]:

$$4Fe^{3+} + 3NaBH_4 + 9H_2O \rightarrow 4Fe^0 \downarrow + 3NaH_2BO_3 + 12H^+ + 6H_2$$

As alternative to other time-consuming preparation techniques, sonochemical methods have been extensively used to generate novel materials with unusual properties.

Kima *et al.* synthesized Fe_3O_4 NPs by sonochemical and co-precipitation methods [174]. They showed that Fe_3O_4 NPs from sonochemical method had a higher crystallinity and saturation magnetization than those obtained from the co-precipitation method.

The microwave-assisted solution method has become widely used due to its advantages such as rapid volumetric heating, high reaction rate, reduced reaction time, and increased product yield compared to conventional heating methods [175, 176]. Wang reported the synthesis of spinel structured $CoFe_2O_4$ and $MnFe_2O_4$ NPs with diameters less than 10 nm by a fast and simple microwave assisted polyol process [177]. Wang and co-workers also applied microwave heating method for preparation of magnetite (Fe_3O_4) and hematite (α-Fe_2O_3) using $FeCl_3$, polyethylene glycol, and $N_2H_4 \cdot H_2O$ [178]. They found that only Fe_3O_4 phase can be formed at higher amounts of $N_2H_4 \cdot H_2O$, but the product was a mixture of Fe_3O_4 and α-Fe_2O_3 at lower concentration of $N_2H_4 \cdot H_2O$. In addition, they reported that the heating method plays an important role in the shape of nanocrystals: ellipsoidal α-Fe_2O_3 NPs were found by microwave assisted synthesis.

Aerosol technologies, such as spray and laser pyrolysis, are attractive because these technologies are continuous chemical processes allowing for high rate of production [179]. In spray pyrolysis a solution of ferric salts and a reducing agent in an organic solvent is sprayed into a series of reactors, where aerosol solutes condense and solvent evaporates. The resulting dried product consists of particles, whose size depends upon the diameter of the original droplets. This method has been used to prepare colloidal aggregates of super-paramagnetic maghemite NPs in the form of either hollow or dense spheres and with the possibility of having surfaces enriched in silica [180]. Maghemite particles with size ranging from 5–60 nm and different shapes have been obtained using different iron precursor salts in alcoholic solutions [179, 181].

In laser pyrolysis laser light induces pyrolysis of a gaseous mixture of iron precursors in an oxidizing atmosphere to produce small, narrow size, and non-aggregated NPs. It is a well-known method of obtaining maghemite NPs with size smaller than 10 nm [182, 183]. Bomati-Miguel *et al.* reported synthesis of maghemite (γ-Fe_2O_3) NPs by in situ hard and soft laser decomposition of gaseous $Fe(CO)_5$ in air [184].

Julian-Lopez *et al.* have reported the synthesis and characterization of hybrid silica-spinel iron oxide composite microspheres built with super-paramagnetic NPs in a hybrid mesoporous matrix, enabling the transport of bioactive molecules. These multifunctional platforms can be obtained by spray drying a sol of tunable composition, which can control the size and amount of magnetic particles embedded in the matrix [185]. Small size, narrow size distribution, and nearly absence of aggregation characterize these NPs.

Park and co-workers developed a method to synthesize carbon-encapsulated magnetic NPs such as Fe–C, Ni–C and Co–C by irradiating nanosecond laser pulses into a metalocene-xylene solution under room temperature and atmospheric pressure conditions [186]. As a possible growth mechanism, the authors supposed that irradiation of metalocene-xylene solutions led metal ions to agglomerate in NPs. During each period between two laser pulses, carbons are supersaturated on the surface of metal core due to a reduction in carbon solubility around the core because of a rapid NP cooling. Super-saturated carbons stop nuclei growth, resulting in the production of carbon-encapsulated magnetic NPs instead of carbon nanotubes [187].

12.3.3 Characterization of Magnetic Nanoparticles

Magnetic properties and various applications of MNPs depend highly on size, morphology, structure, and surface functional groups of prepared NPs. Size and shape of magnetic NPs can be characterized using transmission electron microscopy [188]. This technique reports total particle size (crystalline and amorphous parts) and gives access to a number-weighted mean value. Furthermore, it provides details on size distribution and shape. However, this technique needs an analysis by image treatment and must be performed on a statistically significant large number of particles.

High-resolution transmission electron microscopy gives access to atomic arrangement. It can be used to study local microstructures (such as lattice vacancies and defects, lattice fringes, glide planes, or screw axes) and surface atomic arrangement of crystalline NPs [189].

Scanning electron microscopy is not a good technique for core/shell characterization of NPs because this technique reports total particle size. Scanning electron microscopy resolution is lower than transmission electron microscopy and it is not efficient for NPs with particles size lower than 20 nm.

Dynamic light scattering is a common and easy technique to obtain NPs size. The determination of diffusion coefficient of NPs in solution gives access to the hydrodynamic radius of a corresponding sphere and polydispersity of the colloidal solution [190, 191].

The X-ray diffraction is used for determining the crystallographic identity of produced material, phase purity and mean particle size based on the broadening of the most prominent peak in the X-ray profile. Average particle diameter, D, can be obtained from Scherrer's equation [192]:

$$D = \frac{K\lambda}{\beta \cos\theta}$$

where θ is the Bragg angle of the peak, β is the line broadening at half the maximum intensity, λ is the X-ray radiation wavelength, and K is the shape factor.

Many techniques are available to measure the magnetic properties of an assembly of MNPs. SQUID magnetometry [193] and vibrating sample magnetometry [194, 195] are powerful tools to measure sample net magnetization.

X-ray photoelectron spectroscopy is a very useful technique to study reaction mechanisms that occur on MNP surfaces. X-ray photoelectron spectroscopy allows one to determine the bonding characteristics of the different elements and to confirm the chemical structure of NPs [196].

FT-IR spectroscopy is a powerful tool for identification of functional groups in any organic molecule. It has been widely used to confirm the attachment of different functional groups in each step of functionalization procedures. For example, Grass *et al.* applied FT-IR spectroscopy for the characterization of carbon-coated cobalt NPs before and after functionalization with a chlorobenzene group [197].

Thermal gravimetric analysis can be performed to confirm coating formation (especially surfactants or polymers) and estimate binding efficiency on the surface of MNPs. Shen *et al.* used this technique to confirm the existence of two distinct populations of surfactants coated on the iron oxide surfaces [198].

Chemical composition can be determined by atomic adsorption spectroscopy [199].

Because of the difficulty of synthesizing monodisperse particulate materials, a technique capable of separating magnetic properties of polydisperse samples is of great importance. Magnetic field flow fractionation separates species on the basis of their magnetic susceptibility and is applicable to materials which have sizes ranging from nanometers to micrometers. Samples are injected into a capillary to interact with an external magnetic field gradient that forces them toward an accumulation wall (i.e. toward higher field strength). Material that interacts strongly with the magnetic field are restricted to the slower flow streams near channel walls, while particles that interact weakly are free to experience the faster flow streams in the channel center [200].

12.3.4   Applications of Magnetic Nanoparticles

Magnetic NPs are interesting from both fundamental study of materials science as well as their applications [201–209]. In particular, magnetic NPs may find very important applications in industrial, environmental, analytical, biological, drug delivery, magnetic resonance imaging, and catalysis fields.

Magnetite and hematite have been used as catalysts for synthesis of NH_3 (Haber process), desulfurization of natural gas, Fisher–Tropsch synthesis for hydrocarbons, and large scale manufacture of butadiene [210].

Magnetic iron oxides are commonly used as synthetic pigments in paints, ceramics, and porcelain [211] because they display a range of colors with pure hues and high tinting strength. Pigments based on hematite are red, those based on maghemite are brown, and magnetite-based pigments are black [212].

Iron NPs technology could provide cost-effective solutions to some of the most challenging environmental cleanup problems [2, 172, 213]. Over the past decade permeable reactive barriers have been developed and used to treat groundwater contaminated by different pollutants [214, 215]. In these barriers, nanoscale zero-valent iron can be used as a reactive material due to its great ability to reduce and stabilize different types of compounds. These materials are being accepted as a versatile tool for the remediation of different types of contaminants in groundwater, soil, and air on both experimental and field scales [216].

In recent years other MNPs have been investigated for the removal of organic and inorganic pollutants. MNPs have a high capacity in removal of high concentrations of organic compounds, mostly dyes [55, 217–220]. Therefore, replacing of conventional adsorbents with MNPs for treatment of textile effluents can be a good platform to be investigated [221].

A very important aspect in metal removal is the preparation of functionalized sorbents for affinity or selective removal of hazardous metal ions from complicated matrices [222–231]. For example, very recently, Lee *et al.* prepared boron dipyrromethene-functionalized magnetic silica NPs for a high affinity and selectivity removal of Pb^{2+} from water and human blood [232]. Their findings may lead to the development of a new type of tailor-made biocompatible system, built by immobilizing appropriate fluorescence receptors onto the surfaces of novel magnetic nanomaterials, for the detection, recovery, and removal of other heavy metals from human body.

Magnetic fluorescent particles, such as polystyrene magnetic beads with entrapped organic dyes/quantum dots [233, 234] or shells of quantum dots [235], iron oxide particles coated with dye-doped silica shells [236], and silica NPs embedded with iron oxide and quantum dots [237, 238], are particular magnetic nanosystems used mostly in biological applications such as cellular imaging [239–241]. Other biomedical applications of magnetic nanoparticles include cellular labeling/cell separation [242], tissue repair [243], drug delivery [244–250], magnetic resonance imaging [74], hyperthermia [251], magnetofection [252], and therapeutic applications [253].

12.4 Titania Nanoparticles in Environmental Photo-Catalysis

12.4.1 Advanced Oxidation Processes

Advanced Oxidation Processes are an alternative route to adsorption techniques in environmental remediation. Every day different types of chemicals are used in many industries and large amounts of pollutants, including dyes, phenol, and pesticides are discharged in the environment. Traditional wastewater treatment techniques such as activated carbon adsorption, chemical oxidation, biological treatment are affected by limitations and disadvantages including only a phase transfer of pollutants without decomposition, low mineralization of compounds, very expensive processes, slow reaction rates, inefficiency at higher levels due to the toxicity on the microorganisms [254], and a need to control pH and temperature. Therefore, in addition to adsorption techniques, degradation of pollutants present in wastewaters has been gradually addressed to more efficient and low energy consuming techniques.

Advanced Oxidation Processes are a set of procedures whose final goal is the oxidative degradation of organic compounds in aqueous solutions to harmless end products by catalytic and photo-chemical techniques [255, 256]. The reactions can proceed to complete degradation of organic pollutants, that is, until their total transformation into inorganic compounds. They can be considered a "green" treatment as sunlight is used to convert toxic pollutants in carbon dioxide and water. These processes are characterized by the same chemical features such as production of hydroxyl radicals, $OH^\bullet$, and superoxide anions, $O_2^{-\bullet}$, which are generated when a semiconductor catalyst, in contact with water and oxygen, adsorbs radiation [257, 258]. The various techniques differ each from the other for the adopted procedures to generate reactive species.

Nanostructured semiconductors have focused the interest of several researchers for their particular physico-chemical properties. Semiconductor metal oxide NPs, such as TiO_2, with average diameter of a few nanometers generate hole/electron pairs under UV irradiation with energy intensity larger than the characteristic band-gap. The hole/electron pairs are responsible for oxidative/reductive processes with organic compounds adsorbed or diffusing on NP surface. Electrons on the surface reduce species such as oxygen, whereas holes oxidize electron donor species.

The mechanism of photo-catalytic degradation of an organic pollutant, *P*, in presence of a semiconductor catalyst, *SEM*, can be sketched as follows:

$$SEM + h\nu \rightarrow e_{cb}^- + h_{vb}^+$$

$$O_2 + e_{cb}^- \rightarrow O_2^{-\bullet}$$

$$H_2O + h_{vb}^+ \rightarrow OH^\bullet + H^+$$

$$O_2^{-\bullet} + H_2O \rightarrow H_2O_2 \rightarrow 2OH^{\bullet}$$

$$OH^{\bullet} + P \rightarrow P_{ox}$$

$$P + e_{cb}^{-} \rightarrow P_{red}$$

Upon UV exposure, electrons are promoted from valence band to conduction band of semiconductor resulting in the formation of electron-holes pairs $(e_{cb}^{-} + h_{vb}^{+})$ in the conduction and valence band, respectively [259–261].

Both electrons and holes can move to the NP surface and establish redox reactions with other adsorbed species. Electrons generally react with oxygen to produce superoxide radical anions of oxygen, $O_2^{-\bullet}$, whereas holes react with bound water to produce $OH^{\bullet}$ radicals (Figure 12.7). $O_2^{-\bullet}$ can react with water and produce supplementary $OH^{\bullet}$ radicals. These radicals can oxidize pollutants, P_{ox}, whereas electrons can reduce them, P_{red} [262]. The main advantage of using nanostructured semiconductors resides in their large surface area to volume ratio, that is, in a large active site number density leading to higher photo-catalytic efficiencies. In fact, reaction rates are strongly affected by the total area of photo-catalyst: the smaller the catalyst particle size the faster the reaction rate for a given mass of photo-catalyst. Photo-catalysis can be either homogeneous or heterogeneous, depending on whether the photo-catalyst exists in the same phase of reactants or acts in a different phase (it is generally bound on a substrate). The efficiency of catalytic suspensions is well known, but they suffer of problems related with particle recovery. As a consequence catalytic substrates bearing photo-active sites are preferred in many applications, even if there is a loss in site number density [263].

Some properties for a good substrate are:

- transparency in the UV region,
- capability to strongly bind NPs without affecting their catalytic efficiency,
- high surface area,
- good adsorption of compounds that one wants to degrade, and
- chemical inactivity.

Examples of catalytic supports are SiO_2, glass, silica gel, quartz fibers, micro-porous cellulose membranes, polyethylene films [264]. A good binding without activity loss is the ideal condition for an optimal casting of catalytic species on supports. In fact, it is important to avoid catalyst separation during the catalytic process and changes in the band-gap structure due to thermal treatments and interactions with supports. The activity

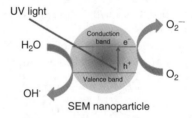

Figure 12.7   *Photo-activity of a semiconductor NP*

can be also reduced by agglomeration phenomena among catalyst particles and their trapping in the micro-porous structure of supports.

The degradation efficiency is defined as $D_E = \dfrac{c(0) - c(t)}{c(0)} 100$, where $c(0)$ and $c(t)$ are the pollutant concentration at time 0 and t, respectively.

The photo-catalytic activity depends on the catalyst ability to create electron–hole pairs, which generate free radicals able to undergo secondary reactions. The process efficiency depends on several parameters including photon adsorption efficiency, charge separation rate, product release from semiconductor surface, redox potentials, long term stability of semiconductor, preparation procedure (e.g. method and amount of doping, calcination temperature, etc.).

Many papers have investigated the relationship between catalyst concentration and process efficiency. It is known that heterogeneous photo-catalytic processes are well described by the following Langmuir–Hinshelwood model:

$$\frac{dc}{dt} = -\frac{k_r k_a c}{1 + k_a c} \cong -k_r k_a c = -k_{app} c$$

where c is the pollutant concentration, k_r is the reaction rate constant for reactant oxidation, k_a is the reactant equilibrium adsorption constant. In the limit of small c values the previous equation can be approximated to apparent first-order kinetics with an apparent reaction rate constant, k_{app} [265].

pH variations change the surface charge of NPs and modify pollutant adsorption equilibrium and, consequently, catalytic activity. In particular, titania surfaces are modified according to the following reactions:

$$TiOH + H^+ \rightarrow TiOH_2^+$$

$$TiOH + OH^- \rightarrow TiO^- + H_2O$$

The higher the catalyst concentration, the faster the degradation rate. Nevertheless turbidity increases and reduces UV penetration with a consequent decrease in the reaction rate [266, 267].

Increasing pollutant amount the degradation percentage decreases as more organic molecules are adsorbed and less photons can impinge onto the catalyst surface.

The addition of ions causes the decrease in degradation percentage due to particular reaction that each ion can establish with reactive species. For example, Cl⁻ plays the role of hole and hydroxyl radical scavenger according to the following reactions [261]:

$$Cl^- + h_{vb}^+ \rightarrow Cl^\bullet$$

$$Cl^- + Cl^\bullet \rightarrow Cl^{\bullet -}$$

$$OH^\bullet + Cl^- \rightarrow HOCl^{\bullet -}$$

$$HOCl^{\bullet -} + H^+ \rightarrow Cl^\bullet + H_2O$$

Ethanol reduces degradation rates as it can quench hydroxyl radicals.

Photo-catalytic degradation is less effective if oxygen is reduced [268–270] as the recombination of hole–electron pairs increases. In fact, oxygen molecules adsorbed on the NP surface trap electrons according to the reaction:

$$O_2 + e_{cb}^- \rightarrow O_2^{\bullet -}$$

An increase in temperature slows down the photo-catalytic activity as recombination of charge carriers and desorption of adsorbed reactants increase [271, 272].

A lot of research has been recently devoted to synthesis and modification of semi-conductor NPs in order to shift their photo-activity to the visible region [273]. Catalysts have been doped in order to shift their band-gap to lower values and activity to visible region of electro-magnetic spectrum. In particular, NPs have been modified with:

1. Noble metals such as Au, Ag, and Pt, which facilitate electron–hole separation, pro-mote electron transfer, decrease energy band-gap, modify surfaces, and attract more pollutant molecules [274, 275]. There exists an optimal metal level doping, above which noble and transition metals act as electron–hole recombination centers lower-ing the photo-activity [276, 277]. Added salts do not affect degradation rates [278].
2. Transition metals, including Fe^{3+}, Mo^{5+}, Ru^{3+}, V^{+4}, Rh^{3+}, which are capable of tuning the electronic structure of semiconductor NPs [259, 279].
3. Lanthanide metals, such as Eu, Ce, Nd, Er, Pr, Sm, and La, which reduce the energy band-gap with a consequent increase of catalytic efficiency [280–282].
4. Alkaline metals, such as Li, K, and Na [283, 284].
5. CdS, which allows a better charge separation [285, 286].
6. Non-metals, such as N, F, S, B, and C [287, 288]. The doping with non-metal is found to be effective in reducing the recombination of electrons and holes, improving the crystallinity of anatase form in titanium dioxide, and generating free OH⁻ radicals.
7. Dyes, which adsorb visible light and are able to sensitize semiconductor NPs [289] as sketched in Figure 12.8.

12.4.2   TiO₂ Assisted Photo-Catalysis

Amongst various oxide semiconductor photo-catalysts, titanium dioxide has proven to be the most widely used due to its strong oxidizing power, low cost, non-toxicity, and long-term photo-stability [290]. The most relevant and active crystalline phases of titanium

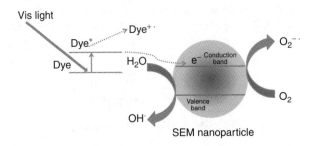

Figure 12.8   *Dye sensitized photo-catalysis*

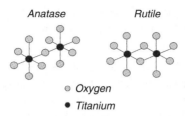

Anatase   Rutile

○ Oxygen
● Titanium

Figure 12.9   *Anatase and rutile structures of TiO$_2$*

dioxide are anatase and rutile (Figure 12.9). TiO$_2$ anatase shows an excellent photo-activity with a band-gap of 3.0–3.2 eV, which allows the formation of hole/electron pairs upon adsorption of photons with a wavelength lower than 400 nm. Titania photo-activity is strongly influenced by structural characteristics (crystal structure, crystallinity degree, surface morphology and area, porosity, sorption characteristics, heat treatment) [259, 291], which depend from preparative conditions.

To address the catalyst retrieval problem TiO$_2$ is generally immobilized on solid supports as bound particles or thin solid films [292–295]. However, in the case of such films the exposed area for photo-catalysis is lower than that of slurries. As a result, a decrease is expected in the overall photo-catalytic performance of thin films compared to slurry solutions [296].

12.4.2.1 TiO$_2$ Assisted Photo-Catalysis of Phenol Compounds

Phenol compounds are important environmental pollutants originating from several industrial (petroleum refinery, coal gasification, metal casting, steel, plastic, dye, and paper manufacturing) and agricultural (chloro- and nitro-phenols are the major products of degradation of some herbicides and pesticides) activities, and processes of chlorine disinfection of drinking water and wastewater [297]. Phenol and phenol compounds present in wastewater industries represent a serious problem for the ecosystem due to their slow biodegradability and antimicrobial activity. In addition, they are endocrine disrupting chemicals with carcinogenic, teratogenic, and mutagenic properties.

Kashif and Ouyang have studied the effects of various parameters such as pH, catalyst concentration, phenol concentration, anions, metal ions, electron acceptors, and surfactants on the anatase assisted photo-catalytic degradation of phenol [298].

The photo-degradation rate of phenol reached a maximum value at pH 5, whereas it was reduced by light shielding effects arising from TiO$_2$ concentration larger than 200 mg L^{-1}. The degradation rate decreased at higher phenol content due to the competitive light adsorption by phenol molecules. Anions such as Cl$^-$, SO$_4^{2-}$, NO$_3^-$, and CO$_3^{2-}$ had different inhibition effects. Fe^{3+} at certain concentration enhanced the degradation rate, whereas metal ions as Ca^{2+}, Cu^{2+}, and Mg^{2+} hindered the degradation process. The presence of oxidants such as BrO$_3^-$, H$_2$O$_2$ and S$_2$O$_8^{2-}$ enhanced the degradation efficiency appreciably due to their electron scavenger properties. Surfactants did not significantly influence the degradation (less than 10%), maybe, due to low concentrations investigated (0.05 mmol L^{-1}).

Similar results were obtained by Zhang *et al.* during the photo-catalytic degradation of phenol by poly-(fluorene-*co*-thiophene) sensitized TiO_2 under visible light irradiation [299]. In particular, they proved:

1. the importance of oxygen, as the reaction was either stopped when the solution was saturated by nitrogen or largely slowed down when NaN_3 (singlet oxygen quencher) was added to the solution;
2. the predominant role played by hydroxyl radicals, $OH^\bullet$, as the addition of alcohols resulted in a decreasing in phenol degradation ratios.

Wang and co-workers have investigated the effect of pH and anion additives in the photo-catalytic degradation of 2-chloro and 2-nitrophenol by titanium dioxide (P25, Evonik) in aqueous solution. The photo-degradation of 2-chlorophenol was faster in acid conditions as low pH values affect the NP surface characteristics and reactive species number [300].

Guo and co-workers detected by gas chromatography/mass spectrometry the intermediates of phenol degradation under UV irradiation. Both in presence and absence of titanium dioxide they found 2-hydroxy-propaldehyde, hydroxy-acetic acid, 3-hydroxy-propyl acid, glycerol, catechol, (E)-2-butenedioic acid, resorcinol, hydroquinone, and 1,2,3-benzenetriol [301].

Ray *et al.* have optimized pollutant concentration, dissolved oxygen, catalyst size and concentration by using a four factor three level Box-Benkhen design [302]. They found that the photo-catalytic degradation rate of phenol by titania could be maximized by using a reaction tube with $40\,mg\,L^{-1}$ phenol concentration, $31\,mg\,L^{-1}$ of dissolved oxygen, a titania particle size of around $9\,nm$, and $1.0\,g\,L^{-1}$ of catalyst. The optimal levels for the selected factors were low, high, medium, and high respectively. Experiments were in agreement with predictions even if the use of $10\,nm$ NPs led to a reduction of about 13% respect to the theoretical catalyst efficiency. In addition, a quantum yield, that is, the number of phenol molecules degraded per unit time over the number of incident photons per unit time, of 35% was gained.

12.4.2.2   *TiO₂ Assisted Photo-Catalysis of Dyes*

Dyes represent another class of pollutants that are found in wastewater industries. About 12% of organic dyes annually produced worldwide are dispersed into the environment during manufacture and application processes [303]. Because of dye toxicity several methods, including biodegradation [304], coagulation [305], adsorption [306], membrane [307], and *advanced oxidation processes* have been proposed for dye recovery and degradation.

Liu and Chiou [308] have investigated the photo-catalytic process of degradation of Reactive Red 239 in a batch photo-reactor as a function of four independent parameters, that is, UV light intensity, TiO_2 concentration, initial pH, and stirring speed, and optimized these values by a 2^4 full-factorial central composite design.

Ao *et al.* have prepared nitrogen doped titania hollow NPs using carbon nanospheres as template and N-atom different content. NP average diameter was $280\,nm$ and the nitrogen shell thickness was about $30\,nm$. Irradiation was performed with visible light (wavelength higher than $400\,nm$) in order to prove that the N-doping has really narrowed titania band-gap. The resulting visible photo-catalysts were more efficient (up to 17 times

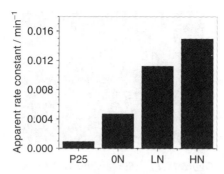

Figure 12.10   *Apparent rate constants in the degradation of Reactive Brilliant Red dye. P25, commercially available NPs; 0N, blank hollow NPs; LN, titania hollow NPs doped with a low nitrogen amount; HN, titania hollow NPs doped with a high nitrogen amount. Reproduced from Journal of Hazardous Materials, 167/1–3, Ao, Xu, Fu and Yuan, A simple method to prepare N-doped titania hollow spheres with hight photocatylitic activity under visible light, August 2009, with permission from Elsevier.*

for the higher N doped catalyst) than commercial available P25 titania in the degradation of Reactive Brilliant Red dye (Figure 12.10) [309].

12.4.2.3   Some Examples of TiO₂ Assisted Photo-Catalysis

As a practical application, the photo-catalytic activity of titanium dioxide NPs in the degradation of some organic pollutants dispersed in water solutions is reported. In particular, films with TiO_2 NPs have been used to photo-degrade Methylene Blue, phenol, and Carbofuran solutions. Carbofuran, 2,3-dihydro-2,2-dimethyl-7-benzofuranol *N*-methylcarbamate, is an insecticide and nematicide used in cultivation of vegetables, sunflowers, cotton, coffee, tobacco. Minimal amounts of this compound are toxic due to its high resistance and inhibition of cholinesterase. The maximum tolerable level of such contaminant in aqueous matrices is $0.04\,mg\,L^{-1}$.

Briefly, for the synthesis 520 mL of water were added drop by drop to 200 g of $TiCl_4$ in a round bottomed flask at 0 °C ($TiCl_4$ tends to react vigorously with air at r.t.) under vigorous stirring and N_2 flow:

$$TiCl_4 + H_2O \rightarrow TiOCl_2 + 2HCl \uparrow$$

100 mL of this solution (0.185 mol of Ti^{+4}) were diluted with 200 mL of water and 20 mL of a water solution (15 wt.%) of $TiOSO_4$ in order to favor the formation of anatase form. Then, other 50 mL of water were added and the solution was stirred at 50 °C for 3 days. The solid $Ti(OH)_4$ was extracted by centrifugation at 2000 rpm, washed with a NaCl solution 0.1 M and isopropyl alcohol. The obtained NPs show a spherical shape with an average size of 20–30 nm, even if they tend to form larger aggregates due to the strong interactions among the OH groups on their surfaces (Figure 12.11).

The volume percentage of anatase, VP_A, was found to be 80% by X-ray diffraction using the following formula [310]:

$$VP_A = \frac{8I_A}{8I_A + 13I_R} \cdot 100$$

Figure 12.11   *SEM picture of TiO₂ NPs*

where I_A and I_R are the intensity of the anatase (101) and rutile (110) peaks, respectively.

An amount of 3 g of TiO_2 aggregates were disrupted in a mortar by addition of 3 mL of water, 0.2 mL of acetyl acetone and some drops of Triton X100. Glass supports were dipped in a hot solution of NaOH (1M@50 °C) for 5 minutes and, then, in a hot acid solution (H_2SO_4 : HNO_3 : H_2O = 1 : 1 : 2 vol.%@50 °C) for 2 min. Finally, they were washed with distilled water and isopropyl alcohol. The TiO_2 suspension was cast on supports by a spin coater at 1500 rpm for 20 s. They were baked in a furnace at 450 °C for 30 min (average thickness of TiO_2 film ≈ 400 nm).

Such films were used to catalyze the photo-degradation of Methylene Blue, phenol, and Carbofuran. Results are shown in Figures 12.12, 12.13, and 12.14. The catalytic efficiency is on average reduced by 18% after 10 days irradiation aging of TiO_2 supports.

12.4.3   Developments in TiO₂ Assisted Photo-Catalysis

As future prospects, Song *et al.* have recently investigated the photo-catalytic degradation of selected dyes by titania thin films characterized by different nanostructures [311].

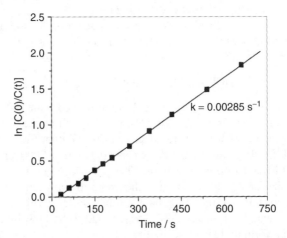

Figure 12.12   *Photo-degradation of Methylene Blue in heterogeneous phase. The line is a first order exponential decay fit*

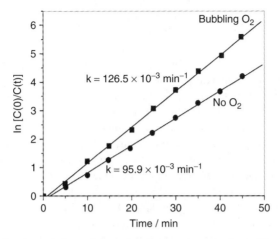

Figure 12.13 *Photo-degradation of phenol in heterogeneous phase. The lines are first order exponential decay fits. The process is faster under oxygen bubbling*

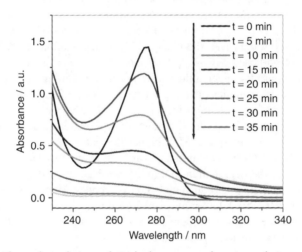

Figure 12.14 *Photo-degradation of Carbofuran as a function of time. After 30 min of irradiation the photo-degradation is almost completed*

The films were fabricated by direct oxidation of metallic titanium in hot H_2O_2 solution (quasi aligned titania nanorods), anodic oxidation of metallic titanium in HF aqueous solution (well aligned titania nanotubes), sol–gel spin coating (NP aggregates), and dip-coating in a P25 NP slurry. Catalytic efficiencies of different films were evaluated with the photo-degradation of Rhodamine B, Methylene Blue, and Methyl Orange dyes under UV irradiation. Titania nanorods and nanotubes were found to be more efficient to assist dye photo-degradation.

Hollow titania NPs and their surface modifications have recently attracted the interest of many researchers for their high surface area and large efficiencies [309, 312].

Wang *et al.* have prepared titania-coated magnetic porous silica and investigated its photo-catalytic activity on Reactive Brilliant Red X-3B under both UV and visible light [313]. They combined the different properties of main components, that is, the high photo-catalytic activity of titania, the large surface area of silica porous NPs, and magnetic separation of iron oxide, to obtain an efficient photo-catalyst without slurry recovery problems or film activity reduction. First, the authors impregnated iron oxide NPs (10 and 15 wt.%) in porous silica; then they reduced Fe^{3+} to Fe^{2+} by ethylene glycol at 450 °C, and, finally, they coated magnetic silica NPs with a shell of titania by sol–gel method under mild conditions. Under both UV and visible light irradiation titania-coated magnetic porous silica with higher iron oxide content was characterized by the highest photo-catalytic activity in the degradation of dye X-3B. Due to the conduction and valence bands of iron oxides, the generated electrons and holes in titania are easily transferred to the lower lying conduction band and upper lying valence band of the iron oxides reducing the recombination processes.

Yu and co-workers have recently proposed a sonochemical-hydrothermal method for preparing fluorinated mesoporous TiO_2 microspheres [314]. The formation and doping of microparticles was achieved by a facile and efficient combination of ultrasound and thermal treatment of a solution containing titanium isopropoxide, template, and sodium fluoride. No calcination or extraction treatment was required for template removal. Despite of particle average diameter, which was around 1.5 μm (aggregates of 9–10 nm crystallites), fluorination remarkably improved the photo-catalytic degradation of Methylene Blue.

Wang *et al.* have proposed a two-step hydrolysis-calcination procedure for fabricating nitrogen doped titania NPs under mild and less time consuming conditions [315]. The authors prepared several samples differing for the calcination temperature and compared N-doped TiO_2 activity in visible light induced photo-decomposition of Methylene Blue. Pure anatase phase was obtained for all samples except that calcinated at lower temperature (250 °C), which was amorphous. The higher the calcination temperature, the higher the crystal size. All N-TiO_2 NPs exhibited a visible photo-activity larger than commercial available P25 NPs in the degradation of Methylene Blue. In particular, the amorphous sample showed the largest photo-decomposition percentage (more than 92% over three runs) as it was affected by a lower nitrogen loss during the calcination process. This result confirms that the nitrogen content plays a more important role than crystallinity in the photo-degradation of Methylene Blue.

12.5   Future Prospects: Is Nano Really Good for the Environment?

Since the discovery of fullerenes and carbon nanotubes in 1990s many metal- and polymer-based NPs have been synthesized and applied. The nanoworld includes materials in which at least one of their dimensions is less than 100 nm and they can be classified according to their dimension, morphology, composition, and state uniformity. The very low dimensions of nanomaterials confer them physico-chemical properties which differ from those of the same bulky materials.

The development of a new technology generally involves the creation of new problems and nanotechnology is associated to the risk of release in the environment of NPs or NP

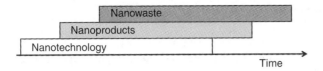

Figure 12.15   *Milestone diagram of Amara's law*

containing materials, creating a new type of waste called *nanowaste*. Nanotoxicology is a new science that deals with the health impacts associated with use of NPs and effects of nanowaste on living organism. According to Environmental Protection Agency nanowaste can be based on carbon (carbon nanotubes, fullerenes, etc.), metals (quantum dots, nanocrystals, metal oxides, etc.), dendrimers, and composites (in which nanomaterials are dispersed). NPs are generally linked or incorporated in a matrix where they exert their action without any particular risk. Nevertheless NP degradation, end-of-life of NP containing materials, decontamination of sewage sludges or wastewaters may disperse nanowaste in the environment (atmosphere, soil, surface waters) where it can be taken up by biological organisms and bioaccumulated in the food chain [316–319].

At the present time there are many doubts about the risks to human health from nanowaste. The nanomaterials contamination pathways include inhalation, ingestion and dermal adsorption. This last can favor ingestion by hands, eyes, and wounds. The scientific literature on nanotoxicology reports about the uptake and accumulation of NPs and their cytotoxic effects in cells, aquatic invertebrates, mice brain cells [4, 320–327]. Oberdörster and co-workers found that nanomaterials, if inhaled, can pass from respiratory into vascular system and disperse in the organism [328]. The cytotoxic effects are dose- and time-dependent [329] and can be attributed to intrinsic toxicity (arising from chemical composition), surface properties, size and shape of nanomaterials. Absorption, distribution, metabolism, and/or breakdown determine nanomaterial fate in the organism. Contamination may cause irritation at the contact site, inflammatory response, and oxidative stress with consequent cell injury. The toxic effects depend on five Ds: dose, dimension and durability of nanowaste, deposition point, and defense mechanisms adopted by the body [330].

No nanowaste is currently considered as hazardous waste. Nevertheless nanotoxicological studies could suggest the need of regulatory actions to the environmental and workplace standards. In fact, according to Amara's law there is the tendency in any society to overestimate the short term effects of a technology and underestimate its long term effects, that is, the real impact of nanotechnology on environment and health (Figure 12.15).

The consideration of regulatory actions requiring basic information from companies producing engineered nanomaterials, expected life-time of the products containing NPs, investigations of parameters as level and intensity of release of NPs from waste under influence of water and moisture, the potential toxicity of the released NPs, the need for reference materials for nanoecotoxicological studies before a massive contamination of the environment have all been recently suggested [331].

12.6   Conclusions

This chapter has reviewed the recent progresses in remediation by nanomaterials. Different classes of nanoparticles such as silica, titania, and magnetic systems, and related synthetic procedures have been described. Their applications for the removal of dyes and organic pollutants from wastewaters have been examined. The advantages of using nanoparticles in remediation have been discussed and, in particular, the attractive opportunity to increase extraction efficiencies and selectivity by changing nanoparticles surface properties has been shown. Applications of some particular nanosystems in biotechnology, chemistry, and biology have been reported. Nanotechnology is a relatively new science with daily new nanodevices that will certainly have a considerable role in the future. However, more research should be devoted to monitor the fate and impact on the environment and health of end-of-life products from nanotechnology.

List of Abbreviations

A-300 HDK®:	hydrophilic nanosilica by Wacker Silicones
AB1:	Acid Black 1
AB25:	Acid Blue 25
AR14:	Acid Red 14
c(0):	pollutant initial concentration
c, c(t):	pollutant concentration at time t
D_E:	degradation efficiency
fcc:	faced-centered cubic unit cell
$Fe(acac)_3$:	iron(III) acetylacetonate
FT-IR spectroscopy:	Fourier Transform – Infra Red spectroscopy
h:	Planck's constant
hcp:	hexagonal close-packed unit cell
$HMS-NH_2$:	aminopropyl-modified hexagonal mesoporous silica
HMS-CD:	monochloro-triazinyl β-cyclodextrin-modified hexagonal mesoporous silica
k_a:	reactant equilibrium adsorption constant
k_{app}:	apparent reaction rate constant
k_r:	reaction rate constant for the oxidation of reactant
$M(acac)_n$:	metal acetylacetonate
MCM41:	a type of mesoporous silica
$M_x cup_x$:	metal cupferronate
MNP:	magnetic nanoparticle
NP:	nanoparticle
P, P_{ox}, P_{red}:	pollutant, oxidized pollutant, reduced pollutant
P25:	titanium dioxide nanoparticles (average diameter 25 nm) commercially available from Evonik
PDDA:	poly (diallydimethylammonium chloride)

Pluronic P123:	triblock copolymer consisting of a central block of polypropylene glycol flanked by two blocks of polyethylene glycol
SEM:	semiconductor
SQUID:	superconducting quantum interference device
UV:	ultraviolet
e_{cb}^{-}:	electron in the conduction band
h_{vb}^{+}:	hole in the valence band
υ:	photon frequency

References

[1] A. Nel, T. Xia, L. Madler and N. Li (2006), "Toxic potential of materials at the nanolevel", *Science* **311**, 622–627.

[2] P.G. Tratnyek and R.L. Johnson (2006), "Nanotechnologies for environmental cleanup", *Nanotoday* **1**, 44–48.

[3] N.C. Mueller and B. Nowack (2010), "Nanoparticles for remediation: solving big problems with little particles", *Elements* **6**, 395–400.

[4] G. Oberdörster, V. Stone and K. Donaldson (2007), "Toxicology of nanoparticles: a historical perspective", *Nanotoxicology* **1**, 2–25.

[5] T.K. Barik, B. Sahu and V. Swain (2008), "Nanosilica - from medicine to pest control", *Parasitol. Res.* **103**, 253–258.

[6] R.K. Iler (1979), "The chemistry of silica", John Wiley & Sons, Inc., New York.

[7] S. Che, A.E.G. Bennett, T. Yokoi, *et al.* (2003), "A novel anionic surfactant templating route for synthesizing mesoporous silica with unique structure", *Nat. Mater.* **2**, 801–805.

[8] A. Salleo, S.T. Taylor, M.C. Martin, *et al.* (2003), "Laser driven phase transformations in amorphous silica", *Nat. Mater.* **2**, 796–800.

[9] K. Suzuki, K. Ikari and H. Imai (2004), "Synthesis of silica nanoparticles having a well-ordered mesostructure using a double surfactant system", *J. Am. Chem. Soc.* **126**, 462–463.

[10] M.T. Alyushin and M.N. Astakhova (1971), "Aerosil and its application in pharmaceutical practice", *Pharmacy* **6**, 73–77.

[11] A.A. Chuiko (ed.) (2003), *Medical Chemistry and Clinical Application of Silicon Dioxide*, Nukova Dumka, Kiev.

[12] J. Kreuter (ed.) (1994), *Colloidal Drug Delivery Systems*, Marcel Dekker, New York.

[13] M. Yokoyoma and T. Okano (1996), "Targetable drug carriers: present status and a future perspective", *Adv. Drug Deliv. Rev.* **21**, 77–80.

[14] M.L. Brongersma (2003) "Nanoscale photonics: nanoshells: gifts in a gold wrapper", *Nat. Mater.* **2**, 296–297.

[15] L.R. Hirsch, J.B. Jackson, A. Lee, *et al.* (2003), "A whole blood immunoassay using gold nanoshells", *Anal. Chem.* **75**, 2377–2381.

[16] L.R. Hirsch, R.J. Stafford, J.A. Bankson *et al.* (2003), "Nanoshell-mediated near-infrared thermal therapy of tumors under magnetic resonance guidance", *Proc. Natl. Acad. Sci. U.S.A.* **100**, 13549–13554.

[17] M. Sliwka-Kaszynska, K. Jaszczolt, A. Kolodziejczyk and J. Rachon (2006), "1,3-alternate 25,27-dibenzoiloxy-26,28-bis-[3-propyloxy]-calix[4]arene bonded silica gel as a new type of HPLC stationary phase", *Talanta* **68**, 1560–1566.

[18] Y.C. Sharma, V. Srivastava, V.K. Singh, *et al.* (2009), "Nano-adsorbents for the removal of metallic pollutants from water and wastewater", *Environ. Technol.* **30**, 583–609.

[19] A. Andrzejewska, A. Krysztafkiewicz and T. Jesionowski (2007), "Treatment of textile dye wastewater using modified silica", *Dyes Pigment.* **75**, 116–124.

[20] Wacker Silicones (2009), Available at http://www.wacker.com/cms/media/publications/downloads/6174_EN.pdf (accessed February 18, 2013).

[21] H.E. Bergna (1994), *Colloid Chemistry of Silica*, American Chemical Society, Washington.

[22] A.P. Legrand (1998), *The Surface Properties of Silicas*, John Wiley & Sons, Inc., New York.

[23] J. Wang, Z.S. Deng, J. Shen and L.Y. Chen (2000), "Silylation of polydiethoxysiloxane derived silica aerogels", *J. Non-Cryst. Solids* **271**, 100–105.

[24] C.A. Morris, D.R. Rolison, K.E. Swider-Lyons, *et al.* (2001), "Modifying nanoscale silica with itself: a method to control surface properties of silica aerogels independently of bulk structure", *J. Non-Cryst. Solids* **285**, 29–36.

[25] H.K. Kammler, L. Madler and S.E. Pratsinis (2001), "Flame synthesis of nanoparticles", *Chem. Eng. Technol.* **24**, 583–596.

[26] J. Chrusciel and L. Slusarski (2003), "Synthesis of nanosilica by the sol–gel method and its activity towards polymers", *Mater. Sci.* **21**, 461–469.

[27] M.T. Swihart (2003), "Vapor-phase synthesis of nanoparticles", *Curr. Opin. Colloid Interface Sci.* **8**, 127–133.

[28] C.J. Brinker and W.J. Scherer (1990), *Sol–gel Science*, Academic Press, New York.

[29] R. Kornak, D. Niznansky, K. Haimann, *et al.* (2005), Synthesis of magnetic nanoparticles via the sol–gel technique', *Mater. Sci.* **23**, 87–92.

[30] F. Banat, S. Al-Asheh and L. Al-Makhadmeh (2003), "Evaluation of the use of raw and activated date pits as potential adsorbents for dye containing waters", *Process Biochem.* **39**, 193–202.

[31] I.M. Banat, P. Nigam, D. Singh and R. Marchant (1996), "Microbial decolorization of textile-dye-containing effluents: a review", *Bioresour. Technol.* **58**, 217–227.

[32] T. Robinson, G. McMullan, R. Marchant and P. Nigam (2001), "Remediation of dyes in textile effluent: a critical review on current treatment technologies with a proposed alternative", *Bioresour. Technol.* **77**, 247–255.

[33] G. Crini (2006), "Non-conventional low-cost adsorbents for dye removal: a review", *Bioresour. Technol.* **97**, 1061–1085.

[34] C.I. Pearce, J.R. Lloyd and J.T. Guthrie (2003), "The removal of colour from textile wastewater using whole bacterial cells: a review", *Dyes Pigment.* **58**, 179–196.

[35] P.C. Vandevivere, R. Bianchi and W. Verstraete (1998), "Treatment and reuse of wastewater from the textile wet-processing industry: review of emerging technologies", *J. Chem. Technol. Biotechnol.* **72**, 289–302.

[36] C. O'Neill, F.R. Hawkes, D.L. Hawkes, *et al.* (1999), "Colour in textile effluents-sources, measurement, discharge consents and simulation: a review", *J. Chem. Technol. Biotechnol.* **74**, 1009–1018.

[37] S.M. Ghoreishi and R. Haghighi (2003), "Chemical catalytic reaction and biological oxidation for treatment of non-biodegradable textile effluent", *Chem. Eng. J.* **95**, 163–169.

[38] A. Dabrowski (2001), "Adsorption, from theory to practice". *Adv. Colloid Interface Sci.* **93**, 135–224.

[39] S.J. Allen and B. Koumanova (2005), "Decolourisation of water/wastewater using adsorption (review)", *J. Univ. Chem. Technol. Metall.* **40**, 175–192.

[40] N.M. Mahmoodi, S. Khorramfar and F. Najafi (2011), "Amine-functionalized silica nanoparticle: preparation, characterization and anionic dye removal ability", *Desalination* **279**, 61–68.

[41] A.R. Cestari, E.F.S. Vieira, G.S. Vieira and L.E. Almeida (2006), "The removal of anionic dyes from aqueous solutions in the presence of anionic surfactant using aminopropylsilica – a kinetic study", *J. Hazard. Mater.* **138**, 133–141.

[42] A.R. Cestari, E.F.S. Vieira and E.S. Silva (2006), "Interactions of anionic dyes with silica-aminopropyl. 1. A quantitative multivariate analysis of equilibrium adsorption and adsorption Gibbs energies", *J. Colloid Interface Sci.* **297**, 22–30.

[43] A.R. Cestari, E.F.S. Vieira, G.S. Vieira and L.E. Almeida (2007), "Aggregation and adsorption of reactive dyes in the presence of an anionic surfactant on mesoporous aminopropyl silica", *J. Colloid Interface Sci.* **309**, 402–411.

[44] A.R. Cestari, E.F.S. Vieira, G.S. Vieira, *et al.* (2009), "The removal of reactive dyes from aqueous solutions using chemically modified mesoporous silica in the presence of anionic surfactant – the temperature dependence and a thermodynamic multivariate analysis", *J. Hazard. Mater.* **161**, 307–316.

[45] E.F.S. Vieira, A.R. Cestari, J.A. Simoni and C. Airoldi (2003), "Thermochemical data for interaction of some primary amines with complexed mercury on mercapto-modified silica gel", *Thermochim. Acta* **399**, 121–126.

[46] M.A. Hassanien and K.S. Abou-El-Sherbini (2006), "Synthesis and characterization of morin-functionalised silica gel for the enrichment of some precious metal ions", *Talanta* **68**, 1550–1559.

[47] A.M. Donia, A.A. Atia, W.A. Al-Amrani and A.M. El-Nahas (2009), "Effect of structural properties of acid dyes on their adsorption behaviour from aqueous solutions by amine modified silica", *J. Hazard. Mater.* **161**, 1544–1550.

[48] H. Yang and Q. Feng (2010), "Direct synthesis of pore-expanded amino-functionalized mesoporous silicas with dimethyldecylamine and the effect of expander dosage on their characterization and decolorization of sulphonated azo dyes", *Microporous Mesoporous Mat.* **135**, 124–130.

[49] T.N.T. Phan, M. Bacquet and M. Morcellet (2000), "Synthesis and characterisation of silica gels functionalized with monochlorotriazinyl β-cyclodextrin and their sorption capacities towards organic compounds", *J. Incl. Phenom. Macrocycl. Chem.* **38**, 343–359.

[50] D.D. Asouhidou, K.S. Triantafyllidis, N.K. Lazaridis and K.A. Matis (2009), "Adsorption of Remazol Red 3BS from aqueous solutions using APTES- and cyclodextrin-modified HMS-type mesoporous silicas", *Colloid Surf. A-Physicochem. Eng. Asp.* **346**, 83–90.

[51] J.B. Joo, J. Park and J. Yi (2009), "Preparation of polyelectrolyte-functionalized mesoporous silicas for the selective adsorption of anionic dye in an aqueous solution", *J. Hazard. Mater.* **168**, 102–107.

[52] T. Jesionowski (2005), "Characterisation of pigments obtained by adsorption of C.I. Basic Blue and C.I. Acid Orange 52 dyes onto silica particles precipitated via the emulsion route", *Dyes Pigment.* **67**, 81–92.

[53] T. Jesionowski, S. Binkowski and A. Krysztafkiewicz (2005), "Adsorption of the selected organic dyes on the functionalized surface of precipitated silica via emulsion route", *Dyes Pigment.* **65**, 267–279.

[54] T. Jesionowski, A. Przybylska, B. Kurc and F. Ciesielczyk (2011), "Hybrid pigments preparation via adsorption of C.I. Mordant Red 3 on both unmodified and aminosilane functionalised silica supports", *Dyes Pigment.* **65**, 267–279.

[55] A.A. Atia, A.M. Donia and W.A. Al-Amrani (2009), "Adsorption/desorption behavior of acid orange 10 on magnetic silica modified with amine groups", *Chem. Eng. J.* **150**, 55–62.

[56] J. Wang, C. Zheng, S. Ding, *et al.* (2011), "Behaviors and mechanisms of tannic acid adsorption on an amino-functionalized magnetic nanoadsorbent", *Desalination* **273**, 285–291.

[57] X. Fu, X. Chen, J. Wang and J. Liu (2011), "Fabrication of carboxylic functionalized superparamagnetic mesoporous silica microspheres and their application for removal basic dye pollutants from water", *Microporous Mesoporous Mat.* **139**, 8–15.

[58] Y.-P. Chang, C.L. Ren, Q. Yang, *et al.* (2011), "Preparation and characterization of hexadecyl functionalized magnetic silica nanoparticles and its application in Rhodamine 6G removal", *Appl. Surf. Sci.* **257**, 8610–8616.

[59] L. Jarup (2003), "Hazards of heavy metal contamination", *Br. Med. Bull.* **68**, 167–182.

[60] X. Huang, K. Caihuan and W. Wen-Xiong (2008), "Bioaccumulation of silver, cadmium and mercury in the abalone Haliotis diversicolor from water and food sources", *Aquaculture* **283**, 194–202.

[61] A. Kouba, M. Buric and P. Kozac (2010), "Bioaccumulation and effects of heavy metals in crayfish: a review", *Water Air Soil Pollut.* **211**, 5–16.

[62] L.D. Benefield, J.F. Judkins and B.L. Weand (1999), *Process Chemistry for Water and Wastewater Treatment*, Prentice-Hall Inc., Englewood Cliffs, NJ.

[63] J. Hu, G. Chen and I.C.M. Lo (2006), "Selective removal of heavy metals from industrial waste water using maghemite nanoparticles: performance and mechanisms", *J. Environ. Eng.-ASCE* **132**, 709–715.

[64] A. Heidari, H. Younesi and Z. Mehraban (2009), "Removal of Ni(II), Cd(II), and Pb(II) from a ternary aqueous solution by amino functionalized mesoporous and nano mesoporous silica", *Chem. Eng. J.* **153**, 70–79.

[65] K.F. Lam, K.L. Yeung and G. McKay (2007), "Efficient approach for Cd^{2+} and Ni^{2+} removal and recovery using mesoporous adsorbent with tunable selectivity", *Environ. Sci. Technol.* **41**, 3329–3334.

[66] S. Pacheco, J. Tapia, M. Medina and R. Rodriguez (2006), "Cadmium ion adsorption in simulated waste water using structured alumina-silica nanoparticles", *J. Non-Cryst. Solids* **352**, 5475–5481.

[67] M. Medina, J. Tapia, S. Pacheco, *et al.* (2006), "Adsorption of lead ions in aqueous solution using silica-alumina nanoparticles", *J. Non-Cryst. Solids* **356**, 383–387.

[68] Y. Wang, G. Wang, H. Wang, *et al.* (2009), "Template-induced synthesis of hierarchical $SiO_2@\gamma$-AlOOH spheres and their application in Cr(VI) removal", *Nanotechnology* **20**, 155604.

[69] C.H. Zong, K.L. Ai, G. Zhang *et al.* (2011) "Dual-emission fluorescent silica nanoparticle-based probe for ultrasensitive detection of Cu^{2+}", *Anal. Chem.* **83**, 3126–3132.

[70] G. Li, Z. Zhao, J. Liu and G. Jiang (2011), "Effective heavy metal removal from aqueous systems by thiol functionalized magnetic mesoporous silica", *J. Hazard. Mater.* **192**, 277–283.

[71] H. Hu, Z. Wang and L. Pan (2010), "Synthesis of monodisperse $Fe_3O_4@$silica core–shell microspheres and their application for removal of heavy metal ions from water", *J. Alloy. Compd.* **492**, 656–661.

[72] Y.-F. Lin, H.-W. Chen, P.S. Chien, *et al.* (2011), "Application of bifunctional magnetic adsorbent to adsorb metal cations and anionic dyes in aqueous solution", *J. Hazard. Mater.* **185**, 1124–1130.

[73] D.G. Rancourt (2001), "Magnetism of earth, planetary, and environmental nano-materials", *Rev. Mineral. Geochem.* **44**, 217–292.

[74] S. Conroy, S.H. Jerry Lee and M. Zhang (2008), "Magnetic nanoparticles in MR imaging and drug delivery", *Adv. Drug Deliv. Rev.* **60**, 1252–1265.

[75] S.P. Gubin, A. Koksharov Yu, G.B. Khomutov, *et al.* (2005), "Magnetic nanoparticles: preparation, structure and properties", *Russ. Chem. Rev.* **74**, 489–520.

[76] A.-H. Lu, E.L. Salabas and F. Schuth (2007), "Magnetic nanoparticles: synthesis, protection, functionalization, and application", *Angew. Chem.-Int. Edit.* **46**, 1222–1244.

[77] M. Mohapatra and S. Anand (2010), "Synthesis and applications of nano-structured iron oxides/hydroxides – a review", *Int. J. Eng. Sci. Technol.* **2**, 127–146.

[78] F. Grasset, N. Labhsetwar, D. Li, *et al.* (2002), "Synthesis and magnetic characterization of zinc ferrite nanoparticles with different environments: powder, colloidal solution, and zinc ferrite–silica core–shell nanoparticles", *Langmuir* **18**, 8209–8216.

[79] S. Neveu, A. Bee, M. Robineau and D. Talbot (2002), "Size-selective chemical synthesis of tartrate stabilized cobalt ferrite ionic magnetic fluid", *J. Colloid Interface Sci.* **255**, 293–298.

[80] S. Sun and H. Zeng (2002), "Size-controlled synthesis of magnetite nanoparticles", *J. Am. Chem. Soc.* **124**, 8204–8205.

[81] S.-J. Park, S. Kim, S. Lee, *et al.* (2000), "Synthesis and magnetic studies of uniform iron nanorods and nanospheres", *J. Am. Chem. Soc.* **122**, 8581–8582.

[82] V.F. Puntes, K.M. Krishan and A.P. Alivisatos (2001), "Colloidal nanocrystal shape and size control: The case of cobalt." *Science* **291**, 2115–2117.

[83] Q. Chen, A.J. Rondinone, B.C. Chakoumakos and Z.J. Zhang (1999), "Synthesis of superparamagnetic $MgFe_2O_4$ nanoparticles by coprecipitation" *J. Magn. Magn. Mater.* **194**, 1–7.

[84] J. Park, K. An, Hwang *et al.* (2004), "Ultra-large-scale syntheses of monodisperse nanocrystals", *Nat. Mater.* **3**, 891–895.

[85] S. Sun, C.B. Murray, Weller *et al.* (2000), "Monodisperse FePt nanoparticles and ferromagnetic FePt nanocrystal superlattices", *Science* **287**, 1989–1992.

[86] E.V. Shevchenko, D.V. Talapin, A.L. Rogach, *et al.* (2002), "Colloidal synthesis and self-assembly of $CoPt_3$ nanocrystals", *J. Am. Chem. Soc.* **124**, 11480–11485.

[87] H. Iida, K. Takayanagi, T. Nakanishi and T. Osaka (2007) "Synthesis of Fe_3O_4 nanoparticles with various sizes and magnetic properties by controlled hydrolysis", *J. Colloid Interface Sci.* **314**, 274–280.

[88] J. Lee, T. Isobe and M. Senna (1996), "Magnetic properties of ultrafine magnetite particles and their slurries prepared via in-situ precipitation", *Colloids Surf. A-Physicochem. Eng. Asp.* **109**, 121–127.

[89] A. Bee, R. Massart and S. Neveu (1995), "Synthesis of very fine maghemite particles", *J. Magn. Magn. Mater.* **149**, 6–9.

[90] T. Ishikawa, T. Takeda and K. Kandori (1992), "Effects of amines on the formation of β-ferric oxide hydroxide", *J. Mater. Sci.* **27**, 4531–4535.

[91] K. Kandori, Y. Kawashima and T. Ishikawa (1992), "Effects of citrate ions on the formation of monodispersed cubic hematite particles", *J. Colloid Interface Sci.* **152**, 284–288.

[92] T. Ishikawa, S. Kataoka and K. Kandori (1993), "The influence of carboxylate ions on the growth of β-FeOOH nanoparticles", *J. Mater. Sci.* **28**, 2693–2698.

[93] L. Cushing, V.L. Kolesnichenko and C.J. O'Connor (2004), "Recent advances in the liquid-phase syntheses of inorganic nanoparticles", *Chem. Rev.* **104**, 3893–3946.

[94] L. Willis, N.J. Turro and S. O'Brien (2005), "Spectroscopic characterization of the surface of iron oxide nanocrystals", *Chem. Mat.* **17**, 5970–5975.

[95] M.S. Wu, Y.H. Ou and Y.-P. Lin (2011), "Iron oxide nanosheets and nanoparticles synthesized by a facile single-step coprecipitation method for lithium-ion batteries" *J. Electrochem. Soc.* **158**, A231–A236.

[96] F.X. Redl, C.T. Black, G.C. Papaefthymiou, *et al.* (2004), "Magnetic, electronic and structural characterization of non-stoichiometric iron oxides at the nanoscale", *J. Am. Chem. Soc.* **126**, 14583–14589.

[97] S. Sun, H. Zeng, D.B. Robinson, *et al.* (2004), "Monodisperse MFe_2O_4 (M = Fe, Co, Mn) nanoparticles", *J. Am. Chem. Soc.* **126**, 273–279.

[98] J. Rockenberger, E.C. Scher and A.P. Alivisatos (1999), "A new nonhydrolytic single-precursor approach to surfactant-capped nanocrystals of transition metal oxides", *J. Am. Chem. Soc.* **121**, 11595–11596.

[99] D. Farrell, S.A. Majetich and J.P. Wilcoxon (2003) "Preparation and characterization of monodisperse Fe nanoparticles", *J. Phys. Chem. B* **107**, 11022–11030.

[100] N.R. Jana, Y. Chen and X. Peng (2004), "Size- and shape-controlled magnetic (Cr, Mn, Fe, Co, Ni) oxide nanocrystals via a simple and general approach", *Chem. Mat.* **16**, 3931–3935.

[101] A.C.S. Samia, K. Hyzer, J.A. Schlueter, *et al.* (2005), "Ligand effects on the growth and digestion of Co nanocrystals", *J. Am. Chem. Soc.* **127**, 4126–4127.

[102] Y. Li, M. Afzaal and P. O'Brien (2006) "The synthesis of amine-capped magnetic (Fe, Mn, Co, Ni) oxide nanocrystals and their surface modification for aqueous dispersibility", *J. Mater. Chem.* **16**, 2175–2180.

[103] K. Butter, A.P. Philipse and G.J. Vroege (2002), "Synthesis and properties of iron ferrofluids", *J. Magn. Magn. Mater.* **2002**, 252, 1–3.

[104] F. Dumestre, B. Chaudret, C. Amiens, *et al.* (2004), "Superlattices of iron nanocubes synthesized from Fe [N(SiMe$_3$)$_2$]$_2$" *Science* **303**, 821–823.

[105] Q. Song and Z.J. Zhang (2004), "Shape control and associated magnetic properties of spinel cobalt ferrite nanocrystals", *J. Am. Chem. Soc.* **126**, 6164–6168.

[106] V.F. Puntes, D. Zanchet, C.K. Erdonmez and A.P. Alivisatos (2002), "Synthesis of hcp-Co nanodisks", *J. Am. Chem. Soc.* **124**, 12874–12880.

[107] F. Dumestre, B. Chaudret, C. Amiens, *et al.* (2002), "Shape control of thermo-dynamically stable cobalt nanorods through organometallic chemistry", *Angew. Chem.-Int. Edit.* **41**, 4286–4289.

[108] F. Dumestre, B. Chaudret, C. Amiens, *et al.* (2003), "Unprecedented crystalline super-lattices of monodisperse cobalt nanorods", *Angew. Chem.-Int. Edit.* **42**, 5213–5216.

[109] N. Cordente, M. Respaud, F. Senocq, *et al.* (2001), "Synthesis and magnetic properties of nickel nanorods", *Nano Lett.* **1**, 565–568.

[110] Y. Chen, D.-L. Peng, D. Lin and X. Luo (2007), "Preparation and magnetic proper-ties of nickel nanoparticles via the thermal decomposition of nickel organometallic precursor in alkylamines", *Nanotechnology* **18**, 505703–505708.

[111] Y. Chen, E. Johnson and X. Peng (2007), "Formation of monodisperse and shape-controlled MnO nanocrystals in non-injection synthesis: self-focusing via ripening", *J. Am. Chem. Soc.* **129**, 10937–10947.

[112] S.G. Kwon, Y. Piao, J. Park, *et al.* (2007), "Kinetics of monodisperse iron oxide nanocrystal formation by "heating-up" process", *J. Am. Chem. Soc.* **129**, 12571–12584.

[113] J. Park, B. Koo, Y. Hwang, *et al.* (2004), "Novel synthesis of magnetic Fe$_2$P nanorods from thermal decomposition of continuously delivered precursors using a syringe pump", *Angew. Chem.-Int. Edit.* **43**, 2282–2285.

[114] Z. Li, Q. Sun and M. Gao (2005), "Preparation of water-soluble magnetite nanocrystals from hydrated ferric salts in 2-pyrrolidone: mechanism leading to Fe$_3$O$_4$" *Angew. Chem.-Int. Edit.* **44**, 123–126.

[115] F.Q. Hu, L. Wei, Z. Zhou, *et al.* (2006), "Preparation of biocompatible magnetite nanocrystals for in vivo magnetic resonance detection of cancer", *Adv. Mater.* **18**, 2553–2556.

[116] S.F. Chin, S.C. Pang and C.H. Tan (2011), "Green synthesis of magnetite nanoparticles (via thermal decomposition method) with controllable size and shape" *J. Mater. Environ. Sci.* **2**, 299–302.

[117] F. Zhao, B. Zhang and L. Feng (2012), "Preparation and magnetic properties of magnetite nanoparticles" *Mater. Lett.* **68**, 112–114.

[118] B. Smith, C.L. Raston and A.N. Sobolev (2005), "Poly(ethyleneglycol) (PEG): a versatile reaction medium in gaining access to 4′-(pyridyl)-terpyridines" *Green Chem.* **7**, 650–654.

[119] N.M. Smith, C.L. Raston C.B. Smith and A.N. Sobolev (2007), "PEG mediated synthesis of amino-functionalised 2,4,6-triarylpyridines" *Green Chem.* **9**, 1185–1190.

[120] M. Kidwai, D. Bhatnagar and N.K. Mishra (2010), "Polyethylene glycol (PEG) mediated green synthesis of 2,5-disubstituted 1,3,4-oxadiazoles catalyzed by ceric ammonium nitrate (CAN)" *Green Chem.* **3**, 55–59.

[121] Y. Yamada, T. Suzuki and E.N. Abarra (1998), "Magnetic properties of electron beam evaporated CoPt alloy thin films", *IEEE Trans. Magn.* **34**, 343–345.

[122] I. Robinson, S. Zacchini, L.D. Tung, *et al.* (2009), "Synthesis and characterization of magnetic nanoalloys from bimetallic carbonyl clusters", *Chem. Mat.* **21**, 3021–3026.

[123] O. Tegus, E. Brvck, K.H.J. Buschow and F.R. de Boer (2002), "Transition-metal-based magnetic refrigerants for room-temperature applications", *Nature* **415**, 150–152.

[124] F. Luo, H.-L Su, W. Song, *et al.* (2004), "Magnetic and magnetotransport properties of Fe_2P nanocrystallites *via* a solvothermal route", *J. Mater. Chem.* **14**, 111–115.

[125] S.C. Perera, G. Tsoi, L.E. Wenger and S.L. Brock (2003), "Synthesis of MnP nanocrystals by treatment of metal carbonyl complexes with phosphines: a new, versatile route to nanoscale transition metal phosphides", *J. Am. Chem. Soc.* **125**, 13960–13961.

[126] K.L. Stamm, J.C. Garno, G.-Y. Liu and S.L. Brock (2003), "A general methodology for the synthesis of transition metal pnictide nanoparticles from pnictate precursors and its application to iron–phosphorus phases", *J. Am. Chem. Soc.* **125**, 4038–4039.

[127] C. Qian, F. Kim, L. Ma, *et al.* (2004), "Solution-phase synthesis of single-crystalline iron phosphide nanorods/nanowires", *J. Am. Chem. Soc.* **126**, 1195–1198.

[128] S. Giri, S. Samanta, S. Maji, *et al.* (2005), "Magnetic properties of α-Fe_2O_3 nanoparticle synthesized by a new hydrothermal method", *J. Magn. Magn. Mater.* **285**, 296–302.

[129] B. Mao, Z. Kang, E. Wang, *et al.* (2006), "Synthesis of magnetite octahedrons from iron powders through a mild hydrothermal method", *Mater. Res. Bull.* **41**, 2226–2231.

[130] H. Zhu, D. Yang and L. Zhu (2007), "Hydrothermal growth and characterization of magnetite (Fe_3O_4) thin films", *Surf. Coat. Technol.* **201**, 5870–5874.

[131] F. Gözüak, Y. Köseoğlu, A. Baykal H. Kavas (2009), "Synthesis and characterization of $Co_xZn_{1-x}Fe_2O_4$ magnetic nanoparticles via a PEG-assisted route", *J. Magn. Magn. Mater.* **321**, 2170–2177.

[132] J. Wang, F. Ren, R. Yi, *et al.* (2009), "Solvothermal synthesis and magnetic properties of size-controlled nickel ferrite nanoparticles", *J. Alloys Compd.* **479**, 791–796.

[133] X. Wang, J. Zhuanq, Q. Peng and Y. Li (2005), "A general strategy for nanocrystal synthesis", *Nature* **437**, 121–124.

[134] H. Deng, X. Peng, Q. Li, *et al.* (2005), "Monodisperse magnetic single-crystal ferrite microspheres", *Angew. Chem.-Int. Edit.* **44**, 2782–2785.

[135] M.-T. Liang, S.-H. Wang, Y.-L. Chang, *et al.* (2010), "Iron oxide synthesis using a continuous hydrothermal and solvothermal system", *Ceram. Int.* **36**, 1131–1135.

[136] Y-H Zheng., Y. Cheng, F. Bao and Y-S. Wang (2006), "Synthesis and magnetic properties of Fe_3O_4 nanoparticles", *Mater. Res. Bull.* **41**, 525–529.

[137] N. Mizutani, T. Iwasaki, S. Watano, *et al.* (2010), "Size control of magnetite nanoparticles in Tomohiro hydrothermal synthesis by coexistence of lactate and sulfate ions", *Current App. Phys.* **10**, 801–806.

[138] X. Sun, C. Zheng, F. Zhang, *et al.* (2009), "Size-controlled synthesis of magnetite (Fe_3O_4) nanoparticles coated with glucose and gluconic acid from a single Fe(III) precursor by a sucrose bifunctional hydrothermal method", *J. Phys. Chem.* **113**, 16002–16008.

[139] Y. Dong, H. Yang, R. Raoand A. Zhang (2009), "Selective synthesis of α-FeOOH and α-Fe_2O_3 nanorods via a temperature controlled process", *J. Nanosci. Nanotechnol.* **9**, 4774–4779.

[140] S. Ge, S. Xiangyang, S. Kai, *et al.* (2009), "Facile hydrothermal synthesis of iron oxide nanoparticles with tunable magnetic properties", *J. Phys. Chem. C* **113**, 13593–13599.

[141] X.-L. Fang, Y. Li, C. Chen, *et al.* (2010), "pH-induced simultaneous synthesis and self-assembly of 3D layered β-FeOOH nanorods", *Langmuir* **26**, 2745–2750.

[142] D. Langevin (1992), "Micelles and microemulsions", *Annu. Rev. Phys. Chem.* **43**, 341–369.

[143] B.K. Paul and S.P. Moulik (2001), "Uses and applications of microemulsions", *Curr. Sci.* **80**, 990–1001.

[144] A.K. Gupta and M. Gupta (2005), "Synthesis and surface engineering of iron oxide nanoparticles for biomedical applications", *Biomaterials* **26**, 3995–4021.

[145] E.E. Carpenter, C.T. Seip and C.J. O'Connor (1999) "Magnetism of nanophase metal and metal alloy particles formed in ordered phases", *J. Appl. Phys.* **85**, 5184–5186.

[146] C. Liu, B. Zou, A.J. Rondinone and Z.J. Zhang (2000), "Reverse micelle synthesis and characterization of superparamagnetic $MnFe_2O_4$ spinel ferrite nanocrystallites", *J. Phys. Chem. B* **104**, 1141–1145.

[147] N.R. Panchal and R.B. Jotania (2010), "Cobalt ferrite nano particles by microemulsion route" *Nanotechnol. Nanosci.* **1**, 7–18.

[148] M. Gobe, K. Kon-No, K. Kandori and A. Kitahara (1983), "Preparation and characterization of monodisperse magnetite sols in W/O microemulsion", *J. Colloid Interface Sci.* **93**, 293–295.

[149] S. Bandow, K. Kimura, K. Kon-No and A. Kitahara (1987), "Magnetic properties of magnetite ultrafine particles prepared by W/O microemulsion method", *Jpn. J. Appl. Phys.* **26**, 713–717.

[150] K.M. Lee, C.M. Sorensen, K.J. Klabunde and G.C. Hadjipanayis (1992), "Synthesis and characterization of stable colloidal Fe_3O_4 particles in water-in-oil microemulsions.", *IEEE Trans. Magn.* **28**, 3180–3182.

[151] L. Liz, M.A. Lopez-Quintela, J. Mira and J. Rivas (1994), "Preparation of colloidal Fe_3O_4 ultrafine particles in microemulsions", *J. Mater. Sci.* **29**, 3797–3801.

[152] J.A. Lopez-Perez, M.A. Lopez-Quintela, J. Mira and J. Rivas (1997), "Preparation of magnetic fluids with particles obtained in microemulsions", *IEEE Trans. Magn.* **33**, 4359–4362.

[153] P.A. Dresco, V.S. Zaitsev, R.J. Gambino and B. Chu (1999), "Preparation and properties of magnetite and polymer magnetite nanoparticles", *Langmuir* **15**, 1945–1951.

[154] H.S. Lee, W.C. Lee and T. Furubayashi (1999), "A comparison of coprecipitation with microemulsion methods in the preparation of magnetite", *J. Appl. Phys.* **85**, 5231–5283.

[155] S. Santra, R. Tapec, N. Theodoropoulou, *et al.* (2001), "Synthesis and characterization of silica-coated iron oxide nanoparticles in microemulsion: the effect of nonionic surfactants", *Langmuir* **17**, 2900–2906.

[156] Z.L. Liu, X. Wang, K.L. Yao, *et al.* (2004), "Synthesis of magnetite nanoparticles in W/O microemulsion", *J. Mater. Sci.* **39**, 2633–2636.

[157] Y. Lee, J. Lee, C.J. Bae, *et al.* (2005), "Large-scale synthesis of uniform and crystalline magnetite nanoparticles using reverse micelles as nanoreactors under reflux conditions", *Adv. Funct. Mater.* **15**, 503–509.

[158] T. Koutzarova, S. Koulev, C. Ghelev, *et al.* (2006), "Microstructural study and size control of iron oxide nanoparticles produced by microemulsion technique", *Phys. Status Solidi C* **3**, 1302–1307.

[159] J. Esquivel, I.A. Facundo, M.E. Trevino and R.G. Lopez (2007), "A novel method to prepare magnetic nanoparticles: precipitation in bicontinuous microemulsions", *J. Mat. Sci.* **42**, 9015–9020.

[160] A.L. Loo, M.G. Pineda, H. Saade, *et al.* (2008), "Synthesis of magnetic nanoparticles in bicontinuous microemulsions. Effect of surfactant concentration", *J. Mater. Sci.* **43**, 3649–3654.

[161] C. Solans, P. Izquierdo, J. Nolla, *et al.* (2005), "Nano emulsions", *Curr. Opin. Colloid Interface Sci.* **10**, 102–110.

[162] F. Geng, Z. Zhao, H. Cong, *et al.* (2006), "An environment-friendly microemulsion approach to α-FeOOH nanorods at room temperature", *Mater. Res. Bull.* **41**, 2238–2243.

[163] A.B. Chin and I.I. Yaacob (2007), "Synthesis and characterization of magnetic iron oxide nanoparticles via w/o microemulsion and Massart's procedure", *J. Mater. Process. Technol.* **191**, 235–237.

[164] F.A. Tourinho, R. Franck and R. Massart (1990), "Aqueous ferrofluids based on manganese and cobalt ferrites", *J. Mater. Sci.* **25**, 3249–3254.

[165] D.E. Zhang, Z.W. Tong, S.Z. Li, *et al.* (2008), "Fabrication and characterization of hollow Fe_3O_4 nanospheres in a microemulsion" *Mater. Lett.* **62**, 4053–4055.

[166] Y.X. Yang, M.L. Liu, H. Zhu, *et al.* (2008), "Preparation, characterization, magnetic property, and Mossbauer spectra of the β-FeOOH nanoparticles modified by nonionic surfactant", *J. Magn. Magn. Mater.* **320**, L132–L136.

[167] Z. Jia, W. Yujun, L. Yangcheng, *et al.* (2006), "In situ preparation of magnetic chitosan/Fe$_3$O$_4$ composite nanoparticles in tiny pools of water-in-oil microemulsion", *React. Funct. Polym.* **66**, 1552–1558.

[168] K. Aslam (2008), "Preparation and characterization of magnetic nanoparticles embedded in microgels", *Mater. Lett.* **62**, 898–902.

[169] L. Zhu, J. Ma, N. Jia, *et al.* (2009), "Chitosan-coated magnetic nanoparticles as carriers of 5-Fluorouracil: Preparation, characterization and cytotoxicity studies.", *Colloid SurfB: Biointerfaces* **68**, 1–6.

[170] M. Green and P. O'Brien (2001), "The preparation of organically functionalised chromium and nickel nanoparticles", *Chem. Commun.* **1912–1913**.

[171] Y. Hou and S. Gao (2003), "Monodisperse nickel nanoparticles prepared from a monosurfactant system and their magnetic properties", *J. Mater. Chem.* **13**, 1510–1512.

[172] Y.-P. Sun, X.-Q. Li, J. Cao, *et al.* (2006), "Characterization of zero-valent iron nano-particles", *Adv. Colloid Interface Sci.* **120**, 47–56.

[173] Y.-P. Sun, X.-Q. Li, W.-X. Zhang and H.P. Wang (2007), "A method for the preparation of stable dispersion of zero-valent iron nanoparticles", *Colloid Surf. A -Physicochem. Eng. Asp.* **308**, 60–66.

[174] E.H. Kima, H.S. Lee, B.K. Kwak and B.-K. Kim (2005), "Synthesis of ferrofluid with magnetic nanoparticles by sonochemical method for MRI contrast agent", *J. Magn. Magn. Mater.* **289**, 328–330.

[175] V.V. Namboodiri and R.S. Varma (2001), "Microwave-accelerated Suzuki cross-coupling reaction in polyethylene glycol (PEG)", *Green Chem.* **3**, 146–148.

[176] S. Komarneni (2003), "Nanophase materials by hydrothermal, microwave-hydrothermal and microwave-solvothermal methods", *Curr. Sci.* **85**, 1730–1734.

[177] W.-W. Wang (2008), "Microwave-induced polyol-process synthesis of M^{II}Fe$_2$O$_4$ (M = Mn, Co) nanoparticles and magnetic property", *Mater. Chem. Phys.* **108**, 227–231.

[178] W.-W. Wang, Y.-J. Zhu and M.-L. Ruan (2007), "Microwave-assisted synthesis and magnetic property of magnetite and hematite nanoparticles", *J. Nanopart. Res.* **9**, 419–426.

[179] S. Veintemillas-Verdaguer, O. Bomatí-Miguel and M.P. Morales (2002), "Effect of the process conditions on the structural and magnetic properties of γ-Fe$_2$O$_3$ nanoparticles produced by laser pyrolysis", *Scr. Mater.* **47**, 589–593.

[180] P. Tartaj, M.P. Morales, S. Veintemillas-Verdaguer, *et al.* (2003), "The preparation of magnetic nanoparticles for applications in biomedicine", *J. Phys. D-Appl. Phys.* **36**, R182–R197.

[181] S. Veintemillas-Verdaguer, M.P. Morales and C.J. Serna (2001), "Effect of the oxidation conditions on the maghemites produced by laser pyrolysis", *Appl. Organomet. Chem.* **15**, 365–372.

[182] M.P. Morales, O. Bomati-Miguel, R. Perez de Alejo, *et al.* (2003), "Contrast agents for MRI based on iron oxide nanoparticles prepared by laser pyrolysis", *J. Magn. Magn. Mater.* **266**, 102–109.

[183] S. Veintemillas-Vendaguer, M.P. Morales, O. Bomati-Miguel, *et al.* (2004), "Colloidal dispersions of maghemite nanoparticles produced by laser pyrolysis with applications as NMR contrast agents", *J. Phys. D-Appl. Phys.* **37**, 2054–2059.

[184] O. Bomati-Miguel, L. Mazeina, A. Navrotsky and S. Veintemillas-Verdaguer (2008), "Calorimetric study of maghemite nanoparticles synthesized by laser-induced pyrolysis", *Chem. Mat.* **20**, 591–598.

[185] B. Julian-Lopez, C. Boissiére, C. Chaneac, *et al.* (2007), "Mesoporous maghemite–organosilica microspheres: a promising route towards multifunctional platforms for smart diagnosis and therapy", *J. Mater. Chem.* **17**, 1563–1569.

[186] J.B. Park, S.H. Jeong, M.S. Jeong, *et al.* (2008), "Synthesis of carbon-encapsulated magnetic nanoparticles by pulsed laser irradiation of solution", *Carbon* **46**, 1369–1377.

[187] J. Gavillet, A. Loiseau, C. Journet, *et al.* (2001), "Root-growth mechanism for single-wall carbon nanotubes", *Phys. Rev. Lett.* **87**, 275504–275507.

[188] R.Y. Hong, B. Feng, L.L. Chen, *et al.* (2008), "Synthesis, characterization and MRI application of dextran-coated Fe_3O_4 magnetic nanoparticles", *Biochem. Eng. J.* **42**, 290–300.

[189] S. Brice-Profeta, M.A. Arrio, E. Tronc, *et al.* (2005), "Magnetic order in γ-Fe_2O_3 nanoparticles: a XMCD study", *J. Magn. Magn. Mater.* **288**, 354–365.

[190] N. De Jaeger, H. Demeye, R. Findy *et al.* (1991), "Particle sizing by photon correlation spectroscopy Part I: monodisperse lattices: influence of scattering angle and concentration of dispersed material", *Part. Part. Syst. Charact.* **8**, 179–186.

[191] Y. Zhao, Z. Qiu and J. Huang (2008), "Preparation and analysis of Fe_3O_4 magnetic nanoparticles used as targeted-drug carriers", *Chin. J. Chem. Eng.* **16**, 451–455.

[192] H.P. Klug and L.E. Alexander (1954), *X-Ray Diffraction Procedures*, John Wiley & Sons, Inc., New York.

[193] A. Ney, P. Poulopoulos, M. Farle and K. Baberschke (2000), "Absolute determination of Co magnetic moments: ultrahigh-vacuum high-T_c SQUID magnetometry", *Phys. Rev. B* **62**, 11336–11339.

[194] S. Foner (1959), "Versatile and sensitive vibrating-sample magnetometer", *Rev. Sci. Instrum.* **30**, 548–557.

[195] L.Y. Zhang, X.J. Zhu, H.W. Sun, *et al.* (2010), "Control synthesis of magnetic Fe_3O_4–chitosan nanoparticles under UV irradiation in aqueous system", *Curr. Appl. Phys.* **10**, 828–833.

[196] S. Gao, Y. Shi, S. Zhang, *et al.* (2008), "Biopolymer-assisted green synthesis of iron oxide nanoparticles and their magnetic properties", *J. Phys. Chem. C* **112**, 10398–10401.

[197] R.N. Grass, E.K. Athanassiou and W.J. Stark (2007), "Covalently functionalized cobalt nanoparticles as a platform for magnetic separations in organic synthesis", *Angew. Chem.-Int. Edit.* **46**, 4909–4912.

[198] L. Shen, P.E. Laibinis and T.A. Hatton (1999), Bilayer surfactant stabilized magnetic fluids: synthesis and interactions at interfaces', *Langmuir* **15**, 447–453.

[199] P.C. Morais, R.L. Santos, A.C.M. Pimenta, *et al.* (2006), "Preparation and characterization of ultra-stable biocompatible magnetic fluids using citrate-coated cobalt ferrite nanoparticles", *Thin Solid Films* **515**, 266–270.

[200] A.H. Latham, R.S. Freitas, P. Schiffer and M.E. Williams (2005), "Capillary magnetic field flow fractionation and analysis of magnetic nanoparticles", *Anal. Chem.* **77**, 5055–5062.

[201] S.W. Charles and J. Popplewell (1982), "Properties and applications of magnetic liquids", *Endeavour.* **6**, 153–161.

[202] R.S. Ruoff, D.C. Lorents, B. Chan, *et al.* (1993), "Single crystal metals encapsulated in carbon nanoparticles", *Science* **259**, 346–348.

[203] R.D. Shull (1993), "Magnetocaloric effect of ferromagnetic particles", *IEEE Trans. Magn.* **29**, 2614–2615.

[204] J.H. Scott and S.A. Majetich (1995), "Morphology, structure, and growth of carbon arc nanoparticles", *Phys. Rev. B* **52**, 12564–12571.

[205] M. Todorovic, S. Schultz, J. Wong and A. Scherer (1999), "Writing and reading of single magnetic domain per bit perpendicular patterned media", *Appl. Phys. Lett.* **74**, 2516–2518.

[206] H.E. Horng, C.-Y. Hong, S.Y. Yang and H.C. Yang (2001), "Novel properties and applications in magnetic fluids", *J. Phys. Chem. Solids* **62**, 1749–1764.

[207] R. Zboril, M. Mashlan and D. Petridis (2002), "Iron(III) oxides from thermal processes-synthesis, structural and magnetic properties, Mössbauer spectroscopy characterization, and applications", *Chem. Mat.* **14**, 969–982.

[208] K.A.J. Gschneidner, V.K. Pecharsky and A.O. Tsokol (2005), "Recent developments in magnetocaloric materials", *Rep. Prog. Phys.* **68** 1479–1539.

[209] M. Faraji, Y. Yamini and M. Rezaee (2010), "Magnetic nanoparticles: synthesis, stabilization, functionalization, characterization, and applications", *J. Iran. Chem. Soc.* **7**, 1–37.

[210] A.S. Teja and P.-Y. Koh (2009), "Synthesis, properties, and applications of magnetic iron oxide nanoparticles", *Prog. Cryst. Growth Charact. Mater.* **55**, 22–45.

[211] R.M. Cornell and U. Schwertmann (2003), *The Iron Oxides: Structure, Properties, Reactions, Occurrences and Uses*, 2nd edn, Wiley-VCH Verlag, Weinheim.

[212] U.T. Lam, R. Mammucari, K. Suzuki and N.R. Foster (2008), "Processing of iron oxide nanoparticles by supercritical fluids", *Ind. Eng. Chem. Res.* **47**, 599–614.

[213] W.-X. Zhang (2003), "Nanoscale iron particles for environmental remediation: an overview", *J. Nanopart. Res.* **5**, 323–332.

[214] D.W. Blowes, C.J. Ptacek, S.G. Benner, *et al.* (2000), "Treatment of inorganic contaminants using permeable reactive barriers", *J. Contam. Hydrol.* **45**, 123–137.

[215] J.T. Nurmi, P.G. Tratnyek, V. Sarathy, *et al.* (2005), "Characterization and properties of metallic iron nanoparticles: spectroscopy, electrochemistry, and kinetics", *Environ. Sci. Technol.* **39**, 1221–1230.

[216] L. Li, M. Fan, R.C. Brown, *et al.* (2006), "Synthesis, properties, and environmental applications of nanoscale iron-based materials: a review", *Crit. Rev. Environ. Sci. Technol.* **36**, 405–431.

[217] S.-Y. Mak and D.-H. Chen (2004), "Fast adsorption of Methylene Blue on polyacrylic acid-bound iron oxide magnetic nanoparticles", *Dyes Pigment.* **61**, 93–98.

[218] Y.-C. Chang and D.-H. Chen (2005), "Adsorption kinetics and thermodynamics of acid dyes on a carboxymethylated chitosan-conjugated magnetic nano-adsorbent", *Macromol. Biosci.* **5**, 254–261.

[219] S.-H. Huang, M.-H. Liao and D.-H. Chen (2006), "Fast and efficient recovery of lipase by polyacrylic acid-coated magnetic nano-adsorbent with high activity retention", *Sep. Purif. Technol.* **51**, 113–117.

[220] B. Zargar, H. Parham and A. Hatamie (2009), "Fast removal and recovery of amaranth by modified iron oxide magnetic nanoparticles", *Chemosphere* **76**, 554–557.

[221] P. Li, D.E. Miser, S. Rabiei, *et al.* (2003), "The removal of carbon monoxide by iron oxide nanoparticles", *Appl. Catal. B-Environ.* **43**, 151–162.

[222] P. Wu and Z. Xu (2005), "Silanation of nanostructured mesoporous magnetic particles for heavy metal recovery", *Ind. Eng. Chem. Res.* **44**, 816–824.

[223] A.-F. Ngomsik, A. Bee, J.-M. Siaugue, *et al.* (2006), "Nickel adsorption by magnetic alginate microcapsules containing an extractant", *Water Res.* **40**, 1848–1856.

[224] L. Wang, Z. Yang, J. Gao, *et al.* (2006), "A biocompatible method of decorporation: bisphosphonate modified magnetite nanoparticles to remove uranyl ions from blood", *J. Am. Chem. Soc.* **128**, 13358–13359.

[225] S.S. Banerjee and D.-H. Chen (2007), "Fast removal of copper ions by gum arabic modified magnetic nano-adsorbent", *J. Hazard. Mater.* **147**, 792–799.

[226] J. Hu, I.M.C. Lo and G. Chen (2007), "Performance and mechanism of chromate (VI) by γ-Fe_2O_3 nanoparticles coated with δ-FeOOH", *Sep. Purif. Technol.* **56**, 249–256.

[227] W. Yantasee, C.L. Warner, T. Sangvanich, *et al.* (2007), "Removal of heavy metals from aqueous systems with thiol functionalized superparamagnetic nanoparticles", *Environ. Sci. Technol.* **41**, 5114–5119.

[228] J.-F. Liu, Z.-S. Zhao and G.-B. Jiang (2008), "Coating Fe_3O_4 magnetic nanoparticles with humic acid for high efficient removal of heavy metals in water", *Environ. Sci. Technol.* **42**, 6949–6954.

[229] S.-H. Huang and D.-H. Chen (2009), "Rapid removal of heavy metal cations and anions from aqueous solutions by an amino-functionalized magnetic nano-adsorbent", *J. Hazard. Mater.* **163**, 174–179.

[230] P. Yuan, M. Fan, D. Yang, *et al.* (2009), "Montmorillonite-supported magnetite nanoparticles for the removal of hexavalent chromium [Cr(VI)] from aqueous solutions", *J. Hazard. Mater.* **166**, 821–829.

[231] B.R. White, B.T. Stackhouse and J.A. Holcombe (2009), "Magnetic γ-Fe_2O_3 nanoparticles coated with poly-L-cysteine for chelation of As(III), Cu(II), Cd(II), Ni(II), Pb(II) and Zn(II)", *J. Hazard. Mater.* **161**, 848–853.

[232] H.Y. Lee, D.R. Bae, J.C. Park, *et al.* (2009), "A selective fluoroionophore based on BODIPY-functionalized magnetic silica nanoparticles: removal of Pb^{2+} from human blood", *Angew. Chem.-Int. Edit.* **48**, 1239–1243.

[233] N. Gaponik, I.L. Radtchenko, G.B. Sukhorukov and A.L. Rogach (2004), "Luminescent polymer microcapsules addressable by a magnetic field", *Langmuir* **20**, 1449–1452.

[234] S.P. Mulvaney, H.M. Mattoussi and L.J. Whitman (2004), "Incorporating fluorescent dyes and quantum dots into magnetic microbeads for immunoassays", *Biotechniques* **36**, 602–609.

[235] D.S. Wang, J.B. He, N. Rosenzweig and Z. Rosenzweig (2004), "Superparamagnetic Fe_2O_3 beads-CdSe/ZnS quantum dots core-shell nanocomposite particles for cell separation", *Nano Lett.* **4**, 409–413.

[236] L. Levy, Y. Sahoo, K.S. Kim, *et al.* (2002), "Nanochemistry: synthesis and characterization of multifunctional nanoclinics for biological applications", *Chem. Mat.* **14**, 3715–3721.

[237] V. Salgueirino-Maceira, M.A. Correa-Duarte, M. Spasova, *et al.* (2006), "Composite silica spheres with magnetic and luminescent functionalities", *Adv. Funct. Mater.* **16**, 509–514.

[238] T.R. Sathe, A. Agrawal and S.M. Nie (2006), "Mesoporous silica beads embedded with semiconductor quantum dots and iron oxide nanocrystals: dual-function microcarriers for optical encoding and magnetic separation", *Anal. Chem.* **78**, 5627–5632.

[239] P.S. Eastman, W.M. Ruan, M. Doctolero, *et al.* (2006), "Qdot nanobarcodes for multiplexed gene expression analysis", *Nano Lett.* **6**, 1059–1064.

[240] C. Moser, T. Mayr and I. Klimant (2006), "Microsphere sedimentation arrays for multiplexed bioanalytics", *Anal. Chim. Acta* **558**, 102–109.

[241] M. Nichkova, D. Dosev, S.J. Gee, *et al.* (2007), "Multiplexed immunoassays for proteins using magnetic luminescent nanoparticles for internal calibration", *Anal. Biochem.* **369**, 34–40.

[242] A. Ito, M. Shinkai, H. Honda and T. Kobayashi (2005), "Medical application of functionalized magnetic nanoparticles", *J. Biosci. Bioeng.* **100**, 1–11.

[243] K. Schulze, A. Koch, B. Schopf, *et al.* (2005), "Intraarticular application of superparamagnetic nanoparticles and their uptake by synovial membrane – an experimental study in sheep", *J. Magn. Magn. Mater.* **293**, 419–432.

[244] A. Kumar, B. Sahoo, A. Montpetit, *et al.* (2007), "Development of hyaluronic acid–Fe_2O_3 hybrid magnetic nanoparticles for targeted delivery of peptides", *Nanomed.-Nanotechnol. Biol. Med.* **3**, 132–137.

[245] J.B. Sun, J.H. Duan, S.L. Dai, *et al.* (2007), "In vitro and in vivo antitumor effects of doxorubicin loaded with bacterial magnetosomes (DBMs) on H22 cells: the magnetic bio-nanoparticles as drug carriers", *Cancer Lett.* **258**, 109–117.

[246] J.L. Arias, V. Gallardo, M.A. Ruiz and A.V. Delgado (2008), "Magnetite/poly (alkylcyanoacrylate) (core/shell) nanoparticles as 5-Fluorouracil delivery systems for active targeting", *Eur. J. Pharm. Biopharm.* **69**, 54–63.

[247] G. Huang, J. Diakur, Z. Xu and L.I. Wiebe (2008), "Asialoglycoprotein receptor-targeted superparamagnetic iron oxide nanoparticles", *Int. J. Pharm.* **360**, 197–203.

[248] E. Munnier, S.C. Jonathan, C. Linassier, *et al.* (2008), "Novel method of doxorubicin–SPION reversible association for magnetic drug targeting", *Int. J. Pharm.* **363**, 170–176.

[249] B. Gaihrea, M.S. Khil, D.R. Lee and H.Y. Kim (2009), "Gelatin-coated magnetic iron oxide nanoparticles as carrier system: drug loading and in vitro drug release study", *Int. J. Pharm.* **365**, 180–189.

[250] S. Kayal and R.V. Ramanujan (2010), "Anti-cancer drug loaded iron-gold core-shell nanoparticles (Fe@Au) for magnetic drug targeting", *J. Nanosci. Nanotechnol.* **10**, 5527–5539.

[251] H. Rahn and S. Odenbach (2009), "X-ray microcomputed tomography as a tool for the investigation of the biodistribution of magnetic nanoparticles", *Nanomedicine* **4**, 981–990.

[252] A. Kumar, P.K. Jena, S. Behera, *et al.* (2010), "Multifunctional magnetic nanoparticles for targeted delivery", *Nanomed. Nanotechnol. Biol. Med.* **6**, 64–69.

[253] J.-J. Lin, J.-S. Chen, S.-J. Huang, *et al.* (2009), "Folic acid-Pluronic F127 magnetic nanoparticle clusters for combined targeting, diagnosis, and therapy applications", *Biomaterials* **30**, 5114–5124.

[254] I. Arslan and I.A. Balcioglu (1999), "Degradation of commercial reactive dyestuffs by heterogeneous and homogenous advanced oxidation processes: a comparative study", *Dyes Pigment.* **43**, 95–108.

[255] D.F. Ollis, E. Pelizzetti and N. Serpone (1989), *"Photo-Catalysis Fundamentals and Applications"*, N. Serpone and E. Pelizzetti (eds), John Wiley & Sons, Inc., New York.

[256] J.-M. Herrmann (1999), "Heterogeneous photo-catalysis: fundamental and application to the removal of various types of aqueous pollutants", *Catal. Today* **53**, 115–129.

[257] X. Fu, W.A. Zeltner and M.A. Anderson (1996), *Applications in Photo-Catalytic Purification of Air, Semiconductor Nanoclusters – Physical, Chemical, and Catalytic Aspects*, Book Series: Studies in Surface Science and Catalysis, **103**, 445–461.

[258] C. Hachem, F. Bocquillon, O. Zahraa and M. Bouchy (2001), "Decolourization of textile industry wastewater by the photo-catalytic degradation process", *Dyes Pigment.* **49**, 117–125.

[259] M.R. Hoffmann, S.T. Martin, W.Y. Choi and D.W. Bahnemann (1995), "Environmental applications of semiconductor photo-catalysis", *Chem. Rev.* **95**, 69–96.

[260] C. Minero (1995), "A rigorous kinetic approach to model primary oxidative steps of photo-catalytic degradations", *Sol. Energy Mater. Sol. Cells* **38**, 421–430.

[261] M.A. Rauf, S.B. Bukallah, A. Hammadi, *et al.* (2007), "The effect of operational parameters on the photoinduced decoloration of dyes using a hybrid catalyst V_2O_5/TiO_2", *Chem. Eng. J.* **129**, 167–172.

[262] N.M. Mahmoodi, M. Arami, N.Y. Limaee and N.S. Tabrizi (2006), "Kinetics of heterogeneous photo-catalytic degradation of reactive dyes in an immobilized TiO_2 photo-catalytic reactor", *J. Colloid Interface Sci.* **295**, 159–164.

[263] J. Winkler (2002), "Nano-scaled titanium dioxide – properties and use in coatings with special functionality", *Macromol. Symp.* **187**, 317–324.

[264] Y. Paz, Z. Luo, L. Rabenberg and A. Heller (1995), "Photooxidative self-cleaning transparent titanium dioxide films on glass", *J. Mater. Res.* **10**, 2842–2848.

[265] A.E.H. Machado, J.A. de Miranda, R.F. de Freitas, *et al.* (2003), "Destruction of the organic matter present in effluent from a cellulose and paper industry using photo-catalysis", *J. Photochem. Photobiol. A-Chem.* **155**, 231–241.

[266] L.C. Macedo, D.A.M. Zaia, G.J. Moore and de Santana H. (2007), "Degradation of leather dye on TiO_2: a study of applied experimental parameters on photoelectrocatalysis", *J. Photochem. Photobiol. A-Chem.* **185**, 86–93.

[267] C.C. Wang, C.K. Lee, M.D. Lyu and L.C. Juang (2008), "Photo-catalytic degradation of C.I. Basic Violet 10 using TiO_2 catalysts supported by Y zeolite: an investigation of the effects of operational parameters", *Dyes Pigment.* **76**, 817–824.

[268] I. Ilisz, Z. Laszlo and A. Dombi (1999), "Investigation of the photodecomposition of phenol in near-UV-irradiated aqueous TiO_2 suspensions. I. Effect of charge-trapping species on the degradation kinetics", *Appl. Catal. A-Gen.* **180**, 25–33.

[269] M. Muruganandham and M. Swaminathan (2006), "Photo-catalytic decoloration and degradation of Reactive Orange 4 by TiO_2-UV process", *Dyes Pigment.* **68**, 133–142.

[270] L. Andronic and A. Duta (2007), "TiO_2 thin films for dyes photodegradation", *Thin Solid Films* **515**, 6294–6297.

[271] M. Canle, L.J.A. Santaballa and E. Vulliet (2005), "On the mechanism of TiO_2-photo-catalysed degradation of aniline derivatives", *J. Photochem. Photobiol. A-Chem.* **175**, 192–200.

[272] D. Chatterjee and S. Dasgupta (2005), "Visible light induced photo-catalytic degradation of organic pollutants", *J. Photochem. Photobiol. C-Photochem. Rev.* **6**, 186–205.

[273] F. Han, K.S.R. Rao, M. Srinivasan, *et al.* (2009), "Tailored titanium dioxide photo-catalysts for the degradation of organic dyes in wastewater treatment: a review" *Appl. Catal. A-Gen.* **359**, 25–40.

[274] C.-Y. Wang, C.-Y. Liu, X. Zheng, *et al.* (1998), "The surface chemistry of hybrid nanometer-sized particles - I. Photochemical deposition of gold on ultrafine TiO_2 particles", *Colloid Surf. A-Physicochem. Eng. Asp.* **131**, 271–280.

[275] K.V.S. Rao, B. Lavedrine and P. Boule (2003), "Influence of metallic species on TiO_2 for the photo-catalytic degradation of dyes and dye intermediates", *J. Photochem. Photobiol. A-Chem.* **154**, 189–193.

[276] H.M. Sung-Suh, J.R. Choi, H.J. Hah, *et al.* (2004), "Comparison of Ag deposition effects on the photo-catalytic activity of nanoparticulate TiO_2 under visible and UV light irradiation", *J. Photochem. Photobiol. A-Chem.* **163**, 37–44.

[277] J. Zhu, F. Chen, J. Zhang, *et al.* (2006), "Fe^{3+}-TiO_2 photo-catalysts prepared by combining sol–gel method with hydrothermal treatment and their characterization", *J. Photochem. Photobiol. A-Chem.* **180**, 196–204.

[278] A.K. Gupta, A. Pal and C. Sahoo (2006), "Photo-catalytic degradation of a mixture of Crystal Violet (Basic Violet 3) and Methyl Red dye in aqueous suspensions using Ag^+ doped TiO_2", *Dyes Pigment.* **69**, 224–232.

[279] Y. Wang, Y. Wang, Y. Meng, *et al.* (2008), "A highly efficient visible-light-activated photo-catalyst based on bismuth- and sulfur-codoped TiO_2", *J. Phys. Chem. C* **112**, 6620–6626.

[280] A.W. Xu, Y. Gao and H.Q. Liu (2002), "The preparation, characterization, and their photo-catalytic activities of rare-earth-doped TiO_2 nanoparticles", *J. Catal.* **207**, 151–157.

[281] J.W. Shi, J.T. Zheng, Y. Hu and Y.C. Zhao (2008), "Photo-catalytic degradation of methyl orange in water by samarium-doped TiO_2", *Environ. Eng. Sci.* **25**, 489–496.

[282] W.Y. Su, E.X. Chen, L. Wu, *et al.* (2008), "Visible light photo-catalysis on praseodymium(III)-nitrate-modified TiO_2 prepared by an ultrasound method", *Appl. Catal. B* **77**, 264–271.

[283] J. Grzechulska and A.W. Morawski (2002), "Photo-catalytic decomposition of azo-dye acid black 1 in water over modified titanium dioxide", *Appl. Catal. B-Environ.* **36**, 45–51.

[284] C. Suwanchawalit and S. Wongnawa (2008), "Influence of calcination on the microstructures and photo-catalytic activity of potassium oxalate-doped TiO_2 powders", *Appl. Catal. A-Gen.* **338**, 87–99.

[285] N. Serpone, P. Maruthamuthu, P. Pichat, *et al.* (1995), "Exploiting the inter-particle electron-transfer process in the photo-catalyzed oxidation of phenol, 2-chlorophenol and pentachlorophenol – chemical evidence for electron and hole transfer between coupled semiconductors", *J. Photochem. Photobiol. A-Chem.* **85**, 247–255.

[286] J. Zhu, D. Yang, J. Geng, *et al.* (2008), "Synthesis and characterization of bamboo-like CdS/TiO_2 nanotubes composites with enhanced visible-light photo-catalytic activity", *J. Nanopart. Res.* **10**, 729–736.

[287] I.M. Arabatzis, T. Stergiopoulos, M.C. Bernard, *et al.* (2003), "Silver-modified titanium dioxide thin films for efficient photodegradation of methyl orange", *Appl. Catal. B-Environ.* **42**, 187–201.

[288] P. Periyat, S.C. Pillai, D.E. McCormack, *et al.* (2008), "Improved high-temperature stability and sun-light-driven photo-catalytic activity of sulfur-doped anatase TiO_2", *J. Phys. Chem. C* **112**, 7644–7652.

[289] C.C. Chen, X.Z. Li, W.H. Ma, *et al.* (2002), "Effect of transition metal ions on the TiO_2-assisted photodegradation of dyes under visible irradiation: a probe for the interfacial electron transfer process and reaction mechanism", *J. Phys. Chem. B* **106**, 318–324.

[290] K. Pirkanniemi and M. Sillanpaa (2002), "Heterogeneous water phase catalysis as an environmental application: a review", *Chemosphere* **48**, 1047–1060.

[291] A. Di Paola, M. Addamo, M. Bellardita, *et al.* (2007), "Preparation of photo-catalytic brookite thin films", *Thin Solid Films* **515**, 3527.

[292] K. Tennakone, C.T.K. Tilakaratne and I.R.M. Kottegoda (1997), "Photomineral-ization of carbofuran by TiO_2-supported catalyst", *Water Res.* **31**, 1909–1912.

[293] G.R.R.A. Kumara, F.M. Sultanbawa, V.P.S. Perera, *et al.* (1999), "Continuous flow photochemical reactor for solar decontamination of water using immobilized TiO_2", *Sol. Energy Mater. Sol. Cells* **58**, 167–171.

[294] M. Karches, M. Morstein, P.R. von Rohr, *et al.* (2002), "Plasma-CVD-coated glass beads as photo-catalyst for water decontamination", *Catal. Today* **72**, 267–279.

[295] S. Fukahori, H. Ichiura, T. Kitaoka and H. Tanaka (2003), "Photo-catalytic decomposition of bisphenol A in water using composite TiO_2-zeolite sheets prepared by a papermaking technique", *Environ. Sci. Technol.* **37**, 1048–1051.

[296] I.M. Arabatzis, S. Antonaraki, T. Stergiopoulos, *et al.* (2002), "Preparation, characterization and photo-catalytic activity of nanocrystalline thin film TiO_2 catalysts towards 3,5-dichlorophenol degradation", *J. Photochem. Photobiol. A-Chem.* **149**, 237–245.

[297] S. Canonica, U. Jans, K. Stemmler and J. Hoigne (1995), "Transformation kinetics of phenols in water: photosensitization by dissolved natural organic material and aromatic ketones", *Environ. Sci. Technol.* **29**, 1822–1831.

[298] N. Kashif and F. Ouyang (2009), "Parameters effect on heterogeneous photo-catalysed degradation of phenol in aqueous dispersion of TiO_2", *J. Environ. Sci.* **21**, 527–533.

[299] D. Zhang, R. Qiu, L. Song, *et al.* (2009), "Role of oxygen active species in the photo-catalytic degradation of phenol using polymer sensitized TiO_2 under visible light irradiation", *J. Hazard. Mater.* **163**, 843–847.

[300] K.-H. Wang, Y.-H. Hsieh, M.-Y. Chou and C.-Y. Chang (1999), "Photo-catalytic degradation of 2-chloro and 2-nitrophenol by titanium dioxide suspensions in aqueous solution", *Appl. Catal. B-Environ.* **21**, 1–8.

[301] Z. Guo, R. Ma and G. Li (2006), "Degradation of phenol by nanomaterial TiO_2 in wastewater", *Chem. Eng. J.* **119**, 55–59.

[302] S. Ray, J.A. Lalmanb and N. Biswas (2009), "Using the Box–Benkhen technique to statistically model phenol photo-catalytic degradation by titanium dioxide nanoparticles", *Chem. Eng. J.* **150**, 15–24.

[303] E. Forgacs, T. Cserhati and G. Oros (2004), "Removal of synthetic dyes from wastewater: a review", *Environ. Int.* **30**, 953–971.

[304] M. Derudi, G. Venturini, G. Lombardi, *et al.* (2007), "Biodegradation combined with ozone for the remediation of contaminated soils", *Eur. J. Soil Biol.* **43**, 297–303.

[305] A.L. Ahmad and S.W. Puasa (2007), "Reactive dyes decolourization from an aqueous solution by combined coagulation/micellar-enhanced ultrafiltration process", *Chem. Eng. J.* **132**, 257–265.

[306] M.J. Martin, A. Artola, M.D. Balaguer and M. Rigola (2003), "Activated carbons developed from surplus sewage sludge for the removal of dyes from dilute aqueous solutions", *Chem. Eng. J.* **94**, 231–239.

[307] J.H. Mo, Y.H. Lee, J. Kim, *et al.* (2008), "Treatment of dye aqueous solutions using nanofiltration polyamide composite membranes for the dye wastewater reuse", *Dyes Pigment.* **76**, 429–434.

[308] H.-L. Liu and Y.-R. Chiou (2005), "Optimal decolorization efficiency of Reactive Red 239 by UV/TiO_2 photo-catalytic process coupled with response surface methodology", *Chem. Eng. J.* **112**, 173–179.

[309] Y. Ao, J. Xu, D. Fua and C. Yuan (2009), "A simple method to prepare N-doped titania hollow spheres with high photo-catalytic activity under visible light", *J. Hazard. Mater.* **167**, 413–417.

[310] F.-L. Toma, L.M. Berger, D. Jacquet, *et al.* (2009), "Comparative study on the photo-catalytic behaviour of titanium oxide thermal sprayed coatings from powders and suspensions", *Surf. Coat. Technol.* **203**, 2150–2156.

[311] X.-M. Song, J.-M. Wu and M. Yan (2009), "Photo-catalytic degradation of selected dyes by titania thin films with various nanostructures", *Thin Solid Films* **517**, 4341–4347.

[312] J.G. Yu, S.W. Liu and H.G. Yu (2007), "Microstructures and photo-activity of mesoporous anatase hollow microspheres fabricated by fluoride-mediated self-transformation", *J. Catal.* **249**, 59–66.

[313] C. Wang, Y. Ao, P. Wang, *et al.* (2010), "A facile method for the preparation of titania-coated magnetic porous silica and its photo-catalytic activity under UV or visible light", *Colloid Surf. A-Physicochem. Eng. Asp.* **360**, 184–189.

[314] C. Yu, J.C. Yu and M. Chan (2009), "Sonochemical fabrication of fluorinated mesoporous titanium dioxide microspheres", *J. Solid State Chem.* **182**, 1061–1069.

[315] J. Wang, W. Zhu, Y. Zhang and S. Liu (2007), "An efficient two-step technique for nitrogen-doped titanium dioxide synthesizing: visible-light-induced photodecomposition of Methylene Blue", *J. Phys. Chem. C* **111**, 1010–1014.

[316] V.L. Colvin (2003), "The potential environmental impact of engineered nanomaterials", *Nat. Biotechnol.* **21**, 1166–1170.

[317] K.H. Ek, S. Rauch, G.M. Morrison and P. Lindberg (2004), "Platinum group elements in raptor eggs, faeces, blood, liver and kidney", *Sci. Total Environ.* **334**, 149–159.

[318] S.A. Blaser, M. Scheringer, M. MacLeod and K. Hungerbühler (2008), "Estimation of cumulative aquatic exposure and risk due to silver: contribution of nanofunctionalized plastics and textiles", *Sci. Total Environ.* **390**, 396–409.

[319] W. Hannah and P.B. Thompson (2008), "Nanotechnology, risk and the environment: a review", *J. Environ. Monit.* **10**, 291–300.

[320] J.S. Kim, T.-J. Yoon, K.N. Yu, *et al.* (2006), "Toxicity and tissue distribution of magnetic nanoparticles in mice", *Toxicol. Sci.* **89**, 338–347.

[321] W.T. Liu (2006), "Nanoparticles and their biological and environmental applications", *J. Biosci. Bioeng.* **102**, 1–7.

[322] T.C. Long, N. Saleh, R.D. Tilton, *et al.* (2006), "Titanium dioxide (P25) produces reactive oxygen species in immortalized brain microglia (BV2) implications for nanoparticle neurotoxicity", *Environ. Sci. Technol.* **40**, 4346–4352.

[323] A.T. Marshall, R.G. Haverkamp, C.E. Davies, *et al.* (2007), "Accumulation of gold nanoparticles in Brassic Juncea", *Int. J. Phytoremediat.* **9**, 197–206.

[324] A.E. Porter, M. Gass, K. Muller, *et al.* (2007), "Direct imaging of single-walled carbon nanotubes in cells", *Nat. Nanotechnol.* **2**, 713–717.

[325] K. Unfried, C. Albrecht, L.O. Klotz, *et al.* (2007), "Cellular responses to nanoparticles: target structures and mechanisms", *Nanotoxicology* **1**, 52–71.

[326] A. Baun, N.B. Hartmann, K. Grieger and K.O. Kusk (2008), "Ecotoxicity of engineered nanoparticles to aquatic invertebrates: a brief review and recommendations for future toxicity testing", *Ecotoxicology* **17**, 387–396.

[327] N. Lewinski, V. Colvin and R. Drezek (2008), "Cytotoxicity of nanoparticles", *Small* **4**, 26–49.

[328] G. Oberdörster, E. Oberdörster and J. Oberdörster (2005), "Nanotoxicology: an emerging discipline evolving from ultrafine particles", *Environ. Health Perspect.* **113**, 823–839.

[329] C. Kirchner, T. Liedl, S. Kudera, *et al.* (2005), "Cytotoxicity of colloidal CdSe and CdSe/ZnS nanoparticles", *Nano Lett.* **5**, 331–338.

[330] P.J.A. Borm and W. Kreyling (2004), "Toxicological hazards of inhaled nanoparticles – Potential implications for drug delivery", *J. Nanosci. Nanotechnol.* **4**, 521–531.

[331] G. Bystrzejewska-Piotrowska, J. Golimowski and P.L. Urban (2009), "Nanoparticles: their potential toxicity, waste and environmental management", *Waste Manage.* **29**, 2587–2595.

Index

Sustainable Development in Chemical Engineering – Innovative Technologies, First Edition.
Edited by Vincenzo Piemonte, Marcello De Falco and Angelo Basile.
© 2013 John Wiley & Sons, Ltd. Published 2013 by John Wiley & Sons, Ltd.